Aircraft Communications and Navigation Systems:
Principles, Operation and Maintenance

Mike Tooley and David Wyatt

Routledge
Taylor & Francis Group

LONDON AND NEW YORK

First published by Butterworth-Heinemann

This edition published 2011 by Routledge
2 Park Square, Milton Park, Abingdon, Oxon OX14 4RN
711 Third Avenue, New York, NY 10017, USA

Routledge is an imprint of the Taylor & Francis Group, an informa business

First edition 2007

British Library Cataloguing in Publication Data
A catalogue record for this book is available from the British Library

Library of Congress Cataloging-in-Publication Data
A catalog record for this book is available from the Library of Congress

Typeset by the author

ISBN: 978-0-7506-81377

Contents

Contents

Contents

Preface

The books in this series have been designed for both independent and tutor assisted studies. They are particularly useful to the 'self-starter' and to those wishing to update or upgrade their aircraft maintenance licence. The series also provides a useful source of reference for those taking *ab initio* training programmes in EASA Part 147 and FAR 147 approved organisations as well as those following related programmes in further and higher education institutions.

This book is designed to cover the essential knowledge base required by certifying mechanics, technicians and engineers engaged in engineering maintenance activities on commercial aircraft. In addition, this book should appeal to members of the armed forces and others attending training and educational establishments engaged in aircraft maintenance and related aeronautical engineering programmes (including BTEC National and Higher National units as well as City and Guilds and NVQ courses).

The book provides an introduction to the principles, operation and maintenance of aircraft communications and navigation systems. The aim has been to make the subject material accessible and presented in a form that can be readily assimilated. The book provides syllabus coverage of the communications and navigation section of Module 13 (ATA 23/34). The book assumes a basic understanding of aircraft flight controls as well as an appreciation of electricity and electronics (broadly equivalent to Modules 3 and 4 of the EASA Part-66 syllabus).

It is important to realise that this book is not designed to replace aircraft maintenance manuals. Nor does it attempt to provide the level of detail required by those engaged in the maintenance of specific aircraft types. Instead it has been designed to convey the essential underpinning knowledge required by all aircraft maintenance engineers.

Chapter 1 sets the scene by providing an explanation of electromagnetic wave propagation and the radio frequency spectrum. The chapter also describes the various mechanisms by which radio waves propagate together with a detailed description of the behaviour of the ionosphere and its effect on radio signals.

Antennas are introduced in Chapter 2. This chapter explains the principles of isotropic and directional radiating elements and introduces a number of important concepts including radiation resistance, antenna impedance, radiated power, gain and efficiency. Several practical forms of antenna are described including dipoles, Yagi beam antennas, quarter wave (Marconi) antennas, corner reflectors, horn and parabolic dish radiators. Chapter 2 also provides an introduction to feeders (including coaxial cable and open-wire types), connectors and standing wave ratio (SWR). The chapter concludes with a brief introduction to waveguide systems.

Radio transmitters and receivers are the subject of Chapter 3. This chapter provides readers with an introduction to the operating principles of AM and FM transmitters as well as tuned radio frequency (TRF) and supersonic-heterodyne (superhet) receivers. Selectivity, image channel rejection and automatic gain control (AGC) are important requirements of a modern radio receiver and these topics are introduced before moving on to describe more complex receiving equipment. Modern aircraft radio equipment is increasingly based on the use of digital frequency synthesis and the basic principles of phase-locked loops and digital synthesisers are described and explained.

Very high frequency (VHF) radio has long been the primary means of communication between aircraft and the ground. Chapter 4 describes the principles of VHF communications (both voice and data). The chapter also provides an introduction to the aircraft communication addressing and reporting system (ACARS).

High frequency (HF) radio provides aircraft with an effective means of communicating over long distance oceanic and trans-polar routes. In addition, global data communication has recently been made possible using strategically located HF data link (HFDL) ground stations. Chapter 5 describes the principles of HF radio communication as well as the equipment and technology used.

As well as communication with ground stations, modern passenger aircraft require facilities for local communication within the aircraft. Chapter 6 describes flight-deck audio systems including the interphone system and all-important cockpit voice recorder (CVR) which captures audio signals so that they can be later analysed in the event of a serious malfunction of the aircraft or of any of its systems.

The detection and location of the site of an air crash is vitally important to the search and rescue (SAR) teams and also to potential survivors. Chapter 7 describes the construction and operation of emergency locator transmitters (ELT) fitted to modern passenger aircraft. The chapter also provides a brief introduction to satellite-based location techniques.

Chapter 8 introduces the subject of aircraft navigation; this sets the scene for the remaining chapters of the book. Navigation is the science of conducting journeys over land and/or sea. This chapter reviews some basic features of the earth's geometry as it relates to navigation, and introduces some basic aircraft navigation terminology, e.g. latitude, longitude, dead reckoning etc. The chapter concludes by reviewing a range of navigation systems used on modern transport and military aircraft. Many aircraft navigation systems utilise radio frequency methods to determine a position fix; this links very well into the previous chapters of the book describing fundamental principles of radio transmitters, receivers and antennas.

Radio waves have directional characteristics as described in the early chapters of the book. This is the basis of the automatic direction finder (ADF); one of earliest forms of radio navigation that is still in use today. ADF is a short–medium range (200 nm) navigation system providing directional information. Chapter 9 looks at the historical background to radio navigation, reviews some typical ADF hardware that is fitted to modern commercial transport aircraft, and concludes with some practical aspects associated with the operational use of ADF.

During the late 1940s, it was evident to the aviation world that an accurate and reliable short-range navigation system was needed. Since radio communication systems based on very high frequency (VHF) were being successfully deployed, a decision was made to develop a radio navigation system based on VHF. This system became the VHF omnidirectional range (VOR) system, and is described in Chapter 10. This system is in widespread use throughout the world today. VOR is the basis of the current network of 'airways' that are used in navigation charts.

Chapter 11 develops this theme with a system for measuring distance to a navigation aid. The advent of radar in the 1940s led to the development of a number of navigation aids including distance measuring equipment (DME). This is a short/medium-range navigation system, often used in conjunction with the VOR system to provide accurate navigation fixes. The system is based on secondary radar principles.

ADF, VOR and DME navigation aids are installed at airfields to assist with approaches to those airfields. These navigation aids cannot however be used for precision approaches and landings. The standard approach and landing system installed at airfields around the world is the instrument landing system (ILS). Chapter 12 describes how the ILS can be used for approach through to autoland. The ILS uses a combination of VHF and UHF radio waves and has been in operation since 1946.

Chapter 13 continues with the theme of guided approaches to an airfield. There are a number of shortcomings with ILS; in 1978 the microwave landing system (MLS) was adopted as the long-term replacement. The system is based on the principle of time referenced scanning beams and provides precision navigation guidance for approach and landing. MLS provides three-dimensional approach guidance, i.e. azimuth, elevation and range. The system provides multiple approach angles for both azimuth and elevation guidance. Despite the advantages of MLS, it has not yet been introduced on a worldwide basis for commercial aircraft. Military operators of MLS often use mobile equipment that can be deployed within hours.

Long-range radio navigation systems are described in Chapter 14. These systems are based on hyperbolic navigation; they were introduced in the 1940s to provide en route operations over oceans and unpopulated areas. Several hyperbolic systems have been developed since, including Decca, Omega and Loran. The operational use of Omega and Decca navigation systems ceased in 1997 and 2000 respectively. Loran systems are still available for use today as stand-alone systems; they are also being proposed as a complementary navigation aid for global navigation satellite systems.

Chapter 15 looks at a unique form of dead reckoning navigation system based on radar and a scientific principle called Doppler shift. This system requires no external inputs or references from ground stations. Doppler navigation systems were developed in the mid-1940s and introduced in the mid-1950s as a primary navigation system. Being self-contained, the system can be used for long distance navigation and by helicopters during hover manoeuvres.

The advent of computers, in particular the increasing capabilities of integrated circuits using digital techniques, has led to a number of advances in aircraft navigation. One example of this is the area navigation system (RNAV); this is described in Chapter 16. Area navigation is a means of combining, or filtering, inputs from one or more navigation sensors and defining positions that are not necessarily co-located with ground-based navigation aids.

A major advance in aircraft navigation came with the introduction of the inertial navigation system (INS); this is the subject of Chapter 17. The inertial navigation system is an autonomous dead reckoning system, i.e. it requires no external inputs or references from ground stations. The system was developed in the 1950s for use by the US military and subsequently the space programmes. Inertial navigation systems (INS) were introduced into commercial aircraft service during the early 1970s. The system is able to compute navigation data such as present position, distance to waypoint, heading, ground speed, wind speed, wind direction etc. The system does not need radio navigation inputs and it does not transmit radio frequencies. Being self-contained, the system can be used for long distance navigation over oceans and undeveloped areas of the globe.

Navigation by reference to the stars and planets has been employed since ancient times; aircraft navigators have utilised periscopes to take celestial fixes for long distance navigation. An artificial constellation of navigation aids was initiated in 1973 and referred to as Navstar (navigation system with timing and ranging). This global positioning system (GPS) was developed for use by the US military; it is now widely available for use in many applications including aircraft navigation. Chapter 18 looks at GPS and other global navigation satellite systems that are in use, or planned for future deployment.

The term 'navigation' can be applied in both the lateral and vertical senses for aircraft applications. Vertical navigation is concerned with optimising the performance of the aircraft to reduce operating costs; this is the subject of Chapter 19. During the 1980s, lateral navigation and performance management functions were combined into a single system known as the flight management system (FMS). Various tasks previously routinely performed by the crew can now be automated with the intention of reducing crew workload.

Chapter 20 reviews how the planned journey from A to B could be affected by adverse weather conditions. Radar was introduced onto passenger aircraft during the 1950s to allow pilots to identify weather conditions and subsequently re-route around these conditions for the safety and comfort of passengers. A secondary use of weather radar is the terrain-mapping mode that allows the pilot to identify features of the ground, e.g. rivers, coastlines and mountains.

Increasing traffic density, in particular around airports, means that we need a method of air traffic control (ATC) to manage the flow of traffic and maintain safe separation of aircraft. The ATC system is based on secondary surveillance radar (SSR). Ground controllers use the system to address individual aircraft. An emerging ATC technology is ADS-B, this is also covered in Chapter 21.

With ever increasing air traffic congestion, and the subsequent demands on air traffic control (ATC) resources, the risk of a mid-air collision increases. The need for improved traffic flow led to the introduction of the traffic alert and collision avoidance system (TCAS); this is the subject of

Chapter 22. TCAS is an automatic surveillance system that helps aircrews and ATC to maintain safe separation of aircraft. TCAS is an airborne system based on secondary radar that interrogates and replies directly with aircraft via a high-integrity data link. The system is functionally independent of ground stations, and alerts the crew if another aircraft comes within a predetermined time to a potential collision.

The book concludes with four useful appendices, including a comprehensive list of abbreviations and acronyms used with aircraft communications and navigation systems.

The review questions at the end of each chapter are typical of these used in CAA and other examinations. Further examination practice can be gained from the four revision papers given in Appendix 2. Other features that will be particularly useful if you are an independent learner are the 'key points' and 'test your understanding' questions interspersed throughout the text.

Acknowledgements

The authors would like to thank the following persons and organisations for permission to reproduce photographs and data in this book: Lees Avionics and Wycombe Air Centre for product/cockpit images; Trevor Diamond for the ADF, VOR and DME photographs; CMC Electronics for data and photographs of Doppler and MLS hardware; the International Loran Association (ILA) and US Coast Guard for information and data on both the existing Loran-C infrastructure and their insight into future developments; Kearfott (Guidance & Navigation Corporation) and Northrop Grumman Corporation for permission to reproduce data on their inertial navigation systems and sensors; ARINC for information relating to TCAS; ADS-B Technologies, LLC for their permission to reproduce data on automatic dependent surveillance-broadcast. Finally, thanks also go to Alex Hollingsworth, Lucy Potter and Jonathan Simpson at Elsevier for their patience, encouragement and support.

Online resources

Additional supporting material (including video clips, sound bites and image galleries) for this book are available at **www.66web.co.uk** or **www.key2study.com/66web**

Introduction

Maxwell first suggested the existence of electromagnetic waves in 1864. Later, Heinrich Rudolf Hertz used an arrangement of rudimentary resonators to demonstrate the existence of electromagnetic waves. Hertz's apparatus was extremely simple and comprised two resonant loops, one for transmitting and the other for receiving. Each loop acted both as a tuned circuit and as a resonant antenna (or 'aerial').

Hertz's transmitting loop was excited by means of an induction coil and battery. Some of the energy radiated by the transmitting loop was intercepted by the receiving loop and the received energy was conveyed to a spark gap where it could be released as an arc. The energy radiated by the transmitting loop was in the form of an **electromagnetic wave**—a wave that has both electric and magnetic field components and that travels at the speed of light.

In 1894, Marconi demonstrated the commercial potential of the phenomenon that Maxwell predicted and Hertz actually used in his apparatus. It was also Marconi that made radio a reality by pioneering the development of telegraphy without wires (i.e. 'wireless'). Marconi was able to demonstrate very effectively that information could be exchanged between distant locations without the need for a 'land-line'.

Marconi's system of **wireless telegraphy** proved to be invaluable for maritime communications (ship to ship and ship to shore) and was to be instrumental in saving many lives. The military applications of radio were first exploited during the First World War (1914 to 1918) and, during that period, radio was first used in aircraft.

This first chapter has been designed to set the scene and to provide you with an introduction to the principles of radio communication systems. The various topics are developed more fully in the later chapters but the information provided here is designed to provide you with a starting point for the theory that follows.

1.1 The radio frequency spectrum

Radio frequency signals are generally understood to occupy a frequency range that extends from a few tens of kilohertz (kHz) to several hundred gigahertz (GHz). The lowest part of the radio frequency range that is of practical use (below 30 kHz) is only suitable for narrow-band communication. At this frequency, signals propagate as ground waves (following the curvature of the earth) over very long distances. At the other extreme, the highest frequency range that is of practical importance extends above 30 GHz. At these microwave frequencies, considerable bandwidths are available (sufficient to transmit many television channels using point-to-point links or to permit very high definition radar systems) and signals tend to propagate strictly along line-of-sight paths.

At other frequencies signals may propagate by various means including reflection from ionised layers in the ionosphere. At frequencies between 3 MHz and 30 MHz ionospheric propagation regularly permits intercontinental broadcasting and communications.

For convenience, the radio frequency spectrum is divided into a number of bands (see Table 1.1), each spanning a decade of frequency. The use to which each frequency range is put depends upon a number of factors, paramount amongst which is the propagation characteristics within the band concerned.

Other factors that need to be taken into account include the efficiency of practical aerial systems in the range concerned and the bandwidth available. It is also worth noting that, although it may appear from Figure 1.1 that a great deal of the radio frequency spectrum is not used, it should be stressed that competition for frequency space is fierce and there is, in fact, little vacant space! Frequency allocations are, therefore, ratified by international agreement and the various user services carefully safeguard their own areas of the spectrum.

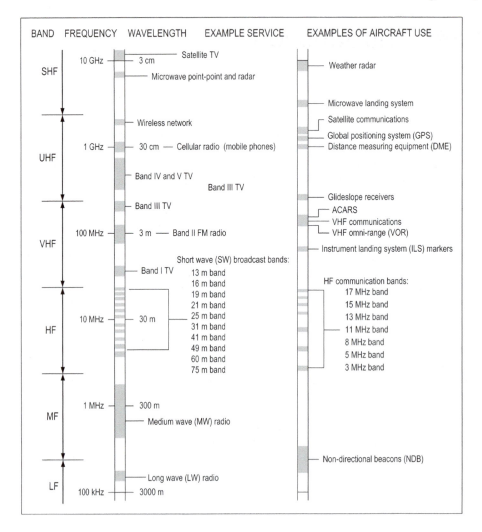

Figure 1.1   Some examples of frequency allocations within the radio frequency spectrum

Table 1.1   Frequency bands

Frequency range	Wavelength	Designation
300 Hz to 3 kHz	1000 km to 100 km	Extremely low frequency (ELF)
3 kHz to 30 kHz	100 km to 10 km	Very low frequency (VLF)
30 kHz to 300 kHz	10 km to 1 km	Low frequency (LF)
300 kHz to 3 MHz	1 km to 100 m	Medium frequency (MF)
3 MHz to 30 MHz	100 m to 10 m	High frequency (HF)
30 MHz to 300 MHz	10 m to 1 m	Very high frequency (VHF)
300 MHz to 3 GHz	1 m to 10 cm	Ultra high frequency (UHF)
3 GHz to 30 GHz	10 cm to 1 cm	Super high frequency (SHF)

1.2 Electromagnetic waves

As with light, radio waves propagate outwards from a source of energy (transmitter) and comprise electric (E) and magnetic (H) fields at right angles to one another. These two components, the **E-field** and the **H-field**, are inseparable. The resulting wave travels away from the source with the E and H lines mutually at right angles to the direction of **propagation**, as shown in Figure 1.2.

Radio waves are said to be **polarised** in the plane of the electric (E) field. Thus, if the E-field is vertical, the signal is said to be vertically polarised whereas, if the E-field is horizontal, the signal is said to be horizontally polarised.

Figure 1.3 shows the electric E-field lines in the space between a transmitter and a receiver. The transmitter aerial (a simple dipole, see page 16) is supplied with a high frequency alternating current. This gives rise to an alternating electric field between the ends of the aerial and an alternating magnetic field around (and at right angles to) it.

The direction of the E-field lines is reversed on each cycle of the signal as the **wavefront** moves outwards from the source. The receiving aerial intercepts the moving field and voltage and current is induced in it as a consequence. This voltage and current is similar (but of smaller amplitude) to that produced by the transmitter.

Note that in Figure 1.3 (where the transmitter and receiver are close together) the field is shown

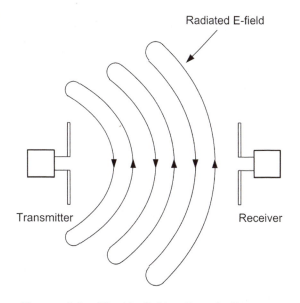

Figure 1.3   Electric field pattern in the near field region between a transmitter and a receiver (the magnetic field has not been shown but is perpendicular to the electric field)

spreading out in a spherical pattern (this is known more correctly as the **near field**). In practice there will be some considerable distance between the transmitter and the receiver and so the wave that reaches the receiving antenna will have a plane wavefront. In this far field region the angular field distribution is essentially independent of the distance from the transmitting antenna.

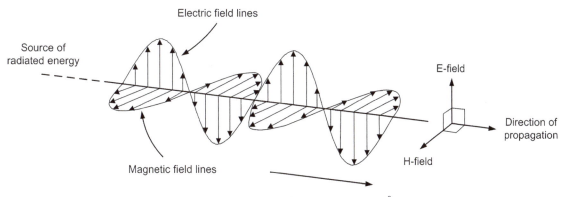

Figure 1.2   An electromagnetic wave

1.3 Frequency and wavelength

Radio waves propagate in air (or space) at the **speed of light** (300 million metres per second). The velocity of propagation, v, wavelength, λ, and frequency, f, of a radio wave are related by the equation:

$$v = f\lambda = 3 \times 10^8 \text{ m/s}$$

This equation can be arranged to make f or λ the subject, as follows:

$$f = \frac{3 \times 10^8}{\lambda} \text{ Hz} \quad \text{and} \quad \lambda = \frac{3 \times 10^8}{f} \text{ m}$$

As an example, a signal at a frequency of 1 MHz will have a wavelength of 300 m whereas a signal at a frequency of 10 MHz will have a wavelength of 30 m.

When a radio wave travels in a cable (rather than in air or 'free space') it usually travels at a speed that is between 60% and 80% of that of the speed of light.

Example 1.3.1

Determine the frequency of a radio signal that has a wavelength of 15 m.

Here we will use the formula $f = \dfrac{3 \times 10^8}{\lambda}$ Hz

Putting $\lambda = 15$ m gives:

$$f = \frac{3 \times 10^8}{15} = \frac{300 \times 10^6}{15} = 20 \times 10^6 \text{ Hz or 20 MHz}$$

Example 1.3.2

Determine the wavelength of a radio signal that has a frequency of 150 MHz.

In this case we will use $\lambda = \dfrac{3 \times 10^8}{f}$ m

Putting $f = 150$ MHz gives:

$$\lambda = \frac{3 \times 10^8}{f} = \frac{3 \times 10^8}{150 \times 10^6} = \frac{300 \times 10^6}{150 \times 10^6} = 2 \text{ m}$$

Example 1.3.3

If the wavelength of a 30 MHz signal in a cable is 8 m, determine the velocity of propagation of the wave in the cable.

Solution

Using the formula $v = f\lambda$, where v is the velocity of propagation in the cable, gives:

$$v = f\lambda = 30 \times 10^6 \times 8 \text{ m} = 240 \times 10^6 = 2.4 \times 10^8 \text{m/s}$$

Test your understanding 1.1

An HF communications signal has a frequency of 25.674 MHz. Determine the wavelength of the signal.

Test your understanding 1.2

A VHF communications link operates at a wavelength of 1.2 m. Determine the frequency at which the link operates.

1.4 The atmosphere

The earth's atmosphere (see Figure 1.4) can be divided into five concentric regions having boundaries that are not clearly defined. These layers, starting with the layer nearest the earth's surface, are known as the troposphere, stratosphere, mesosphere, thermosphere and exosphere.

The boundary between the troposphere and the stratosphere is known as the **tropopause** and this region varies in height above the earth's surface from about 7.5 km at the poles to 18 km at the equator. An average value for the height of the tropopause is around 11 km or 36,000 feet (about the same as the cruising height for most international passenger aircraft).

The thermosphere and the upper parts of the mesosphere are often referred to as the **ionosphere** and it is this region that has a major role to play in the long distance propagation of radio waves, as we shall see later.

The lowest part of the earth's atmosphere is called the **troposphere** and it extends from the surface up to about 10 km (6 miles). The atmosphere above 10 km is called the stratosphere, followed by the mesosphere. It is in the stratosphere that incoming solar radiation creates the ozone layer.

1.5 Radio wave propagation

Depending on a number of complex factors, radio waves can propagate through the atmosphere in various ways, as shown in Figure 1.5. These include:

- ground waves
- ionospheric waves
- space waves
- tropospheric waves.

As their name suggests, **ground waves** (or **surface waves**) travel close to the surface of the earth and propagate for relatively short distances at HF and VHF but for much greater distances at MF and LF. For example, at 100 kHz the range of a ground wave might be in excess of 500 km, whilst at 1 MHz (using the same radiated power) the range might be no more than 150 km and at 10 MHz no more than about 15 km. Ground waves have two basic components; a **direct wave** and a **ground reflected wave** (as shown in Figure 1.6). The **direct path** is that which exists on a

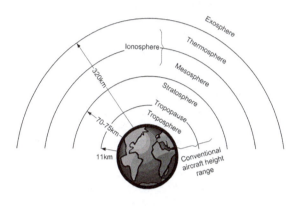

Figure 1.4   Zones of the atmosphere

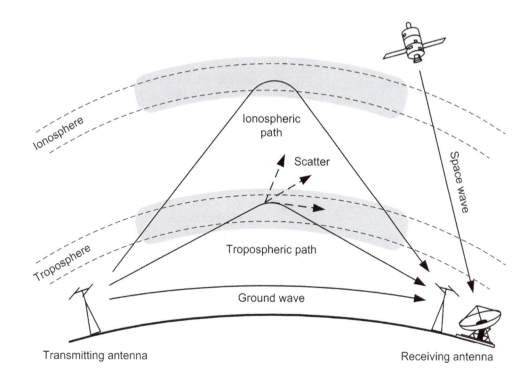

Figure 1.5   Radio wave propagation through the atmosphere

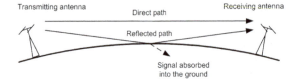

Figure 1.6 Constituents of a ground wave

line-of-sight (**LOS**) basis between the transmitter and receiver. An example of the use of a direct path is that which is used by terrestrial microwave repeater stations which are typically spaced 20 to 30 km apart on a line-of-sight basis. Another example of the direct path is that used for satellite TV reception. In order to receive signals from the satellite the receiving antenna must be able to 'see' the satellite. In this case, and since the wave travels largely undeviated through the atmosphere, the direct wave is often referred to as a **space wave**. Such waves travel over LOS paths at VHF, UHF and beyond.

As shown in Figure 1.6, signals can arrive at a receiving antenna by both the direct path and by means of reflection from the ground. **Ground reflection** depends very much on the quality of the ground with sandy soils being a poor reflector of radio signals and flat marshy ground being an excellent reflecting surface. Note that a proportion of the incident radio signal is absorbed into the ground and not all of it is usefully reflected. An example of the use of a mixture of direct path and ground (or building) reflected radio signals is the reception of FM broadcast signals in a car. It is also worth mentioning that, in many cases, the reflected signals can be stronger than the direct path (or the direct path may not exist at all if the car happens to be in a heavily built-up area).

Ionospheric waves (or **sky waves**) can travel for long distances at MF, HF and exceptionally also at VHF under certain conditions. Such waves are predominant at frequencies below VHF and we shall examine this phenomenon in greater detail a little later but before we do it is worth describing what can happen when waves meet certain types of discontinuity in the atmosphere or when they encounter a physical obstruction. In both cases, the direction of travel can be significantly affected according to the nature and

size of the obstruction or discontinuity. Four different effects can occur (see Figure 1.7) and they are known as:

- reflection
- refraction
- diffraction
- scattering.

Reflection occurs when a plane wave meets a plane object that is large relative to the wavelength of the signal. In such cases the wave is reflected back with minimal distortion and without any change in velocity. The effect is similar to the reflection of a beam of light when it arrives at a mirrored surface.

Refraction occurs when a wave moves from one medium into another in which it travels at a different speed. For example, when moving from a more dense to a less dense medium the wave is bent away from the normal (i.e. an imaginary line constructed at right angles to the boundary). Conversely, when moving from a less dense to a more dense medium, a wave will bend towards the normal. The effect is similar to that experienced by a beam of light when it encounters a glass prism.

Diffraction occurs when a wave meets an edge (i.e. a sudden impenetrable surface discontinuity) which has dimensions that are large relative to the wavelength of the signal. In such cases the wave is bent so that it follows the profile of the discontinuity. Diffraction occurs more readily at lower frequencies (typically VHF and below). An example of diffraction is the bending experienced by VHF broadcast signals when they encounter a sharply defined mountain ridge. Such signals can be received at some distance beyond the 'knife edge' even though they are well beyond the normal LOS range.

Scattering occurs when a wave encounters one or more objects in its path having a size that is a fraction of the wavelength of the signal. When a wave encounters an obstruction of this type it will become fragmented and re-radiated over a wide angle. Scattering occurs more readily at higher frequencies (typically VHF and above) and regularly occurs in the troposphere at UHF and EHF.

Radio signals can also be directed upwards (by suitable choice of antenna) so that signals enter

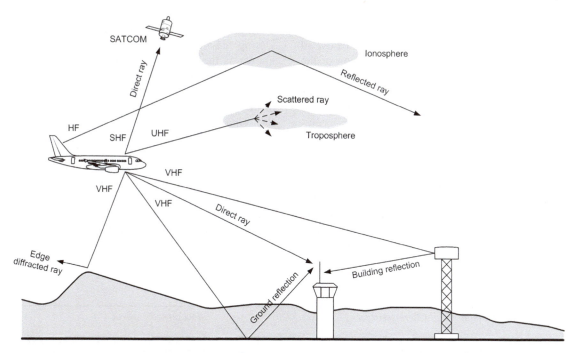

Figure 1.7 Various propagation effects

the troposphere or ionosphere. In the former case, signals can be become scattered (i.e. partially dispersed) in the troposphere so that a small proportion arrives back at the ground. **Tropospheric scatter** requires high power transmitting equipment and high gain antennas but is regularly used for transmission beyond the horizon particularly where conditions in the troposphere (i.e. rapid changes of temperature and humidity with height) can support this mode of communication. Tropospheric scatter of radio waves is analogous to the scattering of a light beam (e.g. a torch or car headlights) when shone into a heavy fog or mist.

In addition to tropospheric scatter there is also **tropospheric ducting** (not shown in Figure 1.7) in which radio signals can become trapped as a result of the change of refractive index at a boundary between air masses having different temperature and humidity. Ducting usually occurs when a large mass of cold air is overrun by warm air (this is referred to as a temperature inversion). Although this condition may occur frequently in certain parts of the world, this mode of propagation is not very predictable and is therefore not used for any practical applications.

1.6 The ionosphere

In 1924, Sir Edward Appleton was one of the first to demonstrate the existence of a reflecting layer at a height of about 100 km (now called the E-layer). This was soon followed by the discovery of another layer at around 250 km (now called the F-layer). This was achieved by broadcasting a continuous signal from one site and receiving the signal at a second site several miles away. By measuring the time difference between the signal received along the ground and the signal reflected from the atmosphere (and knowing the velocity at which the radio wave propagates) it was possible to calculate the height of the atmospheric reflecting layer. Today, the standard technique for detecting the presence of ionised layers (and determining their height above the surface of the earth) is to transmit a very short pulse directed upwards into space and accurately measuring the amplitude and time delay before the arrival back on earth of the reflected pulses. This **ionospheric sounding** is carried out over a range of frequencies.

The ionosphere provides us with a reasonably predictable means of communicating over long distances using HF radio signals. Much of the short and long distance communications below 30 MHz depend on the bending or refraction of the transmitted wave in the earth's ionosphere which are regions of ionisation caused by the sun's ultraviolet radiation and lying about 60 to 200 miles above the earth's surface.

The useful regions of ionisation are the E-layer (at about 70 miles in height for maximum ionisation) and the F-layer (lying at about 175 miles in height at night). During the daylight hours, the F-layer splits into two distinguishable parts: F_1 (lying at a height of about 140 miles) and F_2 (lying at a height of about 200 miles). After sunset the F_1- and F_2-layers recombine into

a single F-layer (see Figures 1.8 and 1.10). During daylight, a lower layer of ionisation known as the D-layer exists in proportion to the sun's height, peaking at local noon and largely dissipating after sunset. This lower layer primarily acts to absorb energy in the low end of the high frequency (HF) band. The F-layer ionisation regions are primarily responsible for long distance communication using **sky waves** at distances of up to several thousand km (greatly in excess of those distances that can be achieved using VHF **direct wave** communication, see Figure 1.9). The characteristics of the ionised layers are summarised in Table 1.2 together with their effect on radio waves.

Table 1.2 Ionospheric layers

Layer	Height (km)	Characteristics	Effect on radio waves
D	50 to 95 km	Develops shortly after sunrise and disappears shortly after sunset. Reaches maximum ionisation when the sun is at its highest point in the sky	Responsible for the absorption of radio waves at lower frequencies (e.g. below 4 MHz) during daylight hours
E	95 to 150 km	Develops shortly after sunrise and disappears a few hours after sunset. The maximum ionisation of this layer occurs at around midday	Reflects waves having frequencies less than 5 MHz but tends to absorb radio signals above this frequency
E_s	80 to 120 km	An intense region of ionisation that sometimes appears in the summer months (peaking in June and July). Usually lasts for only a few hours (often in the late morning and recurring in the early evening of the same day)	Highly reflective at frequencies above 30 MHz and up to 300 MHz on some occasions. Of no practical use other than as a means of long distance VHF communication for radio amateurs
F	250 to 450 km	Appears a few hours after sunset, when the F_1- and F_2-layers (see below) merge to form a single layer	Reflects radio waves up to 20 MHz and occasionally up to 25 MHz
F_1	150 to 200 km	Occurs during daylight hours with maximum ionisation reached at around midday. The F_1-layer merges with the F_2-layer shortly after sunset	Reflects radio waves in the low HF spectrum up to about 10 MHz
F_2	250 to 450 km	Develops just before sunrise as the F-layer begins to divide. Maximum ionisation of the F_2-layer is usually reached one hour after sunrise and it typically remains at this level until shortly after sunset. The intensity of ionisation varies greatly according to the time of day and season and is also greatly affected by solar activity	Capable of reflecting radio waves in the upper HF spectrum with frequencies of up to 30 MHz and beyond during periods of intense solar activity (i.e. at the peak of each 11-year sunspot cycle)

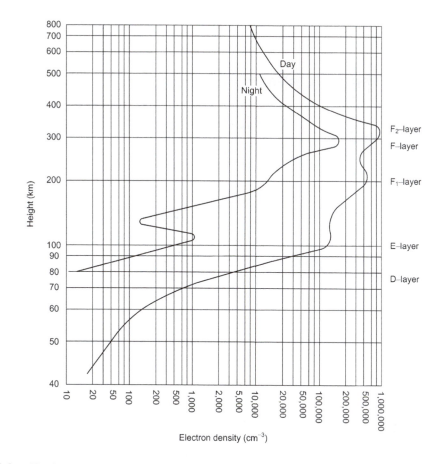

Figure 1.8 Typical variation of electron density versus height (note the use of logarithmic scales for both height and electron density)

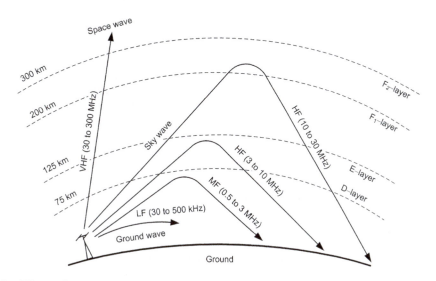
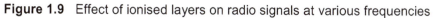

Figure 1.9 Effect of ionised layers on radio signals at various frequencies

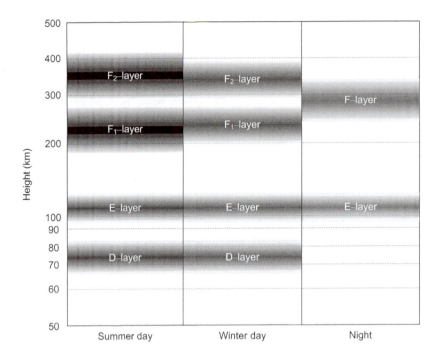

Figure 1.10 Position of ionised layers at day and night

1.7 MUF and LUF

The **maximum usable frequency** (**MUF**) is the highest frequency that will allow communication over a given path at a particular time and on a particular date. MUF varies considerably with the amount of solar activity and is basically a function of the height and intensity of the F-layer. During a period of intense solar activity the MUF can exceed 30 MHz during daylight hours but is often around 16 to 20 MHz by day and around 8 to 10 MHz by night.

The variation of MUF over a 24-hour period for the London to New York path is shown in Figure 1.11. A similar plot for the summer months would be flatter with a more gradual increase in MUF at dawn and a more gradual decline at dusk.

The reason for the significant variation of MUF over any 24-hour period is that the intensity of ionisation in the upper atmosphere is significantly reduced at night and, as a consequence, lower frequencies have to be used to produce the same amount of refractive bending and also to give the same critical angle and skip distance as by day.

Fortunately, the attenuation experienced by lower frequencies travelling in the ionosphere is much reduced at night and this makes it possible to use the lower frequencies required for effective communication. The important fact to remember from this is simply that, for a given path, the frequency used at night is about half that used for daytime communication.

The **lowest usable frequency** (**LUF**) is the lowest frequency that will support communication over a given path at a particular time and on a particular date. LUF is dependent on the amount of absorption experienced by a radio wave. This absorption is worse when the D-layer is most intense (i.e. during daylight). Hence, as with MUF, the LUF rises during the day and falls during the night. A typical value of LUF is 4 to 6 MHz during the day, falling rapidly at sunset to 2 MHz.

The frequency chosen for HF communication must therefore be somewhere above the LUF and below the MUF for a given path, day and time. A typical example might be a working frequency of 5 MHz at a time when the MUF is 10 MHz and the LUF is 2 MHz.

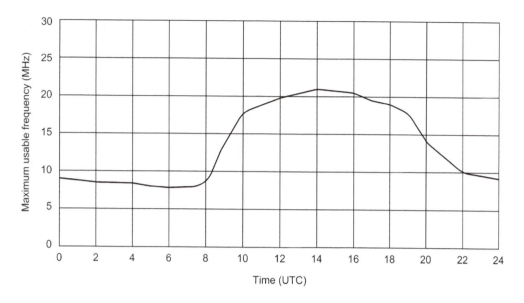

Figure 1.11   Variation of MUF with time for London–New York on 16th October 2006

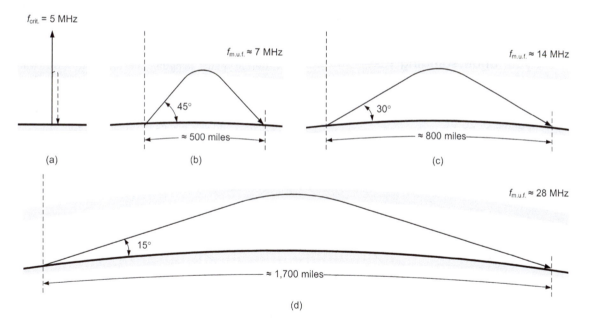

Figure 1.12   Effect of angle of attack on range and MUF

Figure 1.12 shows the typical MUF for various angles of attack together with the corresponding working ranges. This diagram assumes a **critical frequency** of 5 MHz. This is the lowest frequency that would be returned from the ionosphere using a path of vertical incidence (see ionospheric sounding on page 7).

The relationship between the critical frequency, $f_{crit.}$, and electron density, N, is given by:

$$f_{crit.} = 9 \times 10^{-3} \times \sqrt{N}$$

where N is the electron density expressed in cm^3.

The angle of attack, α, is the angle of the transmitted wave relative to the horizon.

The relationship between the MUF, $f_{m.u.f.}$, the critical frequency, $f_{crit.}$, and the angle of attack, α, is given by:

$$f_{m.u.f.} = \frac{f_{crit.}}{\sin \alpha}$$

Example 1.8.1

Given that the electron density in the ionosphere is 5×10^5 electrons per cm^3, determine the critical frequency and the MUF for an angle of attack of 15°.

Now using the relationship $f_{crit.} = 9 \times 10^{-3} \times \sqrt{N}$ gives:

$$f_{crit.} = 9 \times 10^{-3} \times \sqrt{5 \times 10^5} = 6.364 \text{ MHz}$$

The MUF can now be calculated using:

$$f_{m.u.f.} = \frac{f_{crit.}}{\sin \alpha} = \frac{6.364}{\sin 15°} = \frac{6.364}{0.259} = 24.57 \text{ MHz}$$

Test your understanding 1.3

Determine the electron density in the ionosphere when the MUF is 18 MHz for a critical angle of 20°.

1.8 Silent zone and skip distance

The **silent zone** is simply the region that exists between the extent of the coverage of the ground wave signal and the point at which the sky wave returns to earth (see Figure 1.13). Note also that, depending on local topography and soil characteristics, when a signal returns to earth from the ionosphere it is sometimes possible for it to experience a reflection from the ground, as shown in Figure 1.13. The onward reflected signal will suffer attenuation but in some circumstances may be sufficient to provide a further hop and an approximate doubling of the working range. The condition is known as **multi-hop** propagation.

The **skip distance** is simply the distance between the point at which the sky wave is radiated and the point at which it returns to earth (see Figure 1.14).

Note that where signals are received simultaneously by ground wave and sky wave paths, the signals will combine both constructively and destructively due to the different paths lengths and this, in turn, will produce an effect known as **fading**. This effect can often be heard during the early evening on medium wave radio signals as the D-layer weakens and sky waves first begin to appear.

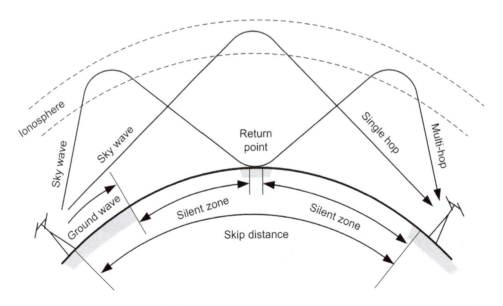

Figure 1.13 Silent zone and skip distance

Table 1.3 See Test your understanding 1.4

Time (UTC)	0	1	2	3	4	5	6	7	8	9	10	11	12
MUF (MHz)	9.7	9.0	8.4	7.9	7.5	9.6	12.3	14.1	15.4	16.4	17.1	17.6	17.9

Time (UTC)	13	14	15	16	17	18	19	20	21	22	23	24
MUF (MHz)	18.1	18.1	17.9	17.7	17.2	16.5	15.6	14.2	12.7	11.5	10.6	9.7

Test your understanding 1.4

Table 1.3 shows corresponding values of time and maximum usable frequency (MUF) for London to Lisbon on 28th August 2006. Plot a graph showing the variation of MUF with time and explain the shape of the graph.

Test your understanding 1.5

Explain the following terms in relation to HF radio propagation:

(a) silent zone
(b) skip distance
(c) multi-hop propagation.

1.9 Multiple choice questions

1. A transmitted radio wave will have a plane wavefront:
   (a) in the near field
   (b) in the far field
   (c) close to the antenna.

2. The lowest layer in the earth's atmosphere is:
   (a) the ionosphere
   (b) the stratosphere
   (c) the troposphere.

3. A radio wave at 115 kHz is most likely to propagate as:
   (a) a ground wave
   (b) a sky wave
   (c) a space wave.

4. The height of the E-layer is approximately:
   (a) 100 km
   (b) 200 km
   (c) 400 km.

5. When a large mass of cold air is overrun by warm air the temperature inversion produced will often result in:
   (a) ionospheric reflection
   (b) stratospheric refraction
   (c) tropospheric ducting.

6. Ionospheric sounding is used to determine:
   (a) the maximum distance that a ground wave will travel
   (b) the presence of temperature inversions in the upper atmosphere
   (c) the critical angle and maximum usable frequency for a given path.

7. The critical frequency is directly proportional to:
   (a) the electron density
   (b) the square of the electron density
   (c) the square root of the electron density.

8. The MF range extends from:
   (a) 300 kHz to 3 MHz
   (b) 3 MHz to 30 MHz
   (c) 30 MHz to 300 MHz.

9. A radio wave has a frequency of 15 MHz. Which one of the following gives the wavelength of the wave?
   (a) 2 m
   (b) 15 m
   (c) 20 m.

10. Which one of the following gives the velocity at which a radio wave propagates?
    (a) 300 m/s
    (b) 3×10^8 m/s
    (c) 3 million m/s.

11. The main cause of ionisation in the upper atmosphere is:
    (a) solar radiation
    (b) ozone
    (c) currents of warm air.

12. The F_2-layer is:
    (a) higher at the equator than at the poles
    (b) lower at the equator than at the poles
    (c) the same height at the equator as at the poles.

13. The free-space path loss experienced by a radio wave:
    (a) increases the frequency but decreases with distance
    (b) decreases with frequency but increases with distance
    (c) increases with both frequency and distance.

14. For a given HF radio path, the MUF changes most rapidly at:
    (a) mid-day
    (b) mid-night
    (c) dawn and dusk.

15. Radio waves tend to propagate mainly as line-of-sight signals in the:
    (a) MF band
    (b) HF band
    (c) VHF band.

16. In the HF band radio waves tend to propagate over long distances as:
    (a) ground waves
    (b) space waves
    (c) ionospheric waves.

17. The maximum distance that can be achieved from a single-hop reflection from the F-layer is in the region:
    (a) 500 to 2,000 km
    (b) 2,000 to 3,500 km
    (c) 3,500 to 5,000 km.

18. The F_1- and F_2-layers combine:
    (a) only at about mid-day
    (b) during the day
    (c) during the night.

19. The path of a VHF or UHF radio wave can be bent by a sharply defined obstruction such as a building or a mountain top. This phenomenon is known as:
    (a) ducting
    (b) reflection
    (c) diffraction.

20. Radio waves at HF can be subject to reflections in ionised regions of the upper atmosphere. This phenomenon is known as:
    (a) ionospheric reflection
    (b) tropospheric scatter
    (c) atmospheric ducting.

21. Radio waves at UHF can sometimes be subject to dispersion over a wide angle in regions of humid air in the atmosphere. This phenomenon is known as:
    (a) ionospheric reflection
    (b) tropospheric scatter
    (c) atmospheric ducting.

22. Radio waves at VHF and UHF can sometimes propagate for long distances in the lower atmosphere due to the presence of a temperature inversion. This phenomenon is known as:
    (a) ionospheric reflection
    (b) tropospheric scatter
    (c) atmospheric ducting.

23. The layer in the atmosphere that is mainly responsible for the absorption of MF radio waves during the day is:
    (a) the D-layer
    (b) the E-layer
    (c) the F-layer.

24. The layer in the atmosphere that is mainly responsible for the reflection of HF radio waves during the day is:
    (a) the D-layer
    (b) the E-layer
    (c) the F-layer.

Chapter 2 Antennas

It may not be apparent from an inspection of the external profile of an aircraft that most large aircraft carry several dozen antennas of different types. To illustrate this point, Figure 2.1 shows just a few of the antennas carried by a Boeing 757. What should be apparent from this is that many of the antennas are of the low profile variety which is essential to reduce drag.

Antennas are used both for transmission and reception. A transmitting antenna converts the high frequency electrical energy supplied to it into electromagnetic energy which is launched or radiated into the space surrounding the antenna. A receiving antenna captures the electromagnetic energy in the surrounding space and converts this into high frequency electrical energy which is then passed on to the receiving system. The **law of reciprocity** indicates that an antenna will have the same gain and directional properties when used for transmission as it does when used for reception.

2.1 The isotropic radiator

The most fundamental form of antenna (which cannot be realised in practice) is the isotropic radiator. This theoretical type of antenna is often used for comparison purposes and as a reference when calculating the gain and directional characteristics of a real antenna.

Isotropic antennas radiate uniformly in all directions. In other words, when placed at the centre of a sphere such an antenna would illuminate the internal surface of the sphere uniformly as shown in Figure 2.2(a). All practical antennas have directional characteristics as illustrated in Figure 2.2(b). Furthermore, such characteristics may be more or less pronounced according to the antenna's application. We shall look at antenna gain and directivity in more detail later on but before we do that we shall introduce you to some common types of antenna.

Figure 2.1   Some of the antennas fitted to a Boeing 757 aircraft. 1, VOR; 2, HF comms.; 3, 5 and 6, VHF comms.; 4, ADF; 7, TCAS (upper); 8, weather radar

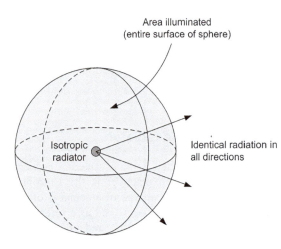

Area illuminated
(entire surface of sphere)

Isotropic
radiator

Identical radiation in
all directions

(a) Isotropic radiator

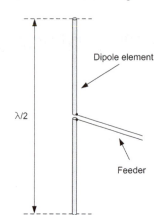

Dipole element

λ/2

Feeder

Figure 2.3   A half-wave dipole antenna

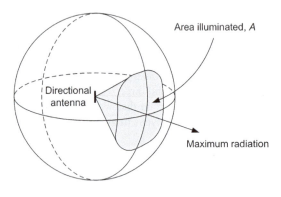

Area illuminated, A

Directional
antenna

Maximum radiation

(b) Directional antenna

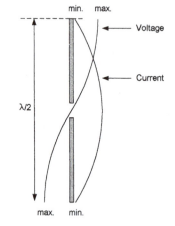

min.   max.

Voltage

Current

λ/2

max.   min.

Figure 2.2   The directional characteristics of isotropic radiators and directional antennas

Figure 2.4   Voltage and current distribution in the half-wave dipole antenna

2.2 The half-wave dipole

The **half-wave dipole** is one of the most fundamental types of antenna. The half-wave dipole antenna (Figure 2.3) consists of a single conductor having a length equal to one-half of the length of the wave being transmitted or received. The conductor is then split in the centre to enable connection to the feeder. In practice, because of the capacitance effects between the ends of the antenna and ground, the antenna is cut a little shorter than a half wavelength.

The length of the antenna (from end to end) is equal to one half wavelength, hence:

$$l = \frac{\lambda}{2}$$

Now since $v = f \times \lambda$ we can conclude that, for a half-wave dipole:

$$l = \frac{v}{2f}$$

Note that l is the **electrical length** of the antenna rather than its actual **physical length**. End effects,

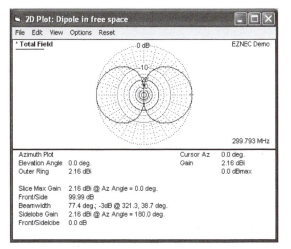

Figure 2.5 E-field polar radiation pattern for a half-wave dipole

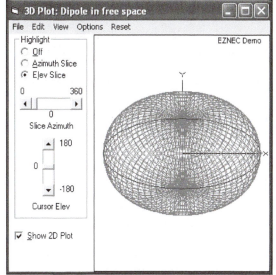

Figure 2.7 3D polar radiation pattern for a half-wave dipole (note the 'doughnut' shape)

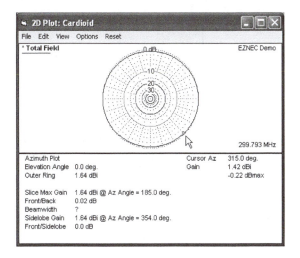

Figure 2.6 H-field polar radiation pattern for a half-wave dipole

or capacitance effects at the ends of the antenna require that we reduce the actual length of the aerial and a 5% reduction in length is typically required for an aerial to be resonant at the centre of its designed tuning range.

Figure 2.4 shows the distribution of current and voltage along the length of a half-wave dipole aerial. The current is maximum at the centre and zero at the ends. The voltage is zero at the centre and maximum at the ends. This

implies that the impedance is not constant along the length of aerial but varies from a maximum at the ends (maximum voltage, minimum current) to a minimum at the centre.

The dipole antenna has directional properties illustrated in Figures 2.5 to 2.7. Figure 2.5 shows the radiation pattern of the antenna in the plane of the antenna's electric field (i.e. the E-field plane) whilst Figure 2.6 shows the radiation pattern in the plane of the antenna's magnetic field (i.e. the H-field plane).

The 3D plot shown in Figure 2.7 combines these two plots into a single 'doughnut' shape. Things to note from these three diagrams are that:

- in the case of Figure 2.5 minimum radiation occurs along the axis of the antenna whilst the two zones of maximum radiation are at 90° (i.e. are 'normal to') the dipole elements
- in the case of Figure 2.6 the antenna radiates uniformly in all directions.

Hence, a vertical dipole will have **omnidirectional** characteristics whilst a horizontal dipole will have a **bi-directional** radiation pattern. This is an important point as we shall see later.

Example 2.1

Determine the length of a half-wave dipole antenna for use at a frequency of 150 MHz.

The length of a half-wave dipole for 150 MHz can be determined from:

$$l = \frac{v}{2f}$$

where $v = 3 \times 10^8$ m/s and $f = 150 \times 10^6$ Hz.

Hence:

$$l = \frac{v}{2f} = \frac{3 \times 10^8}{2 \times 150 \times 10^6} = \frac{3 \times 10^8}{300 \times 10^6} = \frac{3 \times 10^6}{3 \times 10^6} = 1 \text{ m}$$

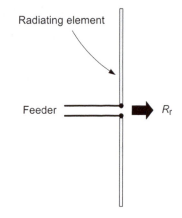

Figure 2.8   Radiation resistance

Test your understanding 2.1

Determine the length of a half-wave dipole for frequencies of (a) 121 MHz and (b) 11.25 MHz.

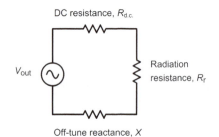

Figure 2.9   Equivalent circuit of an antenna

2.3 Impedance and radiation resistance

Because voltage and current appear in an antenna (a minute voltage and current in the case of a receiving antenna and a much larger voltage and current in the case of a transmitting antenna) an aerial is said to have **impedance**. Here it's worth remembering that impedance is a mixture of resistance, R, and reactance, X, both measured in ohms (Ω). Of these two quantities, X varies with frequency whilst R remains constant. This is an important concept because it explains why antennas are often designed for operation over a restricted range of frequencies.

The impedance, Z, of an aerial is the ratio of the voltage, E, across its terminals to the current, I, flowing in it. Hence:

$$Z = \frac{E}{I} \ \Omega$$

You might infer from Figure 2.7 that the impedance at the centre of the half-wave dipole should be zero. In practice the impedance is usually between 70 Ω and 75 Ω. Furthermore, at resonance the impedance is purely resistive and contains no reactive component (i.e. inductance

and capacitance). In this case X is negligible compared with R. It is also worth noting that the DC resistance (or **ohmic resistance**) of an antenna is usually very small in comparison with its impedance and so it may be ignored. Ignoring the DC resistance of the antenna, the impedance of an antenna may be regarded as its **radiation resistance**, R_r (see Figure 2.8).

Radiation resistance is important because it is through this resistance that electrical power is transformed into radiated electromagnetic energy (in the case of a transmitting antenna) and incident electromagnetic energy is transformed into electrical power (in the case of a receiving aerial).

The equivalent circuit of an antenna is shown in Figure 2.9. The three series-connected components that make up the antenna's impedance are:

- the DC resistance, $R_{d.c.}$
- the radiation resistance, R_r
- the 'off-tune' reactance, X.

Note that when the antenna is operated at a frequency that lies in the centre of its pass-band (i.e. when it is *on-tune*) the off-tune reactance is zero. It is also worth bearing in mind that the radiation resistance of a half-wave dipole varies according to its height above ground. The 70 Ω to 75 Ω impedance normally associated with a half-wave dipole is only realized when the antenna is mounted at an elevation of 0.2 wavelengths, or more.

Test your understanding 2.2

A half-wave dipole is operated at its centre frequency (zero off-tune reactance). If the antenna has a total DC loss resistance of 2.5 Ω and is supplied with a current of 2 A and a voltage of 25 V, determine:

(a) the radiation resistance of the antenna
(b) the power loss in the antenna.

2.4 Radiated power and efficiency

In the case of a transmitting antenna, the radiated power, P_r, produced by the antenna is given by:

$$P_r = I_a^2 \times R_r \quad W$$

where I_a is the antenna current, in amperes, and R_r is the radiation resistance in ohms. In most practical applications it is important to ensure that P_r is maximised and this is achieved by ensuring that R_r is much larger than the DC resistance of the antenna elements.

The efficiency of an antenna is given by the relationship:

$$\text{Radiation efficiency} = \frac{P_r}{P_r + P_{loss}} \times 100\%$$

Where P_{loss} is the power dissipated in the DC resistance present. At this point it is worth stating that whilst efficiency is vitally important in the case of a transmitting antenna it is generally unimportant in the case of a receiving antenna. This explains why a random length of wire can make a good receiving aerial but not a good transmitting antenna!

Example 2.2

An HF transmitting antenna has a radiation resistance of 12 Ω. If a current of 0.5 A is supplied to it, determine the radiated power.

Now:
$$P_r = I_a^2 \times R_r = (0.5)^2 \times 12 = 0.25 \times 12 = 4 \text{ W}$$

Example 2.3

If the aerial in Example 2.2 has a DC resistance of 2 Ω, determine the power loss and the radiation efficiency of the antenna.

From the equivalent circuit shown in Figure 2.9, the same current flows in the DC resistance, $R_{d.c.}$, as flows in the antenna's radiation resistance, R_r.

Hence $I_a = 0.5$ A and $R_{d.c.} = 2$ W

Since $P_{loss} = I_a^2 \times R_{d.c.}$

$$P_{loss} = (0.5)^2 \times 2 = 0.25 \times 2 = 0.5 \text{ W}$$

The radiation efficiency is given by:

$$\text{Radiation efficiency} = \frac{P_r}{P_r + P_{loss}} \times 100\%$$

$$= \frac{4}{4 + 0.5} \times 100\% = \frac{4}{4.5} \times 100\% = 89\%$$

In this example, more than 10% of the power output is actually wasted! It is also worth noting that in order to ensure a high value of radiation efficiency, the loss resistance must be kept very low in comparison with the radiation resistance.

2.5 Antenna gain

The field strength produced by an antenna is proportional to the amount of current flowing in it. However, since different types of antenna produce different values of field strength for the same applied RF power level, we attribute a power gain to the antenna. This power gain is specified in relation to a **reference antenna** (often either a half-wave dipole or a theoretical isotropic radiator) and it is usually specified in decibels (dB)—see Appendix 2.

In order to distinguish between the two types of reference antenna we use subscripts **i** and **d** to denote isotropic and half-wave dipole reference antennas respectively. As an example, an aerial having a gain of 10 dB$_i$ produces ten times power gain when compared with a theoretical isotropic radiator. Similarly, an antenna having a gain of 13 dB$_d$ produces twenty times power gain when compared with a half-wave dipole. Putting this another way, to maintain the same field strength at a given point, you would have to apply 20 W to a half-wave dipole or just 1 W to the antenna in question! Some comparative values of antenna gain are shown on page 28.

2.6 The Yagi beam antenna

Originally invented by two Japanese engineers, Yagi and Uda, the Yagi antenna has remained extremely popular in a wide variety of applications and, in particular, for fixed domestic FM radio and TV receiving aerials. In order to explain in simple terms how the Yagi antenna works we shall use a simple light analogy.

An ordinary filament lamp radiates light in all directions. Just like an antenna, the lamp converts electrical energy into electromagnetic energy. The only real difference is that we can *see* the energy that it produces!

The action of the filament lamp is comparable with our fundamental dipole antenna. In the case of the dipole, electromagnetic radiation will occur all around the dipole elements (in three dimensions the radiation pattern will take on a doughnut shape). In the plane that we have shown in Figure 2.10(c), the directional pattern will be a figure-of-eight that has two lobes of equal size. In order to concentrate the radiation into just one of the radiation lobes we could simply place a reflecting mirror on one side of the filament lamp. The radiation will be reflected (during which the reflected light will undergo a 180° phase change) and this will reinforce the light on one side of the filament lamp. In order to achieve the same effect in our antenna system we need to place a conducting element about one quarter of a wavelength behind the dipole element. This element is referred to as a **reflector** and it is said to be 'parasitic' (i.e. it is not actually connected to

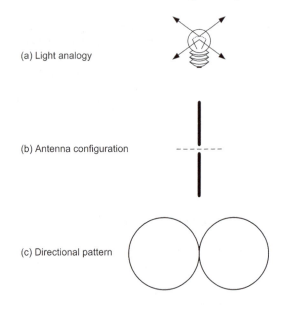

(a) Light analogy

(b) Antenna configuration

(c) Directional pattern

Figure 2.10 Dipole antenna light analogy

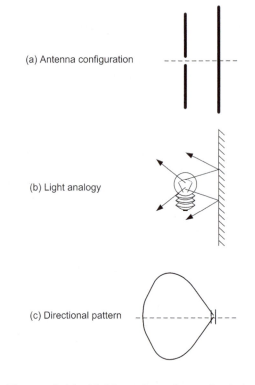

(a) Antenna configuration

(b) Light analogy

(c) Directional pattern

Figure 2.11 Light analogy for a dipole and reflector

the feeder). The reflector needs to be cut slightly longer than the driven dipole element. The resulting directional pattern will now only have one **major lobe** because the energy radiated will be concentrated into just one half of the figure-of-eight pattern that we started with).

Continuing with our optical analogy, in order to further concentrate the light energy into a narrow beam we can add a lens in front of the lamp. This will have the effect of bending the light emerging from the lamp towards the normal line (see Figure 2.12). In order to achieve the same effect in our antenna system we need to place a conducting element, known as a **director**, on the other side of the dipole and about one quarter of a wavelength from it. Once again, this element is parasitic but in this case it needs to be cut slightly shorter than the driven dipole element. The resulting directional pattern will now have a narrower major lobe as the energy becomes concentrated in the normal direction (i.e. at right angles to the dipole elements). The resulting antenna is known as a three-element Yagi aerial, see Figure 2.13.

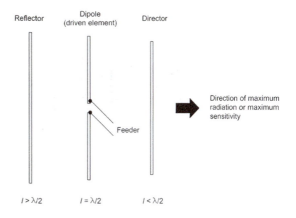

Figure 2.13 A three-element Yagi antenna

If desired, additional directors can be added to further increase the gain and reduce the **beamwidth** (i.e. the angle between the half-power or −3 dB power points on the polar characteristic) of Yagi aerials. Some comparative gain and beamwidth figures are shown on page 28.

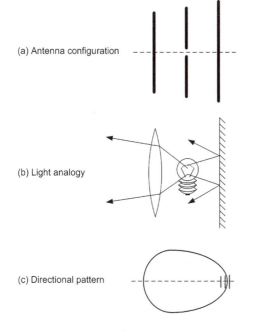

Figure 2.12 Light analogy for a dipole, reflector and director

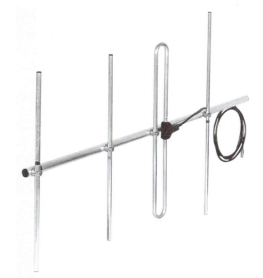

Figure 2.14 A four-element Yagi antenna (note how the dipole element has been 'folded' in order to increase its impedance and provide a better match to the 50 Ω feeder system)

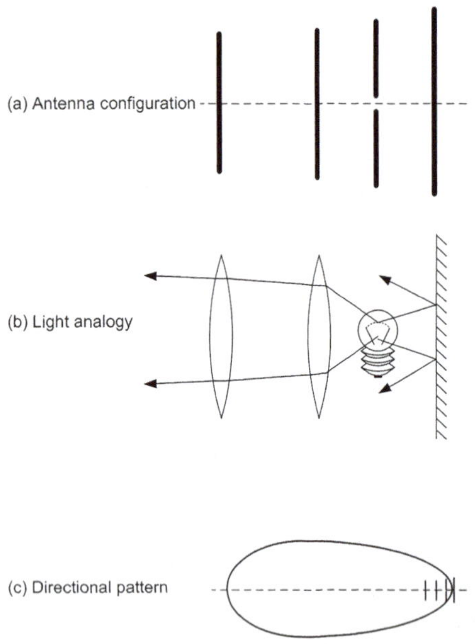

(a) Antenna configuration

(b) Light analogy

(c) Directional pattern

Figure 2.15 Light analogy for the four-element Yagi shown in Figure 2.14

2.7 Directional characteristics

Antenna gain is achieved at the expense of directional response. In other words, as the gain of an antenna increases its radiation pattern becomes more confined. In many cases this is a desirable effect (e.g. in the case of fixed point–point communications). In other cases (e.g. a base station for use with a number of mobile stations) it is clearly undesirable.

The directional characteristics of an antenna are usually presented in the form of a polar response graph. This diagram allows users to determine directions in which maximum and minimum gain can be achieved and allows the antenna to be positioned for optimum effect.

The polar diagram for a horizontal dipole is shown in Figure 2.16. Note that there are two major lobes in the response and two deep nulls. The antenna is thus said to be bi-directional.

Figure 2.17 shows the polar diagram for a dipole plus reflector. The radiation from this antenna is concentrated into a single major lobe and there is a single null in the response at 180° to the direction of maximum radiation.

An alternative to improving the gain but maintaining a reasonably wide beamwidth is that of stacking two antennas one above another (see Figure 2.20). Such an arrangement will usually provide a 3 dB gain over a single antenna but will have the same beamwidth. A disadvantage of stacked arrangements is that they require accurate phasing and matching arrangements.

As a rule of thumb, an increase in gain of 3 dB can be produced each time the number of elements is doubled. Thus a two-element antenna will offer a gain of about 3 dBd, a four-element antenna will produce 6 dBd, an eight-element Yagi will realise 9 dBd, and so on.

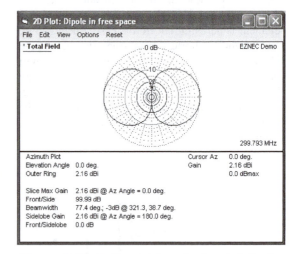

Figure 2.16 Polar plot for a horizontal dipole

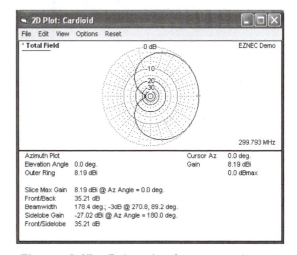

Figure 2.17 Polar plot for a two-element Yagi

Test your understanding 2.3

Identify the antenna shown in Figure 2.19. Sketch a typical horizontal radiation pattern for this antenna.

Test your understanding 2.4

Identify the antenna shown in Figure 2.20. Sketch a typical horizontal radiation pattern for this antenna.

Figure 2.19 See Test your understanding 2.3

Test your understanding 2.5

Figure 2.18 shows the polar response of a Yagi beam antenna (the gain has been specified relative to a standard reference dipole). Use the polar plot to determine:

(a) the gain of the antenna
(b) the beamwidth of the antenna
(c) the size and position of any 'side lobes'
(d) the 'front-to-back' ratio (i.e. the size of the major lobe in comparison to the response of the antenna at 180° to it).

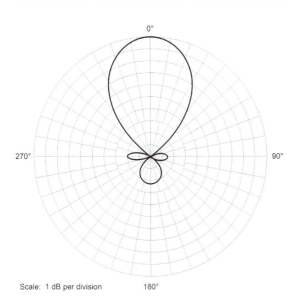

Scale: 1 dB per division   180°

Figure 2.18 Polar diagram for a Yagi beam antenna (see Test your understanding 2.5)

Figure 2.20 See Test your understanding 2.4

2.8 Other practical antennas

Many practical forms of antenna are used in aircraft and aviation-related applications. The following are some of the most common types (several other antennas will be introduced in later chapters).

2.8.1 Vertical quarter-wave antennas

One of the most simple antennas to construct is the quarter-wave antenna (also known as a **Marconi antenna**). Such antennas produce an omnidirectional radiation pattern in the horizontal plane and radiate vertically polarised signals. Practical quarter-wave antennas can be produced for the high-HF and VHF bands but their length is prohibitive for use on the low-HF and LF bands.

In order to produce a reasonably flat radiation pattern (and prevent maximum radiation being directed upwards into space) it is essential to incorporate an effective ground plane. At VHF, this can be achieved using just four quarter-wave radial elements at 90° to the vertical radiating element (see Figure 2.21). All four radials are grounded at the feed-point to the outer screen of the coaxial feeder cable.

A slight improvement on the arrangement in Figure 2.21 can be achieved by sloping the radial elements at about 45° (see Figure 2.22). This arrangement produces a flatter radiation pattern.

At HF rather than VHF, the ground plane can be the earth itself. However, to reduce the earth resistance and increase the efficiency of the antenna, it is usually necessary to incorporate some buried earth radials (see Figure 2.23). These radial wires simply consist of quarter-wave lengths of insulated stranded copper wire grounded to the outer screen of the coaxial feeder at the antenna feed point.

2.8.2 Vertical half-wave antennas

An alternative to the use of a quarter-wave radiating element is that of a half-wave element. This type of antenna must be *voltage fed* (rather than *current fed* as is the case with the quarter-wave antenna). A voltage-fed antenna requires the use of a resonant transformer connected

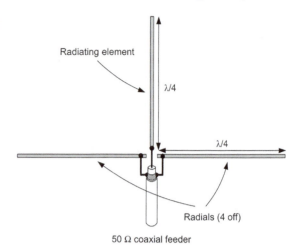

Figure 2.21   Quarter wave vertical antenna

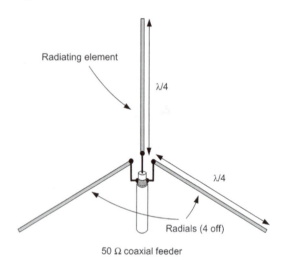

Figure 2.22   Quarter wave vertical antenna with sloping radials

between the low-impedance coaxial feeder and the end of the antenna. Such an arrangement is prone to losses since it requires high-quality, low-loss components. It may also require careful adjustment for optimum results and thus a quarter-wave or three-quarter wave antenna is usually preferred.

2.8.3 5/8th wave vertical antennas

5/8th wave vertical antennas provide a compact solution to the need for an omnidirectional VHF/

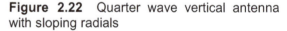

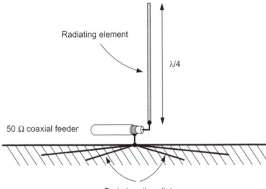

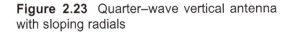

Figure 2.23 Quarter–wave vertical antenna with sloping radials

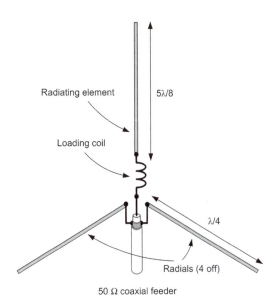

Figure 2.24 5/8th wave vertical antenna with sloping radials

UHF antenna offering some gain over a basic quarter-wave antenna. In fact, a 5/8th wave vertical antenna behaves electrically as a three-quarter wave antenna (i.e. it is current fed from the bottom and there is a voltage maximum at the top). In order to match the antenna, an inductive loading coil is incorporated at the feed-point. A typical 5/8th wave vertical antenna with sloping ground plane is shown in Figure 2.24.

2.8.4 Corner reflectors

An alternative to the Yagi antenna (described earlier) is that of a corner reflecting arrangement like that shown in Figure 2.25. The two reflecting surfaces (which may be solid or perforated to reduce wind resistance) are inclined at an angle of about 90°. This type of aerial is compact in comparison with a Yagi and also relatively unobtrusive.

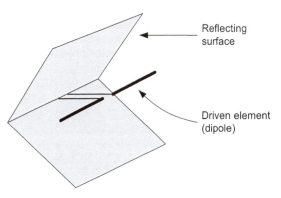

Figure 2.25 High-gain corner reflector antenna with dipole feed

2.8.5 Parabolic reflectors

The need for very high gain coupled with directional response at UHF or microwave frequencies is often satisfied by the use of a parabolic reflector in conjunction with a radiating element positioned at the feed-point of the dish (see Figure 2.26). In order to be efficient, the diameter of a parabolic reflecting surface must be large in comparison with the wavelength of the signal. The gain of such an antenna depends on various factors but is directly proportional to the ratio of diameter to wavelength.

The principle of the parabolic reflector antenna is shown in Figure 2.27. Signals arriving from a distant transmitter will be reflected so that they pass through the focal point of the parabolic surface (as shown). With a conventional parabolic surface, the focal point lies directly on the axis directly in front of the reflecting surface. Placing a radiating element (together with its supporting structure) at the focal point may thus have the

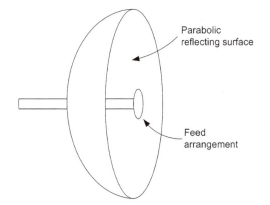

Figure 2.26 Parabolic reflector antenna

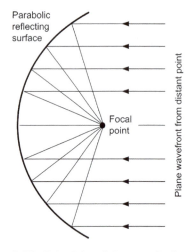

Figure 2.27 Principle of the parabolic reflector

undesirable effect of partially obscuring the parabolic surface! In order to overcome this problem the surface may be modified so that the focus is offset from the central axis.

It is important to realise that the reflecting surface of a parabolic reflector antenna is only part of the story. Equally important (and crucial to the effectiveness of the antenna) is the method of feeding the parabolic surface. What's required here is a means of illuminating or capturing signals from the entire parabolic surface.

Figure 2.28 shows a typical feed arrangement based on a waveguide (see page 38), half-wave dipole and a reflector. The dipole and reflector has a beamwidth of around 90° and this is ideal for illuminating the parabolic surface. The dipole and reflector is placed at the focal point of the

dish. This feed arrangement is often used for **focal plane reflector** antennas where the outer edge of the dish is in the same plane as the half-wave dipole plus reflector feed.

An alternative arrangement using a waveguide and small horn radiator (see page 27) is shown in Figure 2.29. The horn aerial offers some modest gain (usually 6 to 10 dB, or so) and this can be instrumental in increasing the overall gain of the arrangement. Such antennas are generally not focal plane types and the horn feed will usually require support above the parabolic surface.

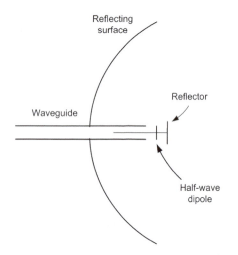

Figure 2.28 Parabolic reflector with half-wave dipole and reflector feed

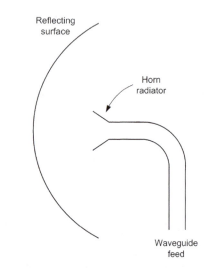

Figure 2.29 Parabolic reflector with horn and waveguide

Figure 2.30 Parabolic reflector antenna with dipole and reflector feed

Figure 2.31 High-gain earth station antenna with parabolic reflector and horn feed

2.8.6 Horn antennas

Like parabolic reflector antennas, horn antennas are commonly used at microwave frequencies. Horn aerials may be used alone or as a means of illuminating a parabolic (or other) reflecting surface. Horn antennas are ideal for use with waveguide feeds; the transition from waveguide (see page 38) to the free space aperture being accomplished over several wavelengths as the waveguide is gradually flared out in both planes. During the transition from waveguide to free space, the impedance changes gradually. The gain of a horn aerial is directly related to the ratio of its aperture (i.e. the size of the horn's opening) and the wavelength. However, as the gain increases, the beamwidth becomes reduced.

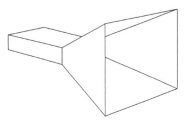

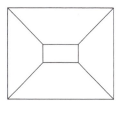

Front view

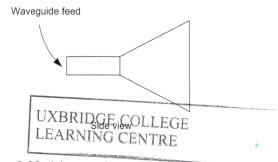

Waveguide feed

Side view

Figure 2.32 A horn antenna

Test your understanding 2.6

Identify an antenna type suitable for use in the following applications. Give reasons for your answers:

(a) an SHF satellite earth station
(b) a low-frequency non-directional beacon
(c) an airfield communication system
(d) a long-range HF communication system
(e) a microwave link between two fixed points.

Table 2.1 Typical characteristics of some common antennas

Application	Gain (dB$_d$)	Beamwidth (degrees)
Vertical half-wave dipole	0	360
Vertical quarter-wave with ground plane	0	360
Four-element Yagi	6	43
UHF corner reflector	9	27
Two stacked vertical half-wave dipoles	3	360
5/8th wave vertical with ground plane	2	360
Small horn antenna for use at 10 GHz	10	20
3 m diameter parabolic antenna for tracking space vehicles at UHF	40	4

Test your understanding 2.7

Identify the antenna shown in Figure 2.33. Sketch a typical horizontal radiation pattern for this antenna.

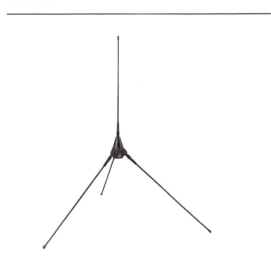

Figure 2.33 See Test your understanding 2.7

2.9 Feeders

The purpose of the feeder line is to convey the power produced by a source to a load which may be some distance away. In the case of a receiver, the source is the receiving antenna whilst the load is the input impedance of the first RF amplifier stage. In the case of a transmitting system, the source is the output stage of the transmitter and the load is the impedance of the transmitting antenna. Ideally, a feeder would have no losses (i.e. no power would be wasted in it) and it would present a perfect match between the impedance of the source to that of the load. In practice, this is seldom the case. This section explains the basic principles and describes the construction of most common types of feeder.

2.9.1 Characteristic impedance

The impedance of a feeder (known as its **characteristic impedance**) is the impedance that would be seen looking into an infinite length of the feeder at the working frequency. The characteristic impedance, Z_0, is a function of the inductance, L, and capacitance, C, of the feeder and may be approximately represented by:

$$Z_0 = \sqrt{\frac{L}{C}} \, \Omega$$

L and C are referred to as the **primary constants** of a feeder. In this respect, L is the loop inductance per unit length whilst C is the shunt capacitance per unit length (see Figure 2.34).

In practice, a small amount of DC resistance will be present in the feeder but this is usually negligible. For the twin open wire shown in Figure 2.21(a), the inductance, L, and capacitance, C, of the line depend on the spacing between the wires and the diameter of the two conductors. For the coaxial cable shown in Figure 2.21(b) the characteristic impedance depends upon the ratio of the diameters of the inner and outer conductors.

Example 2.4

A cable has a loop inductance of 20 nH and a capacitance of 100 pF. Determine the characteristic impedance of the cable.

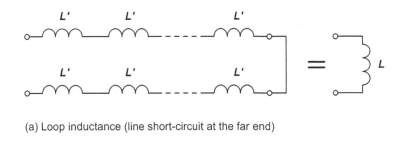

(a) Loop inductance (line short-circuit at the far end)

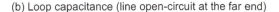

(b) Loop capacitance (line open-circuit at the far end)

Figure 2.34 Loop inductance and loop capacitance

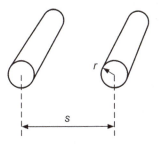

(a) Open wire feeder

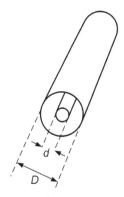

(b) Coaxial cable

Figure 2.35 Dimensions of flat twin feeder and coaxial cables

In this case, $L = 20$ nH $= 20 \times 10^{-9}$ H and $C = 100$ pF $= 100 \times 10^{-12}$ F.

Using $Z_0 = \sqrt{\dfrac{L}{C}}$ Ω gives:

$$Z_0 = \sqrt{\frac{180 \times 10^{-9}}{100 \times 10^{-12}}} = \sqrt{1.8 \times 10^3} = \sqrt{1800} = 42 \ \Omega$$

2.9.2 Coaxial cables

Because they are screened, coaxial cables are used almost exclusively in aircraft applications. The coaxial cable shown in Figure 2.35(b) has a centre conductor (either solid or stranded wire) and an outer conductor that completely shields the inner conductor. The two conductors are concentric and separated by an insulating dielectric that is usually air or some form of polythene. The impedance of such a cable is given by:

$$Z_0 = 138 \log_{10} \left(\frac{D}{d} \right) \ \Omega$$

where Z_0 is the characteristic impedance (in ohms), D is the inside diameter of the outside conductor (in mm), and d is the outside diameter of the inside conductor (in mm).

Example 2.5

A coaxial cable has an inside conductor diameter of 2 mm and an outside conductor diameter of 10 mm. Determine the characteristic impedance of the cable.

In this case, $d = 2$ mm and $D = 10$ mm.

Using $Z_0 = 138 \log_{10}\left(\dfrac{D}{d}\right)$ Ω gives:

$$Z_0 = 138 \log_{10}\left(\frac{10}{2}\right) = 138 \log_{10}(5) = 138 \times 0.7 = 97 \text{ Ω}$$

2.9.3 Two-wire open feeder

The characteristic impedance of the two-wire open feeder shown in Figure 2.35(a) is given by:

$$Z_0 = 276 \log_{10}\left(\frac{s}{r}\right) \text{ Ω}$$

where Z_0 is the characteristic impedance (in ohms), s is the spacing between the wire centres (in mm), and r is the radius of the wire (in mm).

Flat twin **ribbon cable** is a close relative of the two-wire open line (the difference between these two being simply that the former is insulated and the two conductors are separated by a rib of the same insulating material).

When determining the characteristic impedance of ribbon feeder, the formula given above must be modified to allow for the dielectric constant of the insulating material. In practice, however, the difference may be quite small.

Test your understanding 2.8

1. A coaxial cable has an inductance of 30 nH/m and a capacitance of 120 pF/m. Determine the characteristic impedance of the cable.
2. The open wire feeder used with a high-power land-based HF radio transmitter uses wire having a diameter of 2.5 mm and a spacing of 15 mm. Determine the characteristic impedance of the feeder.

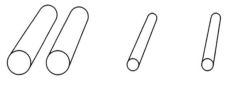

Low impedance   High impedance

(a) Open wire feeder

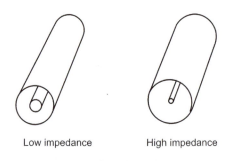

Low impedance   High impedance

(b) Coaxial cable

Figure 2.36 Effect of dimensions on the characteristic impedance of open wire feeder and coaxial cable

2.9.4 Attenuation

The attenuation of a feeder is directly proportional to the DC resistance of the feeder and inversely proportional to the impedance of the line. Obviously, the lower the resistance of the feeder, the smaller will be the power losses. The attenuation is given by:

$$A = 0.143\frac{R}{Z_0} \text{ dB}$$

where A is the attenuation in dB (per metre), R is the resistance in ohms (per metre) and Z is the characteristic impedance (in ohms).

Whilst the attenuation of a feeder remains reasonably constant throughout its specified frequency range, it is usually subject to a progressive increase beyond the upper frequency limit (see Figure 2.38). It is important when choosing a feeder or cable for a particular application to ensure that the operating frequency is within that specified by the manufacturer. As an example, RG178B/U coaxial cable has a loss that increases with frequency from 0.18 dB at 10

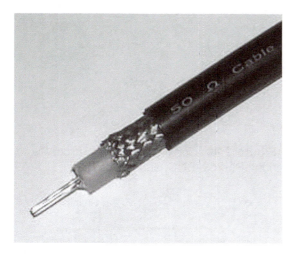

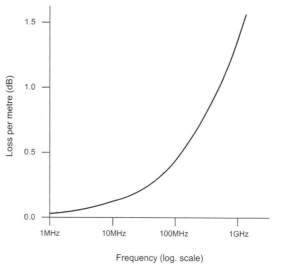

Figure 2.37 Construction of a high-quality coaxial cable (50 Ω impedance)

Figure 2.38 Attenuation of a typical coaxial cable feeder

MHz, to 0.44 dB at 100 MHz, 0.95 dB at 400 MHz, and 1.4 dB at 1 GHz.

Note that, in a 'loss-free' feeder, R and G are both very small and can be ignored (i.e. $R = 0$ and $G = 0$) but with a real feeder both R and G are present.

2.9.5 Primary constants

The two types of feeder that we have already described differ in that one type (the coaxial feeder) is unbalanced whilst the other (the two-wire **transmission line**) is balanced. In order to fully understand the behaviour of a feeder, whether balanced or unbalanced, it is necessary to consider its equivalent circuit in terms of four conventional component values; resistance, inductance, capacitance and conductance, as shown in Figure 2.39. These four parameters are known as primary constants and they are summarised in Table 2.2.

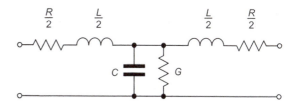

(a) Unbalanced feeder

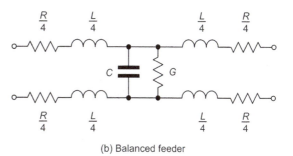

(b) Balanced feeder

Table 2.2 The primary constants of a feeder

Constant	Symbol	Units
Resistance	R	Ohms (Ω)
Inductance	L	Henries (H)
Capacitance	C	Farads (F)
Conductance	G	Siemens (S)

Figure 2.39 Equivalent circuit of balanced and unbalanced feeders

2.9.6 Velocity factor

The velocity of a wave in a feeder is not the same as the velocity of the wave in free space. The ratio of the two (velocity in the feeder compared with the velocity in free space) is known as the **velocity factor**. Obviously, velocity factor must always be less than 1, and in typical feeders it varies from 0.6 to 0.97 (see Table 2.3).

Table 2.3 Velocity factor for various types of feeder

Type of feeder	Velocity factor
Two-wire open line (wire with air dielectric)	0.975
Parallel tubing (air dielectric)	0.95
Coaxial line (air dielectric)	0.85
Coaxial line (solid plastic dielectric)	0.66
Two-wire line (wire with plastic dielectric)	0.68 to 0.82
Twisted-pair line (rubber dielectric)	0.56 to 0.65

2.10 Connectors

Connectors provide a means of linking coaxial cables to transmitting/receiving equipment and antennas. Connectors should be reliable, easy to mate, and sealed to prevent the ingress of moisture and other fluids. They should also be designed to minimise contact resistance and, ideally, they should exhibit a constant impedance which accurately matches that of the system in which they are used (normally 50 Ω for aircraft applications).

Coaxial connectors are available in various format (see Figure 2.40). Of these, the BNC- and N-type connectors are low-loss constant impedance types.

The need for constant impedance connectors (e.g. BNC and N-type connectors) rather than cheaper non-constant impedance connectors (e.g. PL-259) becomes increasingly critical as the frequency increases. As a general rule, constant

Figure 2.40 Coaxial connectors (from left to right: PL-259, BNC, and N-type)

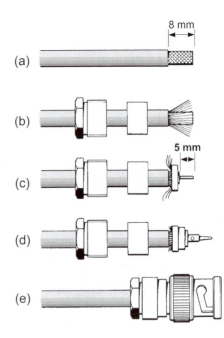

Figure 2.41 Method of fitting a BNC-type connector to a coaxial cable

impedance connectors should be used for applications at frequencies of above 200 MHz. Below this frequency, the loss associated with using non-constant impedance connectors is not usually significant.

Figure 2.41 shows the method of fitting a typical BNC connector to a coaxial cable. Fitting requires careful preparation of the coaxial cable. The outer braided screen is fanned out, as shown in Figure 2.41(b) and Figure 2.41(c) and clamped in place whereas the inner conductor is usually soldered to the centre contact, as shown in Figure 2.41(d).

2.11 Standing wave ratio

Matching a source (such as a transmitter) to a load (such as an aerial) is an important consideration because it allows the maximum transfer of power from one to the other. Ideally, a feeder should present a perfect match between the impedance of the source and the impedance of the load. Unfortunately this is seldom the case and all too often there is some degree of mismatch present. This section explains the consequences of mismatching a source to a load and describes how the effect of a mismatch can be quantified in terms of standing wave ratio (SWR).

Where the impedance of the transmission line or feeder perfectly matches that of the aerial, all of the energy delivered by the line will be transferred to the load (i.e. the aerial). Under these conditions, no energy will be reflected back to the source.

If the match between source and load is imperfect, a proportion of the energy arriving at the load will be reflected back to the source. The result of this is that a standing wave pattern of voltage and current will appear along the feeder (see Fig. 2.42).

The standing wave shown in Figure 2.42 occurs when the wave travelling from the source to the load (i.e. the **forward wave**) interacts with the wave travelling from the load to the source (the **reflected wave**). It is important to note that both the forward and reflected waves are moving but in opposite directions. The standing wave, on the other hand, is stationary.

As indicated in Figure 2.42 when a standing wave is present, at certain points along the feeder the voltage will be a maximum whilst at others it will take a minimum value. The current distribution along the feeder will have a similar pattern (note, however, that the voltage maxima will coincide with the current minima, and vice versa).

Four possible scenarios are shown in Figure 2.43. In Figure 2.43(a) the feeder is perfectly matched to the load. Only the forward wave is present and there is no standing wave. This is the ideal case in which all of the energy generated by the source is absorbed by the load.

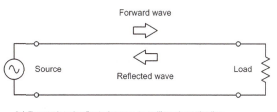

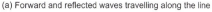

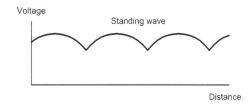

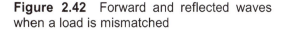

(a) Forward and reflected waves travelling along the line

(b) Voltage standing wave produced

Figure 2.42 Forward and reflected waves when a load is mismatched

In Figure 2.43(b) the load is short-circuit. This represents one of the two worst-case scenarios as the voltage varies from zero to a very high positive value. In this condition, all of the generated power is reflected back to the source.

In Figure 2.43(c) the load is open-circuit. This represents the other worst-case scenario. Here again, the voltage varies from zero to a high positive value and, once again, all of the generated power is reflected back to the source.

In Figure 2.43(d) the feeder is terminated by an impedance that is different from the feeder's characteristic impedance but is neither a short-circuit nor an open-circuit. This condition lies somewhere between the extreme and perfectly matched cases.

The **standing wave ratio** (SWR) of a feeder or transmission line is an indicator of the effectiveness of the impedance match between the transmission line and the antenna. The SWR is the ratio of the maximum to the minimum current along the length of the transmission line, or the ratio of the maximum to the minimum voltage. When the line is absolutely matched the SWR is unity. In other words, we get unity SWR when there is no variation in voltage or current along the transmission line.

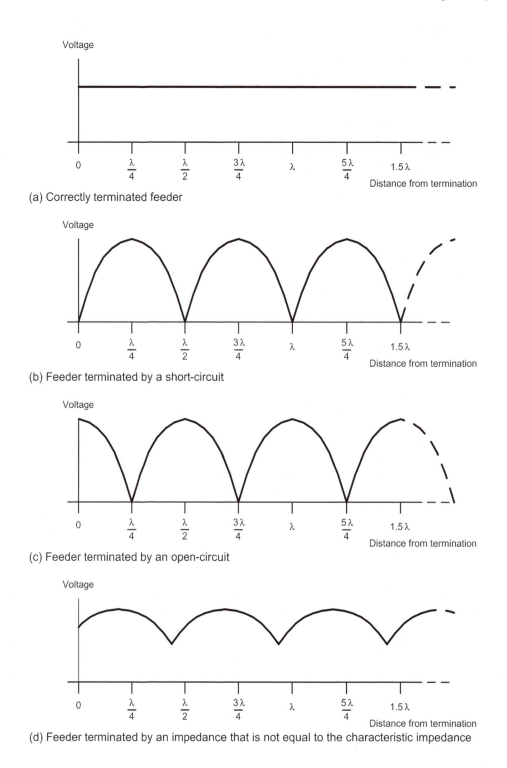

(a) Correctly terminated feeder

(b) Feeder terminated by a short-circuit

(c) Feeder terminated by an open-circuit

(d) Feeder terminated by an impedance that is not equal to the characteristic impedance

Figure 2.43 Effect of different types of mismatch

The greater the number representing SWR, the larger is the mismatch. Also, I^2R losses increase with increasing SWR.

For a *purely resistive* load:

$$SWR = \frac{Z_r}{Z_0} \quad \text{(when } Z_r > Z_0\text{)}$$

and

$$SWR = \frac{Z_0}{Z_r} \quad \text{(when } Z_r < Z_0\text{)}$$

where Z_0 is the characteristic impedance of the transmission line (in ohms) and Z_r is the impedance of the load (also in ohms). Note that, since SWR is a ratio, it has no units.

SWR is optimum (i.e. unity) when Z_r is equal to Z_0. It is unimportant as to which of these terms is in the numerator. Since SWR cannot be less than 1, it makes sense to put whichever is the larger of the two numbers in the numerator.

The average values of RF current and voltage become larger as the SWR increases. This, in turn, results in increased power loss in the series loss resistance and leakage conductance respectively.

For values of SWR of between 1 and 2 this additional feeder loss is not usually significant and is typically less than 0.5 dB. However, when the SWR exceeds 2.5 or 3, the additional loss becomes increasingly significant and steps should be taken to reduce it to a more acceptable value.

2.11.1 SWR measurement

Standing wave ratio is easily measured using an instrument known as an SWR bridge, SWR meter, or a combined SWR/power meter (see Figure 2.44). Despite the different forms of this instrument the operating principle involves sensing the forward and reflected power and displaying the difference between them.

Figure 2.45 shows the scale calibration for the SWR meter circuit shown in Figure 2.46. The instrument comprises a short length of transmission line with two inductively and capacitively coupled secondary lines. Each of these secondary lines is terminated with a matched resistive load and each has a signal

Figure 2.44 A combined RF power and SWR meter

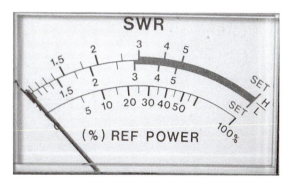

Figure 2.45 A typical SWR meter scale (showing SWR and % reflected power)

detector. Secondary line, L1 (and associated components, D1 and R1) is arranged so that it senses the forward wave whilst secondary line, L2 (and associated components, D2 and R2) is connected so that it senses the reflected wave.

In use, RF power is applied to the system, the meter is switched to indicate the forward power, and VR1 is adjusted for full-scale deflection. Next, the meter is switched to indicate the reflected power and the SWR is read directly from the meter scale. More complex instruments use cross-point meter movements where the two pointers simultaneously indicate forward and reflected power and the point at which they intersect (read from a third scale) gives the value of SWR present.

The point at which the SWR in a system is measured is important. To obtain the most

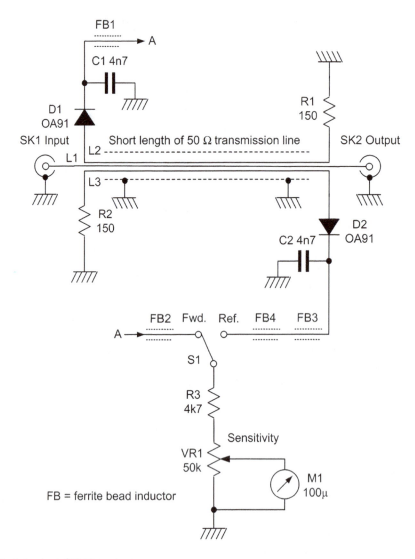

Figure 2.46 A typical SWR meter

meaningful indication of the SWR of an aerial the SWR should ideally be measured at the far end of the feeder (i.e. at the point at which the feeder is connected to the aerial). The measured SWR will actually be *lower* at the other end of the feeder (i.e. at the point at which the feeder is connected to the transmitter). The reason for this apparent anomaly is simply that the loss present in the feeder serves to improve the apparent SWR seen by the transmitter. The more lossy the feeder the better the SWR!

Sudden deterioration of antenna performance and an equally sudden increase in SWR usually points either to mechanical failure of the elements or to electrical failure of the feed-point connection, feeder or RF connectors. Gradual deterioration, on the other hand, is usually associated with corrosion or ingress of fluids into the antenna structure, feeder or antenna termination.

The SWR of virtually all practical aerial/feeder arrangements is liable to some considerable variation with frequency. For this reason, it is advisable to make measurements at the extreme limits of the frequency range as well as at the centre frequency. In the case of a typical transmitting aerial, the SWR can vary from 2:1 at the band edges to 1.2:1 at the centre. Wideband aerials, particularly those designed primarily for receiving applications, often exhibit significantly higher values of SWR. This makes them unsuited to transmitting applications.

2.11.2 A design example

In order to pursue this a little further it's worth taking an example with some measured values to confirm that the SWR of a half-wave dipole (see page 16) really does change in the way that we have predicted. This example further underlines the importance of SWR and the need to have an accurate means of measuring it.

Assume that we are dealing with a simple half-wave dipole aerial that is designed with the following parameters:

Centre frequency: 250 MHz
Feed-line impedance: 75 ohm
Dipole length: 0.564 metres
Element diameter: 5 mm
Bandwidth: 51 MHz
Q-factor: 4.9

The calculated resistance of this aerial varies from about 52 Ω at 235 MHz to 72 Ω at 265 MHz. Over the same frequency range its reactance varies from about −37 Ω (a capacitive reactance) to +38 Ω (an inductive reactance). As predicted, zero reactance at the feed point occurs at a frequency of 250 MHz for the dipole length in question. This relationship is shown in Figure 2.47.

Measurements of SWR show a minimum value of about 1.23 occurring at about 251 MHz and an expected gradual rise either side of this value (see Fig 2.48). This graph shows that the transmitting bandwidth is actually around 33 MHz (extending from 237 MHz to around 270 MHz) for an SWR of 2:1 instead of the intended 51 MHz. Clearly this could be a problem in an application where a transmitter is to be operated with a maximum SWR of 2:1

The bandwidth limitation of a system (comprising transmitter, feeder and aerial) is

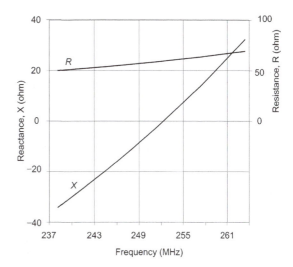

Figure 2.47 Variation of resistance and reactance for the 250 MHz half-wave dipole antenna

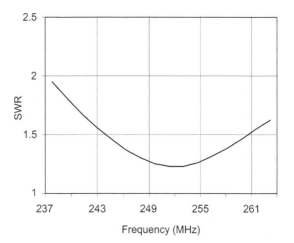

Figure 2.48 Variation of SWR for the 250 MHz half-wave dipole antenna

usually attributable to the inability of a transmitter to operate into a load that has any appreciable amount of reactance present rather than to an inability of the aerial to radiate effectively. Most aerials will radiate happily at frequencies that are some distance away from their resonant frequency—the problem is more one of actually getting the power that the transmitter is capable of delivering into them!

2.12 Waveguide

Conventional coaxial cables are ideal for coupling RF equipment at LF, HF and VHF. However, at microwave frequencies (above 3 GHz, or so), this type of feeder can have significant losses and is also restricted in terms of the peak RF power (voltage and current) that it can handle. Because of this, waveguide feeders are used to replace coaxial cables for SHF and EHF applications, such as weather radar.

A waveguide consists of a rigid or flexible metal tube (usually of rectangular cross-section) in which an electromagnetic wave is launched. The wave travels with very low loss inside the waveguide with its magnetic field component (the H-field) aligning with the broad dimension of the waveguide and the electric field component (the E-field) aligning with the narrow dimension of the waveguide (see Figure 2.49).

A simple waveguide system is shown in Figure 2.50. The SHF signal is applied to a quarter wavelength coaxial probe. The wave launched in the guide is reflected from the plane blanked-off end of the waveguide and travels through sections of waveguide to the load (in this case a horn antenna, see page 27). An example of the use of a waveguide is shown in Figure 2.51. In this application a flexible waveguide is used to feed the weather radar antenna mounted in the nose of a large passenger aircraft. The antenna comprises a flat steerable plate with a large number of radiating slots (each equivalent to a half-wave dipole fed in phase).

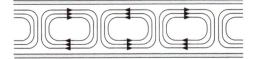

(a) Waveguide (broad dimension) showing H-field

(b) Waveguide (narrow dimension) showing E-field

Figure 2.49 E- and H-fields in a rectangular waveguide

Figure 2.51 Aircraft weather radar with steerable microwave antenna and waveguide

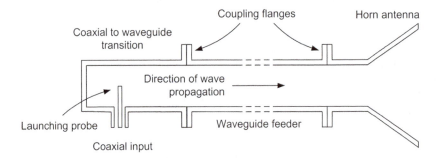

Figure 2.50 A simple waveguide system comprising launcher, waveguide and horn antenna

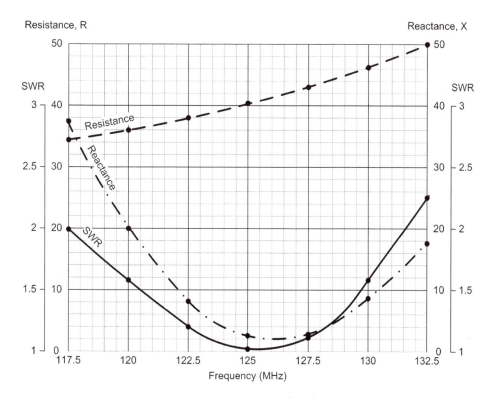

Figure 2.52 See Test your knowledge 2.9

Test your understanding 2.9

Figure 2.52 shows the frequency response of a vertical quarter-wave antenna used for local VHF communications. Use the graph to determine the following:

(a) The frequency at which the SWR is minimum
(b) The 2:1 SWR bandwidth of the antenna
(c) The reactance of the antenna at 120 MHz
(d) The resistance of the antenna at 120 MHz
(e) The frequency at which the reactance of the antenna is a minimum
(f) The frequency at which the resistance of the antenna is 50 Ω.

Test your understanding 2.10

Explain what is meant by standing wave ratio (SWR) and why this is important in determining the performance of an antenna/feeder combination.

2.13 Multiple choice questions

1. An isotropic radiator will radiate:
   (a) only in one direction
   (b) in two main directions
   (c) uniformly in all directions.

2. Another name for a quarter-wave vertical antenna is:
   (a) a Yagi antenna
   (b) a dipole antenna
   (c) a Marconi antenna.

3. A full-wave dipole fed at the centre must be:
   (a) current fed
   (b) voltage fed
   (c) impedance fed.

4. The radiation efficiency of an antenna:
   (a) increases with antenna loss resistance
   (b) decreases with antenna loss resistance
   (c) is unaffected by antenna loss resistance.

5. A vertical quarter-wave antenna will have a
    polar diagram in the horizontal plane which is:
    (a) unidirectional
    (b) omnidirectional
    (c) bi-directional.

6. Which one of the following gives the
    approximate length of a half-wave dipole for
    use at 300 MHz?
    (a) 50 cm
    (b) 1 m
    (c) 2 m.

7. A standing wave ratio of 1:1 indicates:
    (a) that there will be no reflected power
    (b) that the reflected power will be the same
        as the forward power
    (c) that only half of the transmitted power will
        actually be radiated.

8. Which one of the following gives the 2:1
    SWR bandwidth of the antenna whose
    frequency response is shown in Figure 2.53?
    (a) 270 kHz
    (b) 520 kHz
    (c) 11.1 MHz.

9. Which one of the following antenna types
    would be most suitable for a fixed long
    distance HF communications link?
    (a) a corner reflector
    (b) two stacked vertical dipoles
    (c) a three-element horizontal Yagi.

10. What type of antenna is shown in Figure 2.54?
    (a) a folded dipole
    (b) a Yagi
    (c) a corner reflector.

11. When two antennas are vertically stacked the
    combination will have:
    (a) increased gain and decreased beamwidth
    (b) decreased gain and increased beamwidth
    (c) increased gain and unchanged beamwidth.

12. The characteristic impedance of a coaxial
    cable depends on:
    (a) the ratio of inductance to capacitance
    (b) the ratio of resistance to inductance
    (c) the product of the resistance and reactance
        (either capacitive or inductive).

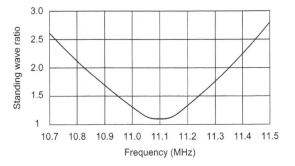

Figure 2.53   See Question 8

Figure 2.54   See Question 10

13. The attenuation of an RF signal in a coaxial
    cable:
    (a) increases with frequency
    (b) decreases with frequency
    (c) stays the same regardless of frequency.

14. If a transmission line is perfectly matched to
    an aerial load it will:
    (a) have no impedance
    (b) be able to carry an infinite current
    (c) appear to be infinitely long.

15. The characteristic impedance of an RF coaxial
    cable is:
    (a) usually between 50 and 75 Ω
    (b) either 300 or 600 Ω
    (c) greater than 600 Ω.

16. The beamwidth of an antenna is measured:
    (a) between the 50% power points
    (b) between the 70% power points
    (c) between the 90% power points.

Chapter 3 · Transmitters and receivers

Transmitters and receivers are used extensively in aircraft communication and navigation systems. In conjunction with one ore more antennas, they are responsible for implementing the vital link between the aircraft, ground stations, other aircraft and satellites. This chapter provides a general introduction to the basic principles and operation of transmitters and receivers. These themes are further developed in Chapters 4 and 5.

3.1 A simple radio system

Figure 3.1 shows a simple radio communication system comprising a **transmitter** and **receiver** for use with **continuous wave** (CW) signals. Communication is achieved by simply switching (or 'keying') the radio frequency signal on and off. Keying can be achieved by interrupting the supply to the power amplifier stage or even the oscillator stage; however, it is normally applied within the driver stage that operates at a more modest power level. Keying the oscillator stage usually results in impaired frequency stability. On the other hand, attempting to interrupt the appreciable currents and/or voltages that appear in the power amplifier stage can also prove to be somewhat problematic.

The simplest form of CW receiver consists of nothing more than a radio frequency amplifier (which provides gain and selectivity) followed by a detector and an audio amplifier. The **detector** stage mixes a locally generated radio frequency signal produced by the **beat frequency oscillator** (BFO) with the incoming signal to produce a signal that lies within the audio frequency range (typically between 300 Hz and 3.4 kHz).

As an example, assume that the incoming signal is at a frequency of 100 kHz and that the BFO is producing a signal at 99 kHz. A signal at the difference between these two frequencies (1 kHz) will appear at the output of the detector stage. This will then be amplified within the audio stage before being fed to the loudspeaker.

Example 3.1.1

A radio wave has a frequency of 162.5 kHz. If a beat frequency of 1.25 kHz is to be obtained, determine the two possible BFO frequencies.

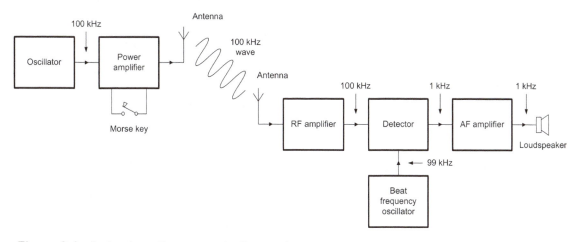

Figure 3.1 A simple radio communication system

The BFO can be above or below the incoming signal frequency by an amount that is equal to the beat frequency (i.e. the audible signal that results from the 'beating' of the two frequencies and which appears at the output of the detector stage).

Hence, $f_{BFO} = f_{RF} \pm f_{AF}$

from which:

$f_{BFO} = (162.5 \pm 1.25)$ kHz = 160.75 or 163.25 kHz

Test your understanding 3.1

An audio frequency signal of 850 Hz is produced when a BFO is set to 455.5 kHz. What is the input signal frequency to the detector?

3.1.1 Morse code

Transmitters and receivers for CW operation are extremely simple but nevertheless they can be extremely efficient. This makes them particularly useful for disaster and emergency communication or for any situation that requires optimum use of low power equipment. Signals are transmitted using the code invented by Samuel Morse (see Figures 3.2 and 3.3).

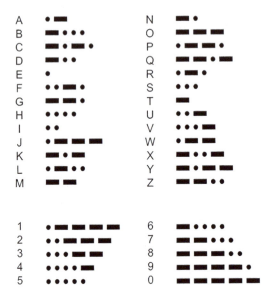

Figure 3.2   Morse code

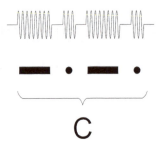

Figure 3.3   Morse code signal for the letter C

3.2 Modulation and demodulation

In order to convey information using a radio frequency carrier, the signal information must be superimposed or 'modulated' onto the carrier. **Modulation** is the name given to the process of changing a particular property of the carrier wave in sympathy with the instantaneous voltage (or current) signal.

The most commonly used methods of modulation are **amplitude modulation** (AM) and **frequency modulation** (FM). In the former case, the carrier amplitude (its peak voltage) varies according to the voltage, at any instant, of the modulating signal. In the latter case, the carrier frequency is varied in accordance with the voltage, at any instant, of the modulating signal.

Figure 3.4 shows the effect of amplitude and frequency modulating a sinusoidal carrier (note that the modulating signal is in this case also sinusoidal). In practice, many more cycles of the RF carrier would occur in the time-span of one cycle of the modulating signal. The process of modulating a carrier is undertaken by a **modulator** circuit, as shown in Figure 3.5. The input and output waveforms for amplitude and frequency modulator circuits are shown in Figure 3.6.

Demodulation is the reverse of modulation and is the means by which the signal information is recovered from the modulated carrier. Demodulation is achieved by means of a **demodulator** (sometimes also called a **detector**). The output of a demodulator consists of a reconstructed version of the original signal information present at the input of the modulator stage within the transmitter. The input and output waveforms for amplitude and frequency

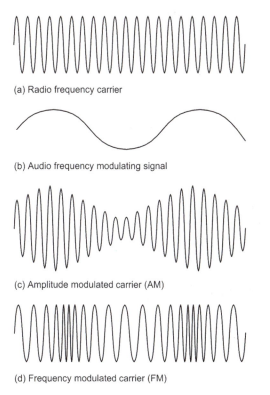

(a) Radio frequency carrier

(b) Audio frequency modulating signal

(c) Amplitude modulated carrier (AM)

(d) Frequency modulated carrier (FM)

Figure 3.4   Modulated waveforms

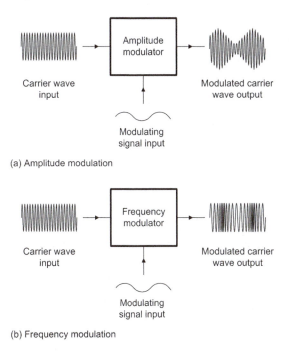

Carrier wave
input

Modulated carrier
wave output

Modulating
signal input

(a) Amplitude modulation

Carrier wave
input

Modulated carrier
wave output

Modulating
signal input

(b) Frequency modulation

Figure 3.5   Action of a modulator

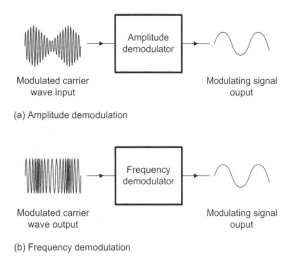

Modulated carrier
wave input

Modulating signal
ouput

(a) Amplitude demodulation

Modulated carrier
wave output

Modulating signal
ouput

(b) Frequency demodulation

Figure 3.6   Action of a demodulator

modulator circuits are shown in Figure 3.6. We shall see how this works a little later.

3.3 AM transmitters

Figure 3.7 shows the block schematic of a simple AM transmitter. An accurate and stable RF oscillator generates the radio frequency **carrier** signal. The output of this stage is then amplified and passed to a modulated RF power amplifier stage. The inclusion of an amplifier between the RF oscillator and the modulated stage also helps to improve frequency stability.

The low-level signal from the microphone is amplified using an AF amplifier before it is passed to an AF power amplifier. The output of the power amplifier is then fed as the supply to the modulated RF power amplifier stage. Increasing and reducing the supply to this stage is instrumental in increasing and reducing the amplitude of its RF output signal.

The modulated RF signal is then passed through an **antenna coupling unit**. This unit matches the antenna to the RF power amplifier and also helps to reduce the level of any unwanted harmonic components that may be present. The AM transmitter shown in Figure 3.7 uses high-level modulation in which the modulating signal is applied to the final RF power amplifier stage.

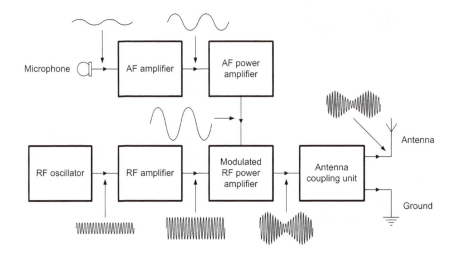

Figure 3.7 An AM transmitter using high-level modulation

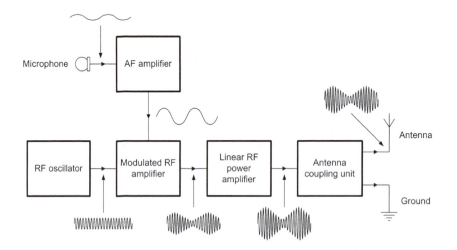

Figure 3.8 An AM transmitter using low-level modulation

An alternative to high-level modulation of the carrier wave is shown in Figure 3.8. In this arrangement the modulation is applied to a low-power RF amplifier stage and the amplitude modulated signal is then further amplified by the final RF power amplifier stage. In order to prevent distortion of the modulated waveform this final stage *must* operate in linear mode (the output waveform must be a faithful replica of the input waveform). Low-level modulation avoids the need for an AF power amplifier.

3.4 FM transmitters

Figure 3.9 shows the block schematic of a simple FM transmitter. Once again, an accurate and stable RF oscillator generates the radio frequency carrier signal. As with the AM transmitter, the output of this stage is amplified and passed to an RF power amplifier stage. Here again, the inclusion of an amplifier between the RF oscillator and the RF power stage helps to improve frequency stability.

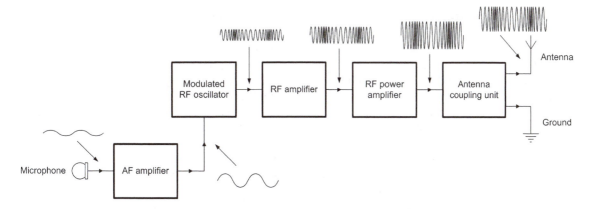

Figure 3.8 An FM transmitter

The low-level signal from the microphone is amplified using an AF amplifier before it is passed to a **variable reactance** element (see Chapter 4) within the RF oscillator tuned circuit. The application of the AF signal to the variable reactance element causes the frequency of the RF oscillator to increase and decrease in sympathy with the AF signal.

As with the AM transmitter, the final RF signal from the power amplifier is passed through an antenna coupling unit that matches the antenna to the RF power amplifier and also helps to reduce the level of any unwanted harmonic components that may be present. Further information on transmitters will be found in Chapters 4 and 5.

3.5 Tuned radio frequency receivers

Tuned radio frequency (TRF) receivers provide a means of receiving local signals using fairly minimal circuitry. The simplified block schematic of a TRF receiver is shown in Figure 3.9.

The signal from the antenna is applied to an RF amplifier stage. This stage provides a moderate amount of gain at the signal frequency. It also provides **selectivity** by incorporating one or more tuned circuits at the signal frequency. This helps the receiver to reject signals that may be present on adjacent channels.

The output of the RF amplifier stage is applied to the demodulator. This stage recovers the audio frequency signal from the modulated RF signal. The demodulator stage may also incorporate a tuned circuit to further improve the selectivity of the receiver.

The output of the demodulator stage is fed to the input of the AF amplifier stage. This stage increases the level of the audio signal from the demodulator so that it is sufficient to drive a loudspeaker.

TRF receivers have a number of limitations with regard to sensitivity and selectivity and this makes this type of radio receiver generally unsuitable for use in commercial radio equipment.

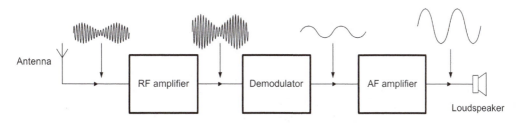

Figure 3.9 A TRF receiver

3.6 Superhet receivers

Superhet receivers provide both improved **sensitivity** (the ability to receive weak signals) and improved **selectivity** (the ability to discriminate signals on adjacent channels) when compared with TRF receivers. Superhet receivers are based on the **supersonic-heterodyne** principle where the wanted input signal is converted to a fixed **intermediate frequency** (IF) at which the majority of the gain and selectivity is applied. The intermediate frequency chosen is generally 455 kHz or 1.6 MHz for AM receivers and 10.7 MHz for communications and FM receivers. The simplified block schematic of a simple superhet receiver is shown in Figure 3.11.

The signal from the antenna is applied to an **RF amplifier** stage. As with the TRF receiver, this stage provides a moderate amount of gain at the signal frequency. The stage also provides selectivity by incorporating one or more tuned circuits at the signal frequency.

The output of the RF amplifier stage is applied to the **mixer** stage. This stage combines the RF signal with the signal derived from the **local oscillator** (LO) stage in order to produce a signal at the **intermediate frequency** (IF). It is worth noting that the output signal produced by the mixer actually contains a number of signal components, including the sum and difference of the signal and local oscillator frequencies as well as the original signals plus harmonic components. The wanted signal (i.e. that which corresponds to the IF) is passed (usually by some form of filter—see page 48) to the IF amplifier stage. This stage provides amplification as well as a high degree of selectivity.

The output of the IF amplifier stage is fed to the demodulator stage. As with the TRF receiver, this stage is used to recover the audio frequency signal from the modulated RF signal.

Finally, the AF signal from the demodulator stage is fed to the AF amplifier. As before, this stage increases the level of the audio signal from the demodulator so that it is sufficient to drive a loudspeaker.

In order to cope with a wide variation in signal amplitude, superhet receivers invariably incorporate some form of **automatic gain control** (AGC). In most circuits the DC level from the AM demodulator (see page 51) is used to control the gain of the IF and RF amplifier stages. As the signal level increases, the DC level from the demodulator stage increases and this is used to reduce the gain of both the RF and IF amplifiers.

The superhet receiver's intermediate frequency f_{IF} is the difference between the signal frequency, f_{RF}, and the local oscillator frequency, f_{LO}. The desired local oscillator frequency can be calculated from the relationship:

$$f_{LO} = f_{RF} \pm f_{IF}$$

Note that in most cases (and in order to simplify tuning arrangements) the local oscillator operates above the signal frequency, i.e. $f_{LO} = f_{RF} + f_{IF}$. So, for example, a superhet receiver with a 1.6 MHz IF tuned to receive a signal at 5.5 MHz will operate with an LO at $(5.5 + 1.6) = 7.1$ MHz.

Figure 3.10 shows the relationship between the frequencies entering and leaving a mixer stage.

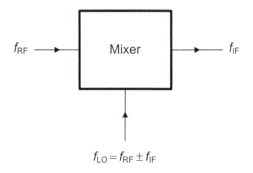

Figure 3.10 Action of a mixer stage in a superhet receiver

Example 3.2

A VHF Band II FM receiver with a 10.7 MHz IF covers the signal frequency range 88 MHz to 108 MHz. Over what frequency range should the local oscillator be tuned?

Using $f_{LO} = f_{RF} + f_{IF}$ when $f_{RF} = 88$ MHz gives $f_{LO} = 88$ MHz + 10.7 MHz = 98.7 MHz

Using $f_{LO} = f_{RF} + f_{IF}$ when $f_{RF} = 108$ MHz gives $f_{LO} = 108$ MHz + 10.7 MHz = 118.7 MHz.

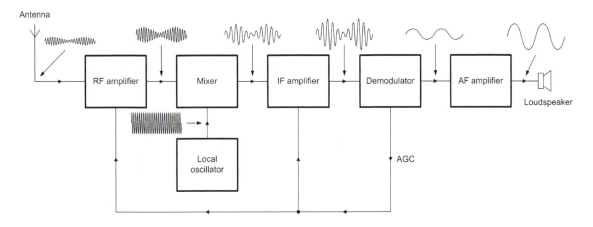

Figure 3.11 A superhet receiver

3.7 Selectivity

Radio receivers use tuned circuits in order to discriminate between incoming signals at different frequencies. Figure 3.12 shows two basic configurations for a tuned circuit; series and parallel. The impedance-frequency characteristics of these circuits are shown in Figure 3.13. It is important to note that the impedance of the series tuned circuit falls to a very low value at the resonant frequency whilst that for a parallel tuned circuit increases to a very high value at

resonance. For this reason, series tuned circuits are sometimes known as **acceptor circuits**. Parallel tuned circuits, on the other hand, are sometimes referred to as **rejector circuits**.

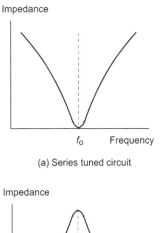

(a) Series tuned circuit

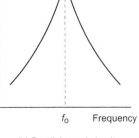

(b) Parallel tuned circuit

(a) Series tuned circuit

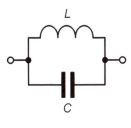

(b) Parallel tuned circuit

Figure 3.12 Series and parallel tuned circuits

Figure 3.13 Frequency response of the tuned circuits shown in Figure 3.12

The frequency response (voltage plotted against frequency) of a parallel tuned circuit is shown in Figure 3.14. This characteristic shows how the signal developed across the circuit reaches a maximum at the resonant frequency (f_o). The range of frequencies accepted by the circuit is normally defined in relation to the half-power (−3dB power) points. These points correspond to 70.7% of the maximum voltage and the frequency range between these points is referred to as the **bandwidth** of the tuned circuit.

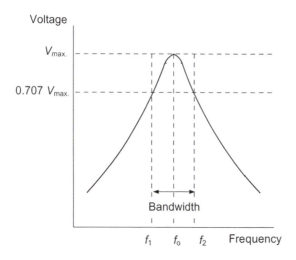

Figure 3.14 Frequency response for a parallel tuned circuit

A perennial problem with the design of the TRF receivers that we met earlier is the lack of selectivity due to the relatively wide bandwidth of the RF tuned circuits. An RF tuned circuit will normally exhibit a quality factor (**Q-factor**) of about 100. The relationship between bandwidth, Δf, Q-factor, Q, and resonant frequency, f_o, for a tuned circuit is given by:

$$\Delta f = \frac{f_o}{Q}$$

As an example, consider a tuned circuit which has a resonant frequency of 10 MHz and a Q-factor of 100. Its bandwidth will be:

$$\Delta f = \frac{f_o}{Q} = \frac{10\ \text{MHz}}{100} = 100\ \text{kHz}$$

Clearly many strong signals will appear within this range and a significant number of them may be stronger than the wanted signal. With only a single tuned circuit at the signal frequency, the receiver will simply be unable to differentiate between the wanted and unwanted signals.

Selectivity can be improved by adding additional tuned circuits at the signal frequency. Unfortunately, the use of multiple tuned circuits brings with it the problem of maintaining accurate tuning of each circuit throughout the tuning range of the receiver. Multiple 'ganged' variable capacitors (or accurately matched variable capacitance diodes) are required.

A **band-pass filter** can be constructed using two parallel tuned circuits coupled inductively (or capacitively), as shown in Figure 3.15. The frequency response of this type of filter depends upon the degree of coupling between the two tuned circuits. Optimum results are obtained with a critical value of coupling (see Figure 3.16). Too great a degree of coupling results in a 'double-humped' response whilst too little coupling results in a single peak in the response curve accompanied by a significant loss in signal. Critical coupling produces a relatively 'flat' pass-band characteristic accompanied by a reasonably steep fall-off either side of the pass-band.

Band-pass filters are often found in the IF stages of superhet receivers where they are used to define and improve the receiver's selectivity. Where necessary, a higher degree of selectivity and **adjacent channel rejection** can be achieved by using a multi-element ceramic, mechanical, or crystal filter. A typical 455 kHz crystal filter (for use with an HF receiver) is shown in Figure 3.18. This filter provides a bandwidth of 9 kHz and a very high degree of attenuation at the two **adjacent channels** on either side of the pass-band.

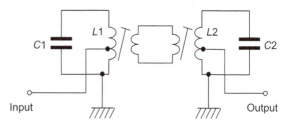

Figure 3.15 A typical band-pass filter

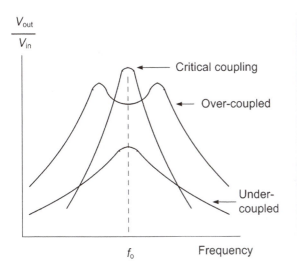

$\dfrac{V_{out}}{V_{in}}$

← Critical coupling

← Over-coupled

← Under-coupled

f_o Frequency

Figure 3.16 Response of coupled tuned circuits

Figure 3.17 Band-pass coupled tuned circuits in the RF stages of a VHF receiver

Test your understanding 3.2

Sketch the block schematic of a superhet receiver and state the function of each of the blocks.

Test your understanding 3.3

An HF communications receiver has an intermediate frequency of 455 kHz. What frequency must the local oscillator operate at when the receiver is tuned to 5.675 MHz?

Test your understanding 3.4

A tuned circuit IF filter is to operate with a centre frequency of 10.7 MHz and a bandwidth of 150 kHz. What Q-factor is required?

Test your understanding 3.5

The ability of a receiver to reject signals on adjacent channels is determined by the selectivity of its IF stages. Explain why this is.

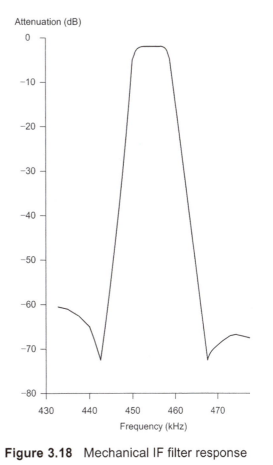

Figure 3.18 Mechanical IF filter response

3.8 Image channel rejection

Earlier we showed that a superhet receiver's intermediate frequency, f_{IF}, is the difference between the signal frequency, f_{RF}, and the local oscillator frequency, f_{LO}. We also derived the following formula for determining the frequency of the local oscillator signal:

$$f_{LO} = f_{RF} \pm f_{IF}$$

The formula can be rearranged to make f_{RF} the subject, as follows:

$$f_{RF} = f_{LO} \pm f_{IF}$$

In other words, there are two potential radio frequency signals that can mix with the local oscillator signal in order to provide the required IF. One of these is the **wanted signal** (i.e. the signal present on the channel to which the receiver is tuned) whilst the other is referred to as the **image channel**.

Being able to reject any signals that may just happen to be present on the image channel of a superhet receiver is an important requirement of any superhet receiver. This can be achieved by making the RF tuned circuits as selective as possible (so that the image channel lies well outside their pass-band). The problem of rejecting the image channel is, however, made easier by selecting a relatively high value of intermediate frequency (note that, in terms of frequency, the image channel is spaced at *twice* the IF away from the wanted signal).

Figure 3.19 shows the relationship that exists between the wanted signal, local oscillator signal, and the image channel for receivers with (a) a 455 kHz IF and (b) a 1.6 MHz IF. A typical response curve for the RF tuned circuits of the receiver (assuming a typical Q-factor) has been

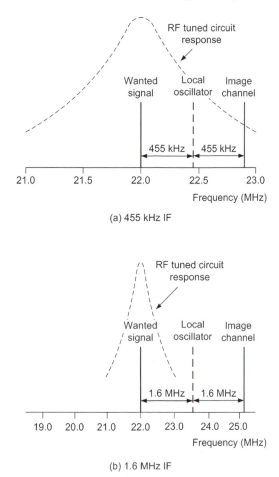

(a) 455 kHz IF

(b) 1.6 MHz IF

Figure 3.19   Image channel rejection

superimposed onto both of the graphs (the same response curve has been used in both cases but the frequency scale has been changed for the two different intermediate frequencies). From Figure 3.19 is should be clear that whilst the image channel for the 455 kHz IF falls inside the RF tuned circuit response, that for the 1.6 MHz IF falls well outside the curve.

3.9 Automatic gain control

The signal levels derived from the antennas fitted to an aircraft can vary from as little as 1 μV to more than 1,000 μV. Unfortunately, this presents us with a problem when signals are to be amplified. The low-level signals benefit from the maximum amount of gain present in a system whilst the larger signals require correspondingly less gain in order to avoid non-linearity and consequent distortion of the signals and modulation. AM, CW and SSB receivers therefore usually incorporate some means of **automatic gain control** (AGC) that progressively reduces the signal gain as the amplitude of the input signal increases (see Figure 3.20).

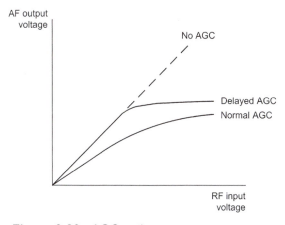

Figure 3.20 AGC action

In simple receivers, the AGC voltage (a DC voltage dependent on signal amplitude) is derived directly from the signal detector and is fed directly to the bias circuitry of the IF stages (see Figure 3.11). In more sophisticated equipment, the AGC voltage is amplified before being applied to the IF and RF stages. There is, in fact, no need to reduce the signal gain for small RF signals. Hence, in more sophisticated equipment, the AGC circuits may be designed to provide a 'delay' so that there is no gain reduction until a predetermined threshold voltage is exceeded. In receivers that feature **delayed AGC** there is no gain reduction until a certain threshold voltage is achieved. Beyond this, there is a progressive reduction in gain (see Figure 3.20).

3.10 Double superhet receivers

The basic superhet receiver shown in Figure 3.11 has an intermediate frequency (IF) of usually either 455 kHz, 1.6 MHz or 10.7 MHz. In order to achieve an acceptable degree of image channel rejection (recall that the image channel is spaced by twice the IF away from the wanted frequency) a 455 kHz IF will generally be satisfactory for the reception of frequencies up to about 5 MHz, whilst an IF of 1.6 MHz (or greater) is often used at frequencies above this. At VHF, intermediate frequencies of 10.7 MHz (or higher) are often used.

Unfortunately, the disadvantage of using a high IF (1.6 MHz or 10.7 MHz) is simply that the bandwidth of conventional tuned circuits is too wide to provide a satisfactory degree of selectivity and thus elaborate (and expensive) IF filters are required.

To avoid this problem and enjoy the best of both worlds, many high-performance receivers make use of *two* separate intermediate frequencies; the first IF provides a high degree of image channel rejection whilst the second IF provides a high degree of selectivity. Such receivers are said to use **dual conversion**.

A typical double superhet receiver is shown in Figure 3.21. The incoming signal frequency (26 MHz in the example) is converted to a first IF at 10.695 MHz by mixing the RF signal with a first local oscillator signal at 36.695 MHz (note that 36.695 MHz − 26 MHz = 10.695 MHz). The first IF signal is then filtered and amplified before it is passed to the second mixer stage.

The input of the second mixer (10.695 MHz) is then mixed with the second local oscillator signal at 10.240 MHz. This produces the second IF at 455 kHz (note that 10.695 MHz − 10.240 MHz = 455 kHz). The second IF signal is then filtered and amplified. It is worth noting that the bulk of the gain is usually achieved in the second IF stages and there will normally be several stages of amplification at this frequency.

In order to tune the receiver, the first local oscillator is either made variable (using conventional tuned circuits) or is synthesised using digital phase-locked loop techniques (see page 53). The second local oscillator is almost invariably crystal controlled in order to ensure good stability and an accurate relationship

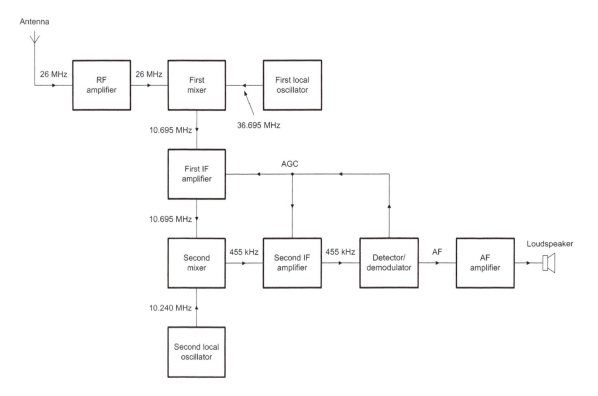

Figure 3.21   A double conversion superhet receiver

between the two intermediate frequencies. Typical IF bandwidths in the receiver shown in Figure 3.21 are 75 kHz at the first IF and a mere 6 kHz in the second IF.

The first IF filter (not shown in Figure 3.21) is connected in the signal path between the first and second mixer. Where a stage of amplification is provided at the first IF, the filter precedes the amplifier stage. The requirements of the filter are not stringent since the ultimate selectivity of the receiver is defined by the second IF filter which operates at the much lower frequency of 455 kHz.

There are, however, some good reasons for using a filter which offers a high degree of rejection of the unwanted second mixer image response which occurs at 9.785 MHz. If this image is present at the input of the second mixer, it will mix with the second mixer injection at 10.240 MHz to produce a second IF component of 455 kHz, as shown in Figure 3.22. The function of the first IF filter is thus best described as **roofing**; bandwidth is a less important

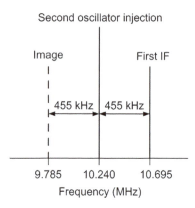

Figure 3.22   Second oscillator signal

Test your understanding 3.8

Explain why AGC is necessary in an HF communications receiver and how it is applied.

3.11 Digital frequency synthesis

The signals used within high-specification radio frequency equipment (both receivers and transmitters) must be both accurate and stable. Where operation is restricted to a single frequency or a limited number of channels, quartz crystals may be used to determine the frequency of operation. However, when a large number of frequencies must be covered, it is necessary to employ digital frequency generating techniques in which a single quartz crystal oscillator is used in conjunction with LSI circuitry to generate a range of discrete frequencies. These frequencies usually have a constant channel spacing (typically 3 kHz, 8.33 kHz, 9 kHz, 12.5 kHz, 25 kHz, etc.). Frequencies are usually selected by means of a rotary switch, push-buttons or a keypad but can also be stored in semiconductor memories.

Digital **phase locked loop** (PLL) circuitry was first used in military communications equipment in the mid-1960s and resulted from the need to generate a very large number of highly accurate and stable frequencies in a multi-channel frequency synthesiser. In this particular application cost was not a primary consideration and highly complex circuit arrangements could be employed involving large numbers of discrete components and integrated circuits.

Phase locked loop techniques did not arrive in mass-produced equipment until the early 1970s.

By comparison with today's equipment such arrangements were crude, employing as many as nine or ten i.c. devices. Complex as they were, these PLL circuits were more cost-effective than their comparable multi-crystal mixing synthesiser counterparts. With the advent of large scale integration in the late 1970s, the frequency generating unit in most radio equipment could be reduced to one, or perhaps two, LSI devices together with a handful of additional discrete components. The cost-effectiveness of this approach is now beyond question and it is unlikely that, at least in the most basic equipment, much further refinement will be made. In the area of more complex receivers and transceivers, however, we are now witnessing a further revolution in the design of synthesised radio equipment with the introduction of dedicated microcomputer controllers which permit keypad programmed channel selection and scanning with pause, search, and lock-out facilities.

The most basic form of PLL consists of a phase detector, filter, DC amplifier and voltage controlled oscillator (VCO), as shown in Figure 3.23. The VCO is designed so that its free-running frequency is at, or near, the reference frequency. The phase detector senses any error between the VCO and reference frequencies. The output of the phase detector is fed, via a suitable filter and amplifier, to the DC control voltage input of the VCO. If there is any discrepancy

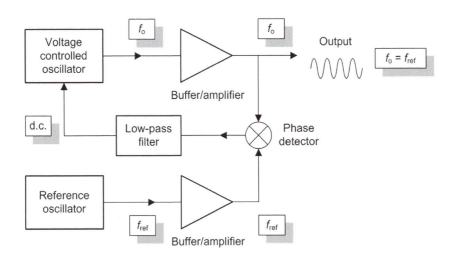

Figure 3.23 A simple phase locked loop

between the VCO output and the reference frequency, an error voltage is produced and this is used to correct the VCO frequency. The VCO thus remains locked to the reference frequency. If the reference frequency changes, so does the VCO. The bandwidth of the system is determined by the time constants of the loop filter. In practice, if the VCO and reference frequencies are very far apart, the PLL may be unable to lock.

The frequency range over which the circuit can achieve lock is known as the **capture range**. It should be noted that a PLL takes a finite time to achieve a locked condition and that the VCO locks to the mean value of the reference frequency.

The basic form of PLL shown in Figure 3.23 is limited in that the reference frequency is the same as that of the VCO and no provision is incorporated for changing it, other than by varying the frequency of the reference oscillator itself. In practice, it is normal for the phase detector to operate at a much lower frequency than that of the VCO output and thus a frequency divider is incorporated in the VCO feedback path (see Figure 3.24). The frequency presented to the phase detector will thus be f_o/n, where n is the **divisor**.

When the loop is locked (i.e. when no phase error exists) we can infer that:

$$f_{ref} = \frac{f_o}{n} \quad \text{or} \quad f_o = n f_{ref}$$

A similar divider arrangement can also be used at the reference input to the phase detector, as shown in Figure 3.25. The frequency appearing at the reference input to the phase detector will be f_{ref}/m and the loop will be locked when:

$$\frac{f_{ref}}{m} = \frac{f_o}{n} \quad \text{or} \quad f_o = \frac{n}{m} f_{ref}$$

Thus if f_{ref}, n and m were respectively 100 kHz, 2,000 and 10, the output frequency, f_{out}, would be:

$$(2,000/10) \times 100 \text{ kHz} = 20 \text{ MHz}$$

If the value of n can be made to change by replacing the fixed divider with a programmable divider, different output frequencies can be generated. If, for example, n was variable from 2,000 to 2,100 in steps of 1, then f_{out} would range from 20 MHz to 21 MHz in 10 kHz steps. Figure 3.26 shows the basic arrangement of a PLL which incorporates a programmable divider driven from the equipment's digital frequency controller (usually a microprocessor).

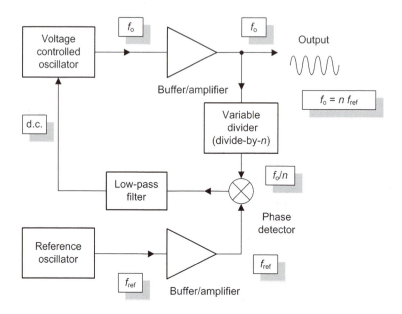

Figure 3.24 A phase locked loop with frequency divider

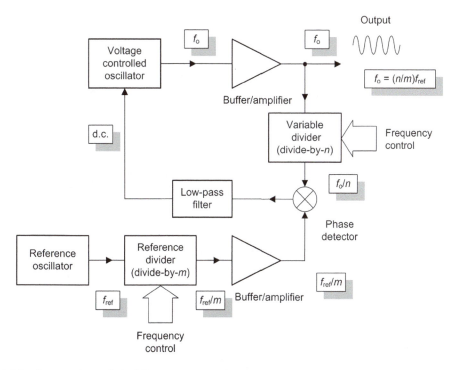

Figure 3.25 A complete digital frequency synthesiser

In practice, problems can sometimes arise in high-frequency synthesisers where the programmable frequency divider, or **divide-by-*n* counter**, has a restricted upper frequency limit. In such cases it will be necessary to mix the high-frequency VCO output with a stable locally generated signal derived from a crystal oscillator. The mixer output (a relatively low difference frequency) will then be within the range of the programmable divider.

3.12 A design example

We shall bring this chapter to a conclusion by providing a design example of a complete HF communications receiver. This receiver was developed by the author for monitoring trans-Atlantic HF communications in the 5.5 MHz aircraft band. The circuit caters for the reception of AM, CW (Morse code) and SSB signals (see Chapter 5). To aid stability, the CIO/BFO frequency is controlled by means of a ceramic resonator. The RF performance is greatly enhanced by the use of dual gate MOSFET

devices in the RF amplifier, mixer, and product detector stages and junction gate FETs in the local oscillator stage. These devices offer high gain with excellent strong-signal handling capability. They also permit simple and effective coupling between stages without the need for complex impedance matching.

The receiver is tunable over the frequency range 5.0 MHz to 6.0 MHz. Used in conjunction with a simple antenna, it offers reception of aircraft signals at distances in excess of 1,000 km. The receiver is based on the single superhet principle operating with an intermediate frequency of 455 kHz. This frequency is low enough to ensure reasonable selectivity with just two stages of IF amplification and with the aid of a low-cost 455 kHz filter. Adequate image rejection is provided by two high-Q ganged RF tuned circuits.

The design uses conventional discrete component circuitry in all stages with the exception of the audio amplifier/output stage and voltage regulator. This approach ensures that the receiver is simple and straightforward to align and does not suffer from the limitations

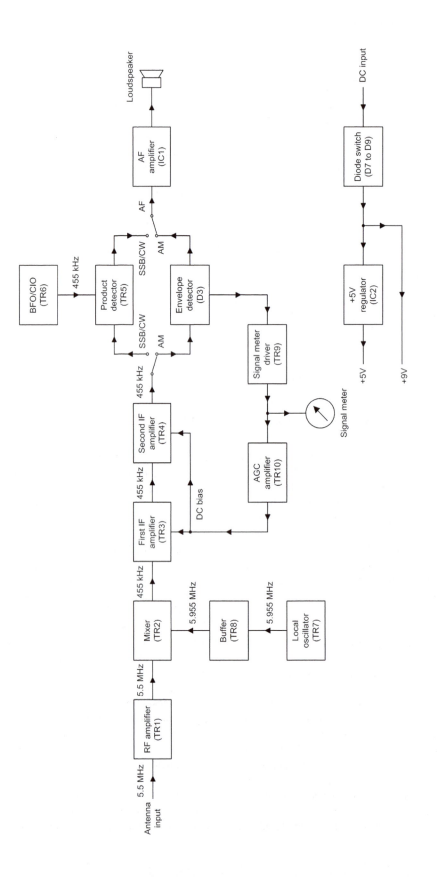

Figure 3.26 Superhet receiver design example

associated with several of the popular integrated circuit IF stages.

The block diagram of the receiver is shown in Figure 3.26. The vast majority of the receiver's gain and selectivity is associated with the two IF stages, TR3 and TR4. These two stages provide over 40 dB of voltage gain and the three IF tuned circuits and filter are instrumental in reducing the IF bandwidth to about 3.4 kHz for SSB reception. The RF stage (TR1) provides a modest amount of RF gain (about 20 dB at the maximum RF gain setting) together with a significant amount of image channel rejection.

The local oscillator stage (TR7) provides the necessary local oscillator signal which tunes from 5.455 MHz to 6.455 MHz. The local oscillator signal is isolated from the mixer stage and the LO output by means of the buffer stage, TR8.

The receiver incorporates two detector stages, one for AM and one for CW and SSB. The AM detector makes use of a simple diode envelope detector (D3) whilst the CW/SSB detector is based on a product detector (TR5). This stage offers excellent performance with both weak and strong CW and SSB signals. The 455 kHz carrier insertion is provided by means of the BFO/CIO

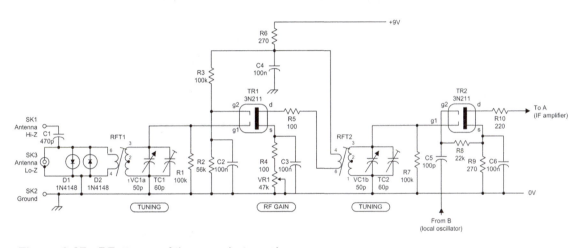

Figure 3.27 RF stages of the superhet receiver

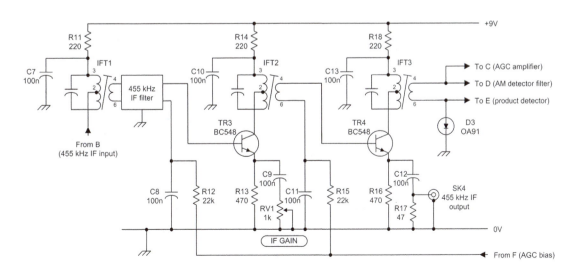

Figure 3.28 IF stages of the superhet receiver

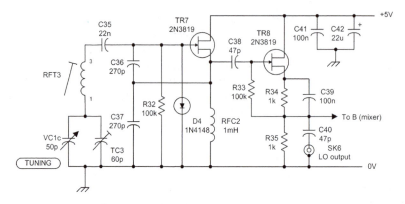

Figure 3.29 Local oscillator and buffer stages of the superhet receiver

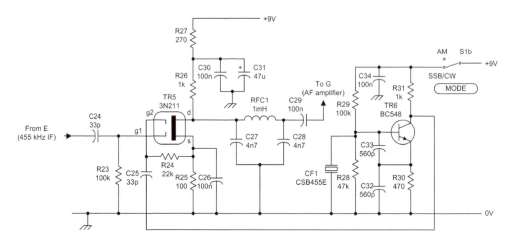

Figure 3.30 Product detector and BFO/CIO stages of the superhet receiver

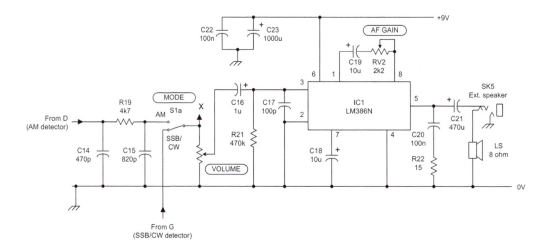

Figure 3.31 AF stages of the superhet receiver

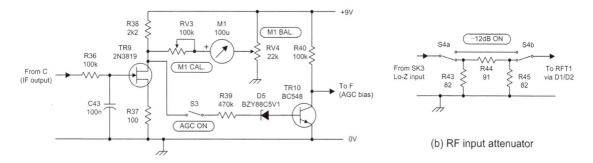

(a) Signal meter and AGC amplifier

(b) RF input attenuator

(c) Noise limiter

(d) Audio filter

Figure 3.32   Signal meter, AGC, RF input attenuator, noise limiter and audio filter stages

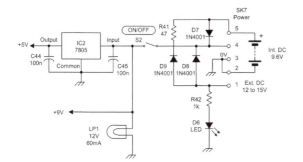

Figure 3.33   Power supply

stage (TR11). Amplified AGC is provided by means of TR9 and TR10.

A conventional integrated circuit audio amplifier stage (IC1) provides the audio gain necessary to drive a small loudspeaker. A 5 V regulator (IC2) is used to provide a stabilised low-voltage DC rail for the local oscillator and buffer stages. Diode switching is used to provide automatic changeover for the external DC or AC supplies. This circuitry also provides charging current for the internal nickel-cadmium (NiCd) battery pack.

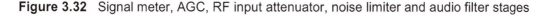

3.13 Multiple choice questions

1. A receiver in which selected signals of any frequency are converted to a single frequency is called a:
   (a) wideband TRF
   (b) multi-channel receiver
   (c) superhet receiver.

2. Delayed AGC:
   (a) maintains receiver sensitivity for very small signals
   (b) increases receiver sensitivity for very large signals
   (c) has no effect on receiver sensitivity.

3. A receiver with a high IF will successfully reject:
   (a) the image frequency
   (b) the adjacent frequency
   (c) the local oscillator frequency.

4. An IF amplifier consists of several stages. These are normally coupled using:
   (a) resistor/capacitor coupling
   (b) pure resistor coupling
   (c) transformer coupling.

5. SSB filters have a typical bandwidth of:
   (a) less than 300 Hz
   (b) 3 kHz to 6 kHz
   (c) more than 10 kHz.

6. The output signal of a diode detector comprises the modulated waveform, a small ripple and a DC component. The DC component is:
   (a) independent of the carrier strength
   (b) proportional to the carrier strength
   (c) inversely proportional to the carrier strength.

7. What is the principal function of the RF stage in a superhet receiver?
   (a) To improve the sensitivity of the receiver
   (b) To reduce second channel interference
   (c) To reduce adjacent channel interference.

8. A receiver having an IF of 1.6 MHz is tuned to a frequency of 12.8 MHz. Which of the following signals could cause image channel interference?
   (a) 11.2 MHz
   (b) 14.5 MHz
   (c) 16.0 MHz.

9. In an FM transmitter, the modulating signal is applied to:
   (a) the final RF amplifier stage
   (b) the antenna coupling unit
   (c) the RF oscillator stage.

10. The response of two coupled tuned circuits appears to be 'double-humped'. This is a result of:
    (a) undercoupling
    (b) overcoupling
    (c) critical coupling.

11. A disadvantage of low-level amplitude modulation is the need for:
    (a) a high-power audio amplifier
    (b) a high-power RF amplifier
    (c) a linear RF power amplifier.

12. The function of an antenna coupling unit in a transmitter is:
    (a) to provide a good match between the RF power amplifier and the antenna
    (b) to increase the harmonic content of the radiated signal
    (c) the reduce the antenna SWR to zero.

13. In order to improve the stability of a local oscillator stage:
    (a) a separate buffer stage should be used
    (b) the output signal should be filtered
    (c) an IF filter should be used.

14. A dual conversion superhet receiver uses:
    (a) a low first IF and a high second IF
    (b) a high first IF and a low second IF
    (c) the same frequency for both first and second IF.

15. The majority of the gain in a superhet receiver is provided by:
    (a) the RF amplifier stage
    (b) the IF amplifier stage
    (c) the AF amplifier stage.

16. Image channel rejection in a superhet receiver is improved by:
    (a) using an IF filter
    (b) using a low IF
    (c) using a high IF.

Chapter 4

VHF communications

Very high frequency (VHF) radio has long been the primary means of communication between aircraft and the ground. The system operates in the frequency range extending from 118 MHz to 137 MHz and supports both voice and data communication (the latter becoming increasingly important). This chapter describes the equipment used and the different modes in which it operates.

VHF communication is used for various purposes including air traffic control (ATC), approach and departure information, transmission of meteorological information, ground handling of aircraft, company communications, and also for the Aircraft Communications and Reporting System (ACARS).

4.1 VHF range and propagation

In the VHF range (30 MHz to 300 MHz) radio waves usually propagate as direct line-of-sight (LOS) waves (see Chapter 1). Sky wave propagation still occurs at the bottom end of the VHF range (up to about 50 MHz depending upon solar activity) but at the frequencies used for aircraft communication, reflection from the ionosphere is exceptionally rare.

Communication by strict line-of-sight (LOS) paths, augmented on occasions by diffraction and reflection, imposes a limit on the working range that can be obtained. It should also be evident that the range will be dependent on the height of an aircraft above the ground; the greater this is the further the range will be.

The maximum line-of-sight (LOS) distance (see Figure 4.1) between an aircraft and a ground station, in nautical miles (nm), is given by the relationship:

$$d = 1.1\sqrt{h}$$

where h is the aircraft's altitude in feet above ground (assumed to be flat terrain).

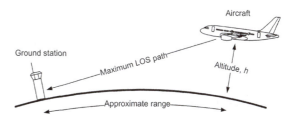

Figure 4.1 VHF line-of-sight range

Example 4.1

Determine the maximum line-of-sight distance when an aircraft is flying at a height of (a) 2,500 feet, and (b) 25,000 feet.

In (a), $h = 2{,}500$ hence:

$$d = 1.1\sqrt{2{,}500} = 1.1 \times 50 = 55 \text{ nm}$$

In (b), $h = 25{,}000$ hence:

$$d = 1.1\sqrt{25{,}000} = 1.1 \times 158 = 174 \text{ nm}$$

The actual range obtained depends not only on the LOS distance but also on several other factors, including aircraft position, transmitter power, and receiver sensitivity. However, the LOS distance usually provides a good approximation of the range that can be obtained between an aircraft and a ground station (see Table 4.1). The situation is slightly more complex when communication is from one aircraft to another; however, in such cases summing the two LOS distances will normally provide a guide as to the maximum range that can be expected.

Test your understanding 4.1

Determine the altitude of an aircraft that would provide a line-of-sight distance to a ground station located at a distance of 125 nm.

Table 4.1 Theoretical LOS range

Altitude (feet)	Approx. LOS range (nm)
100	10
1,000	32
5,000	70
10,000	100
20,000	141

4.2 DSB modulation

Amplitude modulation is used for voice communications as well as several of the **VHF data link** (VDL) modes. The system uses **double sideband** (DSB) modulation and, because this has implications for the bandwidth of modulated signals, it is worth spending a little time explaining how this works before we look at how the available space is divided into channels.

Figure 4.2 shows the frequency spectrum of an RF carrier wave at 124.575 MHz amplitude modulated by a single pure sinusoidal tone with a frequency of 1 kHz. Note how the amplitude modulated waveform comprises three separate components:

- an **RF carrier** at 124.575 MHz
- a **lower side frequency** (LSF) component at 124.574 MHz
- an **upper side frequency** (USF) component at 124.576 MHz.

Note how the LSF and USF are spaced away from the RF carrier by a frequency that is equal to that of the modulating signal (in this case 1 kHz). Note also from Figure 4.2 that the **bandwidth** (i.e. the range of frequencies occupied by the modulated signal) is *twice* the frequency of the modulating signal (i.e. 2 kHz).

Figure 4.3 shows an RF carrier modulated by a speech signal rather than a single sinusoidal tone. The **baseband** signal (i.e. the voice signal itself) typically occupies a frequency range extending from around 300 Hz to 3.4 kHz. Indeed, to improve intelligibility and reduce extraneous noise, the frequency response of the microphone and speech amplifier is invariable designed to

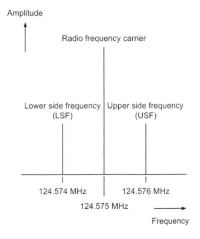

Figure 4.2 Frequency spectrum of an RF carrier using DSB modulation and a pure sinusoidal modulating signal

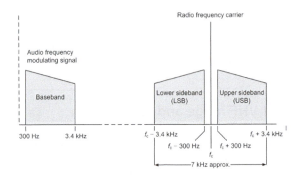

Figure 4.3 Frequency spectrum of a baseband voice signal (left) and the resulting DSB AM RF carrier (note that the bandwidth of the RF signal is approximately twice that of the highest modulating signal frequency)

select this particular range of frequencies and reject any audio signals that lie outside it. From Figure 4.3 it should be noted that the bandwidth of the RF signal is approximately 7 kHz (i.e. twice that of the highest modulating signal).

Test your understanding 4.2

Determine the RF signal frequency components present in a DSB amplitude modulated carrier wave at 118.975 MHz when the modulating signal comprises pure tones at 2 kHz and 5 kHz.

4.3 Channel spacing

VHF aircraft communications take place in a number of allocated channels. These channels were originally spaced at 200 kHz intervals throughout the VHF aircraft band. However, a relentless increase in air traffic coupled with the increasing use of avionic systems for data link communication has placed increasing demands on the available frequency spectrum. In response to this demand, the spacing between adjacent channels in the band 118 MHz to 137 MHz has been successively reduced so as to increase the number of channels available for VHF communication (see Table 4.2).

Figure 4.4 shows the channel spacing for the earlier 25 kHz and current European 8.33 kHz VHF systems. Note how the 8.33 kHz system of channel spacing allows three DSB AM signals to occupy the space that was previously occupied by a single signal.

The disadvantage of narrow channel spacing is that the **guard band** of unused spectrum that previously existed with the 25 kHz system is completely absent and that receivers must be designed so that they have a very high degree of **adjacent channel rejection** (see page 48). Steps must also be taken to ensure that the bandwidth of the transmitted signal does not exceed the 7 kHz, or so, bandwidth required for effective voice communication. The penalty for not restricting the bandwidth is that signals from one channel may 'spill over' into the adjacent channels, causing interference and degrading communication (see Figure 4.7).

Table 4.2 Increase in the number of available VHF channels

Date	Frequency range	Channel spacing	Number of channels
1947	118 MHz to 132 MHz	200 kHz	70
1958	118 MHz to 132 MHz	100 kHz	140
1959	118 MHz to 136 MHz	100 kHz	180
1964	118 MHz to 136 MHz	50 kHz	360
1972	118 MHz to 136 MHz	25 kHz	720
1979	118 MHz to 137 MHz	25 kHz	760
1995	118 MHz to 137 MHz	8.33 kHz	2280

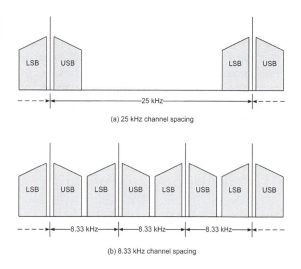

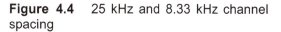

Figure 4.4 25 kHz and 8.33 kHz channel spacing

Test your understanding 4.3

How many channels at a spacing of 12.5 kHz can occupy the band extending from 118 MHz to 125 MHz?

Test your understanding 4.4

A total of 1520 data channels are to be accommodated in a band extending from 316 MHz to 335 MHz. What channel spacing must be used and what range of frequencies can the baseband signal have?

4.4 Depth of modulation

The depth of modulation of an RF carrier wave is usually expressed in terms of **percentage modulation**, as shown in Figure 4.6. Note that the level of modulation can vary between 0% (corresponding to a completely unmodulated carrier) to 100% (corresponding to a fully modulated carrier).

In practice, the intelligibility of a signal (i.e. the ability to recover information from a weak signal that may be adversely affected by noise and other disturbances) increases as the percentage modulation increases and hence there is a need to ensure that a transmitted signal is fully modulated but without the attendant risk of over-modulation (see Fig. 4.6). The result of over-modulation is excessive bandwidth, or 'splatter', causing adjacent channel interference, as shown in Fig. 4.7.

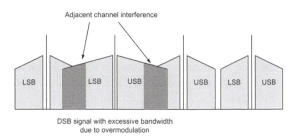

Figure 4.7 Adjacent channel interference caused by overmodulation

4.5 Compression

In order to improve the intelligibility of VHF voice communications, the speech amplifier stage of an aircraft VHF radio is invariably fitted with a **compressor** stage. This stage provides high gain for low amplitude signals and reduced gain for high amplitude signals. The result is an increase in the average modulation depth (see Figure 4.8).

Figure 4.9 shows typical speech amplifier characteristics with and without compression. Note that most aircraft VHF radio equipment provides adjustment both for the level of modulation and for the amount of compression that is applied (see Figure 4.10).

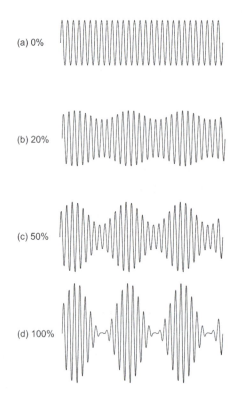

Figure 4.5 Different modulation depths

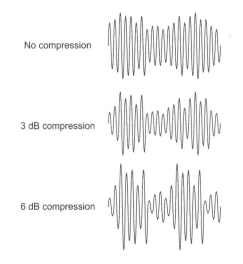

Figure 4.8 Modulated RF carrier showing different amounts of compression applied to the modulating signal

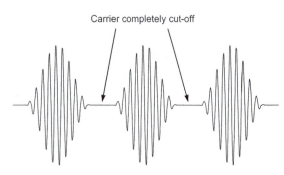

Figure 4.6 Over-modulation

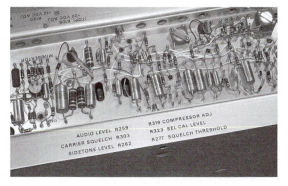

Figure 4.9 Effect of compression on average modulation depth

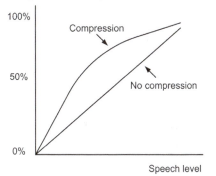

Figure 4.10 VHF radio adjustment points

4.6 Squelch

Aircraft VHF receivers invariably incorporate a system of muting the receiver audio stages in the absence of an incoming signal. This system is designed to eliminate the annoying and distracting background noise that is present when no signals are being received. Such systems are referred to as **squelch** and the threshold at which this operates is adjusted (see Figure 4.10) so that the squelch 'opens' for a weak signal but 'closes' when no signal is present.

Two quite different squelch systems are used but the most common (and easy to implement) system responds to the amplitude of the received carrier and is known as **carrier operated squelch**. The voltage used to inhibit the receiver audio can be derived from the receiver's AGC system and fed to the squelch gate (Figure 4.11).

The alternative (and somewhat superior) squelch system involves sensing the noise present at the output of the receiver's detector stage and using this to develop a control signal which is dependent on the signal-to-noise ratio of the received signal rather than its amplitude. This latter technique, which not only offers better sensitivity but is also less prone to triggering from general background noise and off-channel signals, is often found in FM receivers and is referred to as **noise operated squelch**.

4.7 Data modes

Modern aircraft VHF communications equipment supports both data communication as well as voice communication. The system used for the aircraft data link is known as Aircraft Communications Addressing and Reporting System (ACARS). Currently, aircraft are equipped with three VHF radios, two of which are used for ATC voice communications and one is used for the ACARS data link (also referred to as **airline operational control** communications).

A data link terminal on board the aircraft (see Figure 4.12) generates **downlink** messages and processes **uplink** messages received via the VHF data link. The downlink and uplink ACARS messages are encoded as plain ASCII text. In the Unites States, the ACARS ground stations are operated by ARINC whilst in Europe, Asia and Latin America, the equivalent service is provided by SITA.

Initially each VHF ACARS provider was allocated a single VHF channel. However, as the use of VHF data links (VDL) has grown, the number of channels used in the vicinity of the busiest airports has increased to as many as four and these are often operating at full capacity.

Unfortunately, due to the pressure for additional voice channels, it has not been possible to assign a number of additional VHF channels for ACARS data link operation. As a result, several new data modes have recently been introduced that support higher data rates and make more efficient use of each 25 kHz channel currently assigned for data link purposes.

In addition, the FAA is developing a system that will permit the integration of ATC voice and

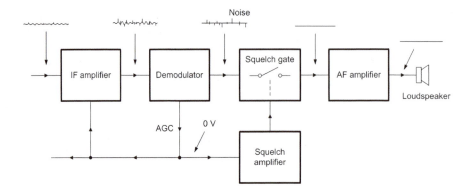

(a) No signal present (squelch gate open)

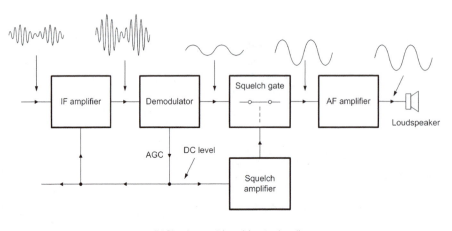

(b) Signal present (squelch gate closed)

Figure 4.11 Action of the squelch system

data communications. This system uses digitally encoded audio rather than conventional analogue voice signals.

When operating in **VDL Mode 0**, the required data link protocols are implemented in the ACARS **management unit** (see Figure 4.11). Data is transferred from the VHF radio to the management unit at a rate of 2400 bits per second (bps) by means of **frequency shift keying** (FSK). The FSK audio signal consists of two sinusoidal tones, one at a 1.2 kHz and one at 2.4 kHz depending on whether the polarity of the information bit being transmitted is the same as that of the previous bit or is different. Note that the phase of the tones varies linearly and that

there is no phase change on the transition between the two tones.

This type of modulation (in which the frequency spacing between the two audio tones is exactly half the data rate) is highly efficient in terms of bandwidth and is thus referred to as **minimum shift keying** (MSK). When data is transmitted, the MSK signal is used to modulate the amplitude of the VHF carrier (in much the same was as the voice signal). The resultant transmitted signal is then a double side-band (DSB) AM signal whose amplitude is modulated at 2400 bps. The RF carrier is then said to use DSB AM MSK modulation.

VHF carrier frequency selection and transmit/

Table 4.3 Summary of voice and data modes

Mode	Modulation	Channel spacing	Access method	Data rate	Type of traffic	Radio interface
Voice	DSB AM	25/8.33 kHz	PTT	Not applicable	Voice	Analogue
Data (Mode 0)	DSB AM MSK	25 kHz	CSMA	2,400 bps	ACARS	Analogue
Data (Mode A)	DSB AM MSK	25 kHz	CSMA	2,400 bps	ACARS	ARINC 429
Data (Mode 2)	D8PSK	25 kHz	CSMA	31,500 bps	ACARS and ATN	ARINC 429

receive control is provided by the ACARS management unit working in conjunction with an ARINC 429 interface to the VHF radio (Figure 4.12). The channel access protocol employed is known as **carrier sense multiple access** (CSMA). It consists of listening for activity on the channel (i.e. transmissions from other users) and transmitting only when the channel is free.

Operation in **VDL Mode A** is similar to Mode 0 except uplink and downlink ACARS data packets are transferred between the VHF radio and the ACARS management unit via a transmit/receive pair of 100 kbps ARINC 429 digital interfaces rather than the analogue audio interface used by Mode 0. The digital data is then used by the VHF radio to modulate the RF carrier at a rate of 2400 bps using the same DSB AM MSK modulation scheme used by VDL Mode 0.

Another difference between VDL Mode 0 and VDL Mode A is that, when using the latter, the VHF radio controls when to access the channel to transmit data using the same CSMA protocol employed by the management unit in VDL Mode 0. However, the selection of the frequency to be used is still controlled by the CMU or ATSU by means of commands issued via the same ARINC 429 interface used for data transfer. Note that, as far as the VHF data link ground stations are concerned, there is no difference in the air/ground VDL Mode 0 or VDL Mode A transmissions.

Operation in **VDL Mode 2** is based on an improved set of data transfer protocols and, as a result, it provides a significant increase in data capacity. VDL Mode 2 has been designed to provide for the future migration of VDL to the aeronautical telecommunications network (ATN). This network will permit more efficient and seamless delivery of data messages and data files between aircraft and the ground computer systems used by airlines and air traffic control facilities.

ATN will be supported by a number of air/ground networks and ground/ground networks. The air/ground and ground/ground networks will be interconnected by means of ATN **routers** that implement the required protocols and will operate in much the same way as the Internet with which you are probably already familiar.

VDL Mode 2 employs a data rate of 31,500 bits per second over the air/ground link using a single 25 kHz channel. The increased utilization of the 25 kHz channel is achieved by employing a system of modulation that is more efficient in terms of its use of bandwidth. This system is known as **differential eight phase shift keying** (D8PSK). In this system, an audio carrier signal is modulated be means of shift in phase that can take one of eight possible phases; 0°, 45°, 90°, 135°, 180°, 225°, 270° or 315°. These phase changes correspond to three bits of digital data as follows: 000, 001, 011, 010, 110, 111, 101, or 100. The D8PSK modulator uses the bits in the data message, in groups of three, to determine the carrier phase change at a rate of 10.5 kHz. Consequently, the bit rate will be three times this value, or 31.5 kbps. D8PSK modulation of the phase of the VHF carrier is accomplished using a **quadrature modulator**. Note that, in D8PSK modulation, groups of three bits are often referred to as **D8PSK symbols**.

VDL Mode 3 offers an alternative to the European solution of reducing the channel spacing to 8.33 kHz. VDL Mode 3 takes a 25 kHz frequency assignment and divides it into 120 ms frames with four 30 ms time slots (each of which constitutes a different channel). Thus Mode 3 employs **time division multiplexing** (TDM) rather than **frequency division multiplexing** (FDM) used in the European system. Note that VDL Mode 3 is the only planned VDL mode that is designed to support voice and data traffic *on the same frequency*.

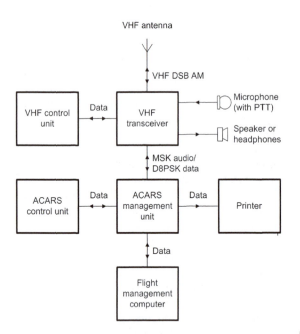

Figure 4.13 VHF radio data management

4.8 ACARS

ACARS (Aircraft Communication Addressing and Reporting System) is a digital data link system transmitted in the VHF range (118 MHz to 136 MHz). ACARS provides a means by which aircraft operators can exchange data with an aircraft without human intervention. This makes it possible for an airline to communicate with the aircraft in their fleet in much the same way as it is possible to exchange data using a land-based digital network. ACARS uses an aircraft's unique identifier and the system has

some features that are similar to those currently used for electronic mail.

The ACARS system was originally specified in the ARINC 597 standard but has been revised as ARINC 724B. A significant feature of ACARS is the ability to provide real-time data on the ground relating to aircraft performance; this has made it possible to identify and plan aircraft maintenance activities.

ACARS communications are automatically directed through a series of ground-based ARINC (Aeronautical Radio Inc.) computers to the relevant aircraft operator. The system helps to reduce the need for mundane HF and VHF voice messages and provides a system which can be logged and tracked. Typical ACARS messages are used to convey routine information such as:

- passenger loads
- departure reports
- arrival reports
- fuel data
- engine performance data.

This information can be requested by the company and retrieved from the aircraft at periodic intervals or on demand. Prior to ACARS this type of information would have been transferred via VHF voice.

ACARS uses a variety of hardware and software components including those that are installed on the ground and those that are present

```
ACARS mode: E Aircraft reg: N27015
Message label: H1 Block id: 3
Msg no: C36C
Flight id: CO0004
Message content:-
#CFBBY ATTITUDE INDICATOR
MSG 2820121 A 0051 06SEP06 CL H PL
DB FUEL QUANTITY PROCESSOR UNIT
MSG 3180141 A 0024 06SEP06 TA I 23
PL
DB DISPLAYS-2 IN LEFT AIMS
MSG 2394201 A 0005 06SEP06 ES H PL
MSG 2717018
```

Figure 4.14 Example of a downlink ACARS message sent from a Boeing 777 aircraft

in the aircraft. The aircraft ACARS components include a **management unit** (see Figure 4.12) which deals with the reception and transmission of messages via the VHF radio transceiver, and the **control unit** which provides the crew interface and consists of a display screen and printer. The ACARS **ground network** comprises the ARINC ACARS remote transmitting/ receiving stations and a network of computers and switching systems. The ACARS **command, control and management subsystem** consists of the ground-based airline operations and associated functions including operations control, maintenance and crew scheduling.

There are two types of ACARS messages; **downlink** messages that originate from the aircraft and **uplink** messages that originate from ground stations (see Figures 4.14 to 4.17). Frequencies used for the transmission and reception of ACARS messages are in the band extending from 129 MHz to 137 MHz (VHF) as shown in Table 4.4. Note that different channels are used in different parts of the world. A typical ACARS message (see Figure 4.14) consists of:

- mode identifier (e.g. 2)
- aircraft identifier (e.g. G-DBCC)
- message label (e.g. 5U—a weather request)
- block identifier (e.g. 4)
- message number (e.g. M55A)
- flight number (e.g. BD01NZ)
- message content (see Figure 4.14).

Table 4.4 ACARS channels

Frequency	ACARS service
129.125 MHz	USA and Canada (additional)
130.025 MHz	USA and Canada (secondary)
130.450 MHz	USA and Canada (additional)
131.125 MHz	USA (additional)
131.475 MHz	Japan (primary)
131.525 MHz	Europe (secondary)
131.550 MHz	USA, Canada, Australia (primary)
131.725 MHz	Europe (primary)
136.900 MHz	Europe (additional)

```
ACARS mode: 2
Aircraft reg: G-DBCC
Message label: 5U
Block id: 4
Msg no: M55A
Flight id: BD01NZ
Message content:-
01 WXRQ 01NZ/05 EGLL/EBBR .G-DBCC
/TYP 4/STA EBBR/STA EBOS/STA EBCI
```

Figure 4.14 Example of an ACARS message (see text)

```
ACARS mode: 2 Aircraft reg: N788UA
Message label: RA Block id: L
Msg. no: QUHD
Flight id: QWDUA~
Message content:-
WEIGHT MANIFEST
UA930 SFOLHR
SFO
ZFW      383485
TOG      559485
MAC      40.1
TRIM     02.8
PSGRS    285
```

Figure 4.15 Example of aircraft transmitted data (in this case, a weight manifest)

```
ACARS mode: X Aircraft reg: N199XX
Message label: H1 Block id: 7
Msg no: F00M
Flight id: GS0000
Message content:-
#CFBER FAULT/WRG [SWPA2]
INTERFACE
TCAS FAIL ADVISORY
TERRAIN 1 FAIL ADVISORY
TERRAIN 1-2 FAIL ADVISORY
THROTTLE QUADRANT 1-2 FAIL ADVISORY
22-10 221009ATA1 OC=1
TQA FAULT [ATA1]
INTERFACE
22-10 221009ATA
```

Figure 4.16 Example of a failure advisory message transmitted from an aircraft

```
ACARS mode: R Aircraft reg: G-EUPR
Message label: 10 Block id: 8
Msg no: M06A
Flight id: BA018Z
Message content:-
FTX01.ABZKOBA
BA1304
WE NEED ENGINEERING TO DO PDC ON
NUMBER 2 IDG
CHEERS
ETL 0740 GMT
```

Figure 4.17 Example of a plain text message sent via ACARS

Figure 4.18 Three VHF radios (on the extreme left) installed in the aircraft's avionic equipment bay

Test your understanding 4.5

Explain the need for (a) speech compression and (b) squelch in an aircraft VHF radio.

Test your understanding 4.6

Explain, with the aid of a block diagram, how data transfer is possible using an aircraft VHF radio.

Figure 4.19 VHF communications frequency selection panel (immediately above the ILS panel)

Test your understanding 4.7

Explain the difference between MSK and D8PSK modulation. Why is the latter superior?

4.9 VHF radio equipment

The typical specification of a modern aircraft VHF data radio is shown in Table 4.5. This radio can be used with analogue voice as well as data in Modes 0, A and 2 (see page 65). Figures 4.18 to 4.20 show typical equipment and control locations in a passenger aircraft whilst Figures 4.21 to 4.24 show internal and external views of a typical VHF radio. Finally, Figure 4.25 shows a typical VHF quarter-wave blade antenna fitted to an Airbus A380 aircraft.

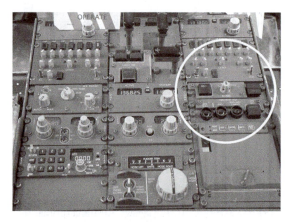

Figure 4.20 ACARS control panel (immediately to the right of the VHF communications frequency selection panel)

Figure 4.21 Aircraft VHF radio removed from its rack mounting

Table 4.5 Aircraft VHF radio specifications

Parameter	Specification
Frequency range	118.00 MHz to 136.99167 MHz
Channel spacing	8.33 kHz or 25 kHz
Operating modes	Analogue voice (ARINC 716); Analogue data 2400 bps AM MSK ACARS (external modem); ARINC 750 Mode A analogue data 2400 bps AM MSK ACARS; Mode 2 data 31.5 kbps D8PSK
Sensitivity	2 μV for 6 dB (S+N)/N
Selectivity (25 kHz channels)	6 dB max. attenuation at ±16 kHz 60 dB min. attenuation at ±34 kHz
Selectivity (8.33 kHz channels)	6 dB max. attenuation at ± 5.5 kHz 60 dB min. attenuation at ±14.7 kHz
Audio power output	Adjustable from less than 50 μW to 50 m into 600 Ω ± 20%
RF output power	25 W min. DSB AM operation 18 W min. D8PSK operation
Frequency stability	±0.005%
Modulation level	0.25 V RMS input at 1 kHz will modulate the transmitter at least 90%
Speech processing	Greater than 20 dB of compression
Mean time between failure	Greater than 40,000 hours

Figure 4.22 Digital frequency synthesiser stages of the VHF radio. The quartz crystal controlled reference oscillator is at the bottom left corner and the frequency divider chain runs from left to right with the screened VCO at the top

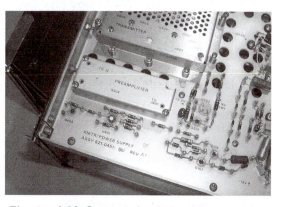

Figure 4.23 Screened receiver pre-amplifier and transmitter power amplifier stages (top)

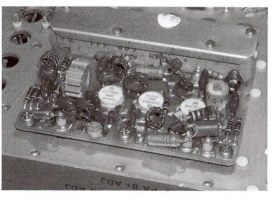

Figure 4.24 RF power amplifier stages with the screening removed. There are three linear power stages and one driver (left)

Figure 4.25 The forward quarter-wave VHF blade antenna on the Airbus A380 (see page 15 for the VHF antenna locations on a Boeing 757)

4.10 Multiple choice questions

1. The angle between successive phase changes of a D8PSK signal is:
   (a) 45°
   (b) 90°
   (a) 180°.

2. The method of modulation currently employed for aircraft VHF voice communication is:
   (a) MSK
   (b) D8PSK
   (c) DSB AM.

3. The channel spacing currently used in Europe for aircraft VHF voice communication is:
   (a) 8.33 kHz and 25 kHz
   (b) 12.5 kHz and 25 kHz
   (c) 25 kHz and 50 kHz.

4. Which one of the following gives the approximate LOS range for an aircraft at an altitude of 15,000 feet?
   (a) 74 nm
   (b) 96 nm
   (c) 135 nm.

5. The function of the compressor stage in an aircraft VHF radio is:
   (a) to reduce the average level of modulation
   (b) to increase the average level of modulation
   (c) to produce 100% modulation at all times.

6. The function of the squelch stage in an aircraft VHF radio is:
   (a) to eliminate noise when no signal is received
   (b) to increase the sensitivity of the receiver for weak signals
   (c) to remove unwanted adjacent channel interference.

7. Large passenger aircraft normally carry:
   (a) two VHF radios
   (b) three VHF radios
   (c) four VHF radios.

8. The typical bandwidth of a DSB AM voice signal is:
   (a) 3.4 kHz
   (b) 7 kHz
   (c) 25 kHz.

9. The disadvantage of narrow channel spacing is:
   (a) the need for increased receiver sensitivity
   (b) the possibility of adjacent channel interference
   (c) large amounts of wasted space between channels.

10. The standard for ACARS is defined in:
    (a) ARINC 429
    (b) ARINC 573
    (c) ARINC 724.

11. The frequency band currently used in Europe for aircraft VHF voice communication is:
    (a) 88 MHz to 108 MHz
    (b) 108 MHz to 134 MHz
    (c) 118 MHz to 137 MHz.

12. The typical output power of an aircraft VHF radio using voice mode is:
    (a) 25 W
    (b) 150 W
    (c) 300 W.

Chapter 5   HF communications

High frequency (HF) radio provides aircraft with an effective means of communication over long distance oceanic and trans-polar routes. In addition, global data communication has recently been made possible using strategically located HF data link (HFDL) ground stations. These provide access to ARINC and SITA airline networks. HF communication is thus no longer restricted to voice and is undergoing a resurgence of interest due to the need to find a means of long distance data communication that will augment existing VHF and SATCOM data links.

An aircraft HF radio system operates on spot frequencies within the HF spectrum. Unlike aircraft VHF radio, the spectrum is not divided into a large number of contiguous channels but aircraft allocations are interspersed with many other services, including short wave broadcasting, fixed point-to-point, marine and land-mobile, government and amateur services. This chapter describes the equipment used and the different modes in which it operates.

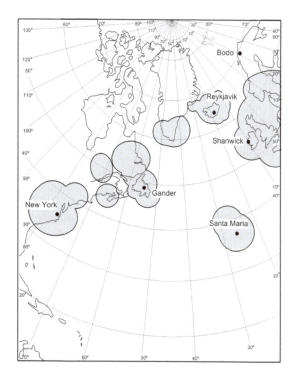

Figure 5.1 VHF aircraft coverage in the North Atlantic area

5.1 HF range and propagation

In the HF range (3 MHz to 30 MHz) radio waves propagate over long distances due to reflection from the ionised layers in the upper atmosphere. Due to variations in height and intensities of the ionised regions, different frequencies must be used at different times of day and night and for different paths. There is also some seasonal variation (particularly between winter and summer). Propagation may also be disturbed and enhanced during periods of intense solar activity. The upshot of this is that HF propagation has considerable vagaries and is far less predictable than propagation at VHF.

Frequencies chosen for a particular radio path are usually set roughly mid-way between the lowest usable frequency (LUF) and the maximum usable frequency (MUF). The daytime LUF is usually between 4 to 6 MHz during the day, falling rapidly after sunset to around 2 MHz. The MUF is dependent on the season and sunspot cycle but is often between 8 MHz and 20 MHz. Hence a typical daytime frequency for aircraft communication might be 8 MHz whilst this might be as low as 3 MHz during the night. Typical ranges are in the region of 500 km to 2500 km and this effectively fills in the gap in VHF coverage (see Figure 5.1).

As an example of the need to change frequencies during a 24-hour period, Figure 5.2

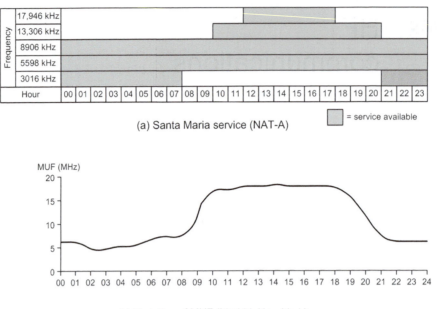

(a) Santa Maria service (NAT-A)   = service available

(b) Variation of MUF (Madrid–New York)

Figure 5.2 Santa Maria oceanic service (NAT-A) showing operational frequencies and times together with typical variation of MUF for a path from Madrid to New York

shows how the service provided by the Santa Maria HF oceanic service makes use of different parts of the HF spectrum at different times of the day and night. Note the correlation between the service availability chart shown in Figure 5.2(a) and the typical variation in maximum usable frequency (MUF) for the radio path between Madrid and New York.

The following HF bands are allocated to the aeronautical service:

- 2850 to 3155 kHz
- 3400 to 3500 kHz
- 4650 to 4750 kHz
- 5480 to 5730 kHz
- 6525 to 6765 kHz
- 8815 to 9040 kHz
- 10,005 to 10,100 kHz
- 11,175 to 11,400 kHz
- 13,200 to 13,360 kHz
- 15,010 to 15,100 kHz
- 17,900 to 18,030 kHz
- 21,870 to 22,000 kHz
- 23,200 to 23,350 kHz.

5.2 SSB modulation

Unfortunately, the spectrum available for aircraft communications at HF is extremely limited. As a result, steps are taken to restrict the bandwidth of transmitted signals, for both voice and data. **Double sideband** (DSB) amplitude modulation requires a bandwidth of at least 7 kHz but this can be reduced by transmitting only one of the two sidebands. Note that either the **upper sideband** (USB) or the **lower sideband** (LSB) can be used because they both contain the same modulating signal information. In addition, it is possible to reduce (or 'suppress') the carrier as this, in itself, does not convey any information.

In order to demodulate a signal transmitted without a carrier it is necessary to reinsert the carrier at the receiving end (this is done in the demodulator stage where a beat frequency oscillator or **carrier insertion oscillator** replaces the missing carrier signal at the final intermediate frequency—see Figure 5.9). The absence of the carrier means that less power is wasted in the transmitter which consequently operates at significantly higher efficiency.

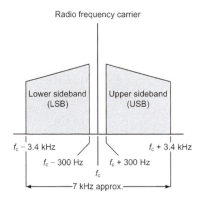

(a) Double sideband (DSB) full-carrier AM

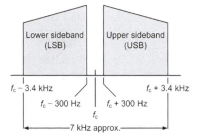

(b) Double sideband suppressed-carrier (DSB-SC)

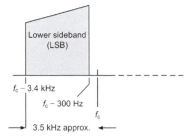

(c) Single sideband suppressed-carrier (SSB-SC)

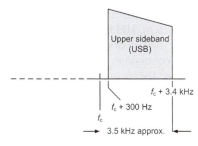

(d) Single sideband suppressed-carrier (SSB-SC)

Figure 5.3 Frequency spectrum of an RF carrier using DSB and SSB modulation

Figure 5.3 shows the frequency spectrum of an RF signal using different types of amplitude modulation, with and without a carrier.

In Figure 5.3(a) the mode of transmission is conventional **double sideband** (DSB) amplitude modulation with full-carrier. This form of modulation is used for VHF aircraft communications and was described earlier in Chapter 4.

Figure 5.3(b) shows the effect of suppressing the carrier. This type of modulation is known as **double sideband suppressed-carrier** (DSB-SC). In practical DSB-SC systems the level of the carrier is typically reduced by 30 dB, or more. The DSB-SC signal has the same overall bandwidth as the DSB full-carrier signal but the reduction in carrier results in improved efficiency as well as reduced susceptibility to heterodyne interference.

Figure 5.3(c) shows the effect of removing both the carrier and the upper sideband. The resulting signal is referred to as **single sideband** (SSB), in this case using only the **lower sideband** (LSB). Note how the overall bandwidth has been reduced to only around 3.5 kHz, i.e. half that of the comparable DSB AM signal shown in Figure 5.3(a).

Finally, Figure 5.3(d) shows the effect of removing the carrier and the lower sideband. Once again, the resulting signal is referred to as single sideband (SSB), but in this case we are using only the **upper sideband** (USB). Here again, the overall bandwidth has been reduced to around 3.5 kHz. Note that aircraft HF communication requires the use of the upper sideband (USB). DSB AM may also be available but is now very rarely used due to the superior performance offered by SSB.

Test your understanding 5.1

1. Explain why HF radio is used on trans-oceanic routes.

2. Explain why different frequencies are used for HF aircraft communications during the day and at night.

3. State TWO advantages of using SSB modulation for aircraft HF communications.

5.3 SELCAL

Selective calling (SELCAL) reduces the burden on the flight crew by alerting them to the need to respond to incoming messages. SELCAL is available at HF and VHF but the system is more used on HF. This is partly due to the intermittent nature of voice communications on long oceanic routes and partly due to the fact that squelch systems are more difficult to operate when using SSB because there is no transmitted carrier to indicate that a signal is present on the channel.

The aircraft SELCAL system is defined in Annex 10 to the Convention on International Civil Aviation (ICAO), Volume 1, 4th edition of 1985 (amended 1987). The system involves the transmission of a short burst of audio tones.

Each transmitted code comprises two consecutive tone pulses, with each pulse containing two simultaneously transmitted tones. The pulses are of 1 second duration separated by an interval of about 0.2 seconds. To ensure proper operation of the SELCAL decoder, the frequency of the transmitted tones must be held to an accuracy of better than ± 0.15%.

SELCAL codes are uniquely allocated to particular aircraft by Air Traffic Control (ATC). As an example, a typical transmitted SELCAL code might consist of a 1 second burst of 312.6 Hz and 977.2 Hz followed by a pause of about 0.2 seconds and a further 1 second burst of tone comprising 346.7 Hz and 977.2 Hz. Table 5.1 indicates that the corresponding transmitted SELCAL code is 'AM-BM' and only the aircraft with this code would then be alerted to the need to respond to an incoming message.

The RF signal transmitted by the ground radio station should contain (within 3 dB) equal amounts of the two modulating tones and the combination of tones should result in a modulation envelope having a nominal modulation percentage as high as possible (and in no case less than 60%).

The transmitted tones are made up from combinations of the tones listed in Table 5.1. Note that the tones have been chosen so that they are not harmonically related (thus avoiding possible confusion within the SELCAL decoder when harmonics of the original tone frequencies might be present in the demodulated waveform).

Table 5.1 SELCAL tone frequencies

Character	Frequency
A	312.6 Hz
B	346.7 Hz
C	384.6 Hz
D	426.6 Hz
E	473.2 Hz
F	524.8 Hz
G	582.1 Hz
H	645.7 Hz
J	716.1 Hz
K	794.3 Hz
L	881.0 Hz
M	977.2 Hz
P	1083.9 Hz
Q	1202.3 Hz
R	1333.5 Hz
S	1479.1 Hz

5.4 HF data link

ARINC's global high frequency data link (HFDL) coverage provides a highly cost-effective data link capability for carriers on remote oceanic routes, as well as the polar routes at high latitudes where SATCOM coverage is unavailable. HFDL is lower in cost than SATCOM and many carriers are using HFDL instead of satellite services, or as a backup system. HFDL is still the only data link technology that works over the North Pole, providing continuous, uninterrupted data link coverage on the popular polar routes between North America and eastern Europe and Asia.

The demand for HFDL has grown steadily since ARINC launched the service in 1998, and today HFDL avionics are offered as original equipment by all the major airframe manufacturers. HFDL offers a cost-effective solution for global data link service. The demand for HFDL service is currently growing by more

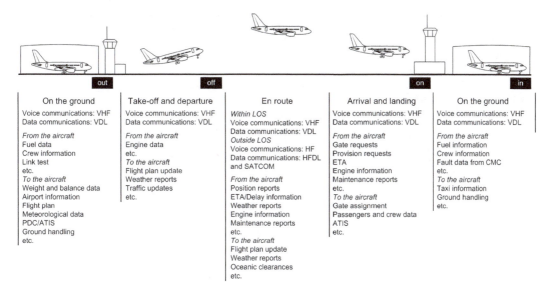

Figure 5.4 Aircraft operational control at various 'out-off-on-in' (OOOI) stages

than several hundred aircraft per year.

Advantages of HFDL can be summarised as:

- wide coverage due to the extremely long range of HF signals
- simultaneous coverage on several bands and frequencies (currently 60)
- multiple ground stations (currently 14) at strategic locations around the globe
- relatively simple avionics using well-tried technology
- rapid network acquisition
- exceptional network availability.

Disadvantages of HFDL are:

- very low data rates (making the system unsuitable for high-speed wideband communications).

As a result of the above, the vast majority of HFDL messages are related to **airline operational control** (AOC) (see Fig.ure 5.4) but HFDL is also expected to play an important part in **future air navigation systems** (FANS) where it will provide a further means of data linking with an aircraft, supplementing VDL, GPS, and SATCOM systems. Note that SATCOM can support much faster data rates but it can also be susceptible to interruptions and may not available at high latitudes.

HFDL uses **phase shift keying** (PSK) at data rates of 300, 600, 1200 and 1800 bps. The rate used is dependent on the prevailing propagation conditions. HFDL is based on **frequency division multiplexing** (FDM) for access to ground station frequencies and **time division multiplexing** (TDM) within individual communication channels. Figure 5.5 shows how the frequency spectrum of a typical HFDL signal at 300 bps compares with an HF voice signal.

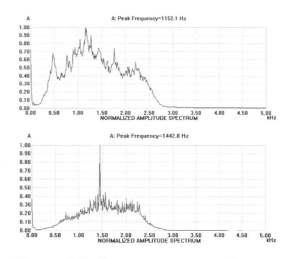
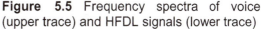

Figure 5.5 Frequency spectra of voice (upper trace) and HFDL signals (lower trace)

```
Preamble 300  bps 1.8 sec Interleaver FREQ ERR 5.398116 Hz Errors 0
[MPDU AIR CRC PASS]
Nr LPDUs = 1 Ground station ID SHANNON - IRELAND SYNCHED
Aircraft ID LOG-ON
Slots Requested medium = 0 Low = 0
Max Bit rate 1800 bps U(R) = 0 UR(R)vect = 0
[LPDU LOG ON DLS REQUEST] ICAO AID 0A123C
[HFNPDU FREQUENCY DATA]
14:45:24  UTC  Flight ID = AB3784  LAT 39 37 10  N  LON 0 21 20  W
07 87 FF 00 04 00 14 85 92 BF 3C 12 0A FF D5      ...............
41 42 33 37 38 34 C8 C2 31 BF FF C2 67 88 8C      A B 3 7 8 4 ..1 ...g ..
00 00 00 00 00 00 00 00 00 00 00 00 00 00 00      ...............
00 00 00 00 00 00 00 00 00 00 00 00 00 00 00      ...............
00 00 00 00 00 00 00                              .......

Preamble 300  bps 1.8 sec Interleaver FREQ ERR -18.868483 Hz Errors 19
[MPDU AIR CRC PASS]
Nr LPDUs = 1 Ground station ID SHANNON - IRELAND SYNCHED
Aircraft ID LOG-ON
Slots Requested medium = 0 Low = 0
Max Bit rate 1200 bps U(R) = 0 UR(R)vect = 0
[LPDU LOG ON DLS REQUEST] ICAO AID 4A8002
[HFNPDU FREQUENCY DATA]
14:45:30  UTC  Flight ID = SU0106  LAT 54 42 16  N  LON 25 50 42  E
07 87 FF 00 03 00 14 80 1E BF 02 80 4A FF D5      ............J ..
53 55 30 31 30 36 6A 6E F2 60 12 C5 67 33 FB      S U 0 1 0 6 j n ....g 3 .
00 00 00 00 00 00 00 00 00 00 00 00 00 00 00      ...............
00 00 00 00 00 00 00 00 00 00 00 00 00 00 00      ...............
00 00 00 00 00 00 00                              .......

Preamble 300  bps 1.8 sec Interleaver FREQ ERR 15.059247 Hz Errors 2
[MPDU AIR CRC PASS]
Nr LPDUs = 1 Ground station ID SHANNON - IRELAND SYNCHED
Aircraft ID AF
Slots Requested medium = 0 Low = 0
Max Bit rate 1200 bps U(R) = 0 UR(R)vect = 0
[LPDU UNNUMBERED DATA]
[HFNPDU PERFORMANCE]
14:45:30  UTC  Flight ID = LH8409  LAT 46 42 34  N  LON 21 22 55  E
07 87 AF 00 03 00 31 4D 1D 0D FF D1 4C 48 38      ......1 M ....L H 8
34 30 39 73 13 82 34 0F C5 67 01 36 03 02 02      4 0 9 s ..4 ..g .6 ...
00 B6 00 00 00 00 00 00 00 00 03 00 00 00 00      ...............
02 00 00 00 00 00 01 00 00 00 01 01 D3 EA 00      ...............
00 00 00 00 00 00 00                              .......

Preamble 300  bps 1.8 sec Interleaver FREQ ERR 8.355845 Hz Errors 0
[MPDU AIR CRC PASS]
Nr LPDUs = 1 Ground station ID SHANNON - IRELAND SYNCHED
Aircraft ID AD
Slots Requested medium = 0 Low = 0
Max Bit rate 1200 bps U(R) = 0 UR(R)vect = 0
[LPDU UNNUMBERED DATA]
[HFNPDU PERFORMANCE]
14:43:30  UTC  Flight ID = LH8393  LAT 52 37 27  N  LON 16 46 41  E
07 87 AD 00 03 00 31 C5 0B 0D FF D1 4C 48 38      ......1 .....L H 8
33 39 33 BF 56 62 EE 0B 89 67 01 8A 07 01 B8      3 9 3 .V b ...g .....
00 7E 00 00 00 00 00 00 00 00 06 0F 00 00 00 00   ...............
2E 00 00 00 00 00 05 00 00 00 05 07 08 27 00      ...............
00 00 00 00 00 00 00                              .......
```

Figure 5.6 Examples of aircraft communication using HFDL

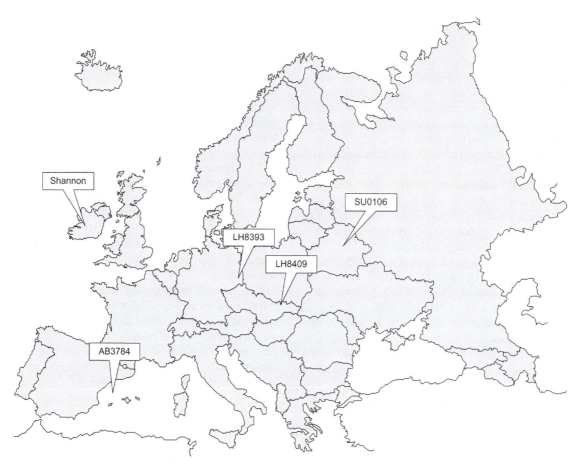

Figure 5.7 Ground station and aircraft locations for the HFDL communications in Figure 5.6

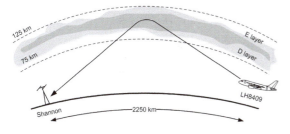

Figure 5.8 Radio path for LH8409

Figure 5.6 shows typical HFDL messages sent from the four aircraft shown in Figure 5.7 to the Shannon HFDL ground station using the same communications channel. The radio path from one of the aircraft (LH8409) is illustrated in Figure 5.8. The first two of the messages shown in Figure 5.6 are **log-on requests** and the maximum bit rate is specified in the header. In each log-on request, the aircraft is identified by its unique 24-bit **ICAO address**. Once logged on, the aircraft is allocated an 8-bit **address code** (AF hex in the case of the third message and AD hex in the case of the fourth message). Each aircraft also transmits its current location data (longitude and latitude).

The system used for HFDL data exchange is specified in ARINC 635. Each ground station transmits a frame called a 'squitter' every 32 seconds. The **squitter frame** informs aircraft of the system status, provides a timing reference and provides protocol control. Each ground station has a time offset for its squitters. This allows aircraft to jump between ground stations finding the best one before logging on. When passing traffic, dedicated TDM time slots are used. This prevents two aircraft transmitting at the same time causing **data collisions**.

5.5 HF radio equipment

The block schematic of a simple HF transmitter/ receiver is shown in Figure 5.9. Note that, whilst this equipment uses a single intermediate frequency (IF), in practice most modern aircraft HF radios are much more complex and use two or three intermediate frequencies.

On transmit mode, the DSB suppressed carrier (Figure 5.2b) is produced   by means of a **balanced modulator** stage. The balanced modulator rejects the carrier and its output just comprises the upper and lower sidebands. The DSB signal is then passed through a multiple-stage crystal or mechanical filter. This filter has a very narrow pass-band (typically 3.4 kHz) at the intermediate frequency (IF) and this rejects the unwanted sideband. The resulting SSB signal is then mixed with a signal from the digital frequency synthesiser to produce a signal on the wanted channel. The output from the mixer is then further amplified before being passed to the output stage. Note that, to avoid distortion, all of the stages must operate in linear mode.

When used on receive mode, the incoming signal frequency is mixed with the output from the digital frequency synthesiser in order to produce the intermediate frequency signal. Unwanted adjacent channel signals are removed by means of another multiple-stage crystal or mechanical filter which has a pass-band similar to that used in the transmitter. The IF signal is then amplified before being passed to the demodulator.

The (missing) carrier is reinserted in the demodulator stage. The carrier signal is derived from an accurate crystal controlled carrier oscillator which operates at the IF frequency. The recovered audio signal from the demodulator is then passed to the audio amplifier where it is amplified to an appropriate level for passing to a loudspeaker.

The typical specification for an aircraft HF radio is shown in Table 5.2. One or two radios of this type are usually fitted to a large commercial aircraft (note that at least one HF radio is a requirement for *any* aircraft following a trans-oceanic route). Figure 5.10 shows the flight deck location of the **HF radio controller**.

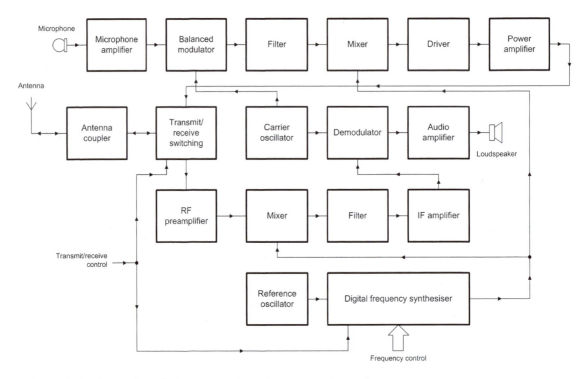

Figure 5.9 A simple SSB transmitter/receiver

Table 5.2 Aircraft HF radio specifications

Parameter	Specification
Frequency range	2.0000 MHz to 29.9999 MHz
Tuning steps	100 Hz
Operating modes	SSB SC analogue voice (ARINC 719) and analogue data (ARINC 753 and ARINC 635) at up to 1800 bps; DSB AM (full carrier)
Sensitivity	1 µV for 10 dB (S+N)/N SSB; 4 µV for 10 dB (S+N)/N AM
Selectivity	6 dB max. attenuation at +2.5 kHz 60 dB min. attenuation at +3.4 kHz
Audio output	50 mW into 600 Ω
SELCAL output	50 mW into 600 Ω
RF output power	200 W pep min. SSB; 50 W min. DSB AM
Frequency stability	±20 Hz
Audio response	350 Hz to 2500 Hz at −6 dB
Mean time between failure	Greater than 50,000 hours

Figure 5.10   HF radio control unit

Test your understanding 5.2

1. Explain how HF data link (HFDL) differs from VHF data link (VDL). Under what circumstances is HFDL used and what advantages does it offer?

2. Explain briefly how an aircraft logs on to the HFDL system. How are data collisions avoided?

5.6 HF antennas and coupling units

External wire antennas were frequently used on early aircraft. Such antennas would usually run from the fuselage to the top of the vertical stabiliser and they were sufficiently long to permit resonant operation on one or more of the aeronautical HF bands. Unfortunately this type of antenna is unreliable and generally unsuitable for use with a modern high-speed passenger aircraft. The use of a large probe antenna is unattractive due to its susceptibility to static discharge and lightning strike. Hence an alternative solution in which the HF antenna is protected within the airframe is highly desirable. Early experiments (see Figure 5.13) showed that the vertical stabiliser (tail fin) would be a suitable location

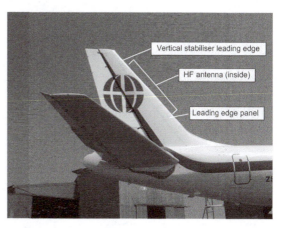

Figure 5.11   HF antenna location

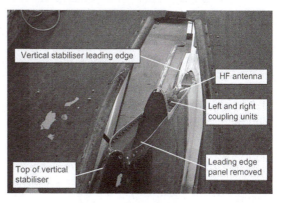

Figure 5.12   View from the top of the vertical stabiliser (leading edge panel removed)

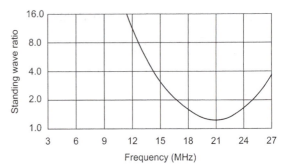

Figure 5.14 Variation of SWR with frequency for an HF notch antenna (note the logarithmic scale used for SWR)

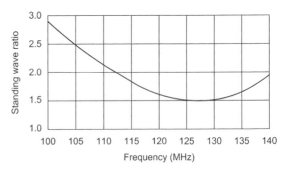

Figure 5.15 Variation of SWR with frequency for a VHF quarter-wave blade antenna (note the linear scale used for SWR)

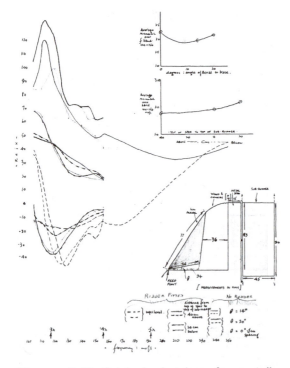

Figure 5.13 Original sketches for a tail-mounted antenna from work carried out by E. H. Tooley in 1944

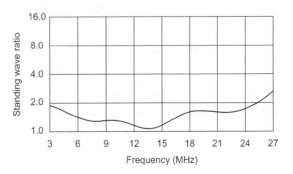

Figure 5.16 Variation of SWR with frequency for an HF notch antenna fitted with an antenna coupling/tuning unit

and is now invariably used to house the HF antenna and its associated coupling unit on most large transport aircraft—see Figures 5.11 and 5.12.

Due to the restriction in available space (which mitigates against the use of a resonant antenna such as a quarter-wave Marconi antenna—see page 24) the HF antenna is based on a notch which uses part of the airframe in order to radiate effectively. The notch itself has a very high-Q factor and its resistance and reactance varies very widely over the operating frequency range (i.e. 3 MHz to 24 MHz). The typical variation of **standing wave ratio** (SWR—see page 33) against frequency for an HF notch antenna is shown in Figure 5.14. For comparison, the variation of SWR with frequency for a typical quarter-wave VHF blade antenna is shown in Figure 5.15.

From Figures 5.14 and 5.15 it should be obvious that the HF antenna, whilst well matched at 21 MHz, would be severely mismatched to a

conventional 50 Ω feeder/transmitter at most other HF frequencies. Because of this, and because the notch antenna is usually voltage fed, it is necessary to use an HF coupling/tuning unit

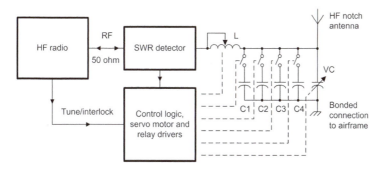

Figure 5.17 Typical feedback control system used in an HF antenna coupler

between the HF radio feeder and the notch antenna. This unit is mounted in close proximity to the antenna, usually close to the top of the vertical stabiliser (see Figure 5.12). Figure 5.16 shows the effect of using a coupling/tuning unit on the SWR-frequency characteristic of the same notch antenna that was used in Figure 5.14. Note how the SWR has been reduced to less than 2:1 for most (if not all) of the HF range.

The tuning adjustment of HF antenna coupler is entirely automatic and only requires a brief signal from the transmitter to retune to a new HF frequency. The HF antenna coupler unit incorporates an SWR bridge (see page 35) and a feedback control system (see Figure 5.17) to adjust a roller coater inductor (L1) and high-voltage vacuum variable capacitor (C1) together with a number of switched high-voltage capacitors (C1 to C4). The internal arrangement of a typical HF antenna coupler is shown in Figures 5.18 and 5.19. The connections required between the HF antenna coupler, HF radio and control unit are shown in Figure 5.20.

Voltages present in the vicinity of the HF antenna (as well as the field radiated by it) can be extremely dangerous. It is therefore **essential** to avoid contact with the antenna and to maintain a safe working distance from it (at least 5 metres) whenever the HF radio system is 'live'.

Test your understanding 5.3

Explain the function of an HF antenna coupler. What safety precautions need to be observed when accessing this unit?

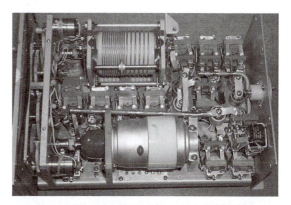

Figure 5.18 Interior view of an HF antenna coupler showing the roller coaster inductor (top) and vacuum variable capacitor (bottom). The high-voltage antenna connector is shown in the extreme right

Figure 5.19 SWR bridge circuit incorporated in the HF antenna coupler. The output from the SWR bridge provides the error signal input to the automatic feedback control system

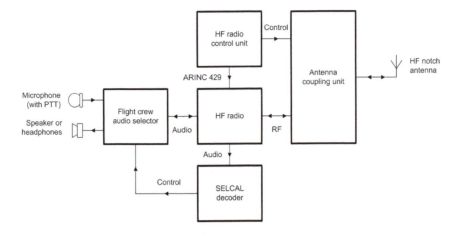

Figure 5.20 Connections to the HF radio, control unit and antenna coupling unit

5.7 Multiple choice questions

1. The typical bandwidth of an aircraft HF SSB
   signal is:
   (a) 3.4 kHz
   (b) 7 kHz
   (c) 25 kHz.

2. The principal advantage of SSB over DSB
   AM is:
   (a) reduced bandwidth
   (b) improved frequency response
   (c) faster data rates can be supported.

3. HF data link uses typical data rates of:
   (a) 300 bps and 600 bps
   (b) 2400 bps and 4800 bps
   (c) 2400 bps and 31,500 bps.

4. The standard for HF data link is defined in:
   (a) ARINC 429
   (b) ARINC 573
   (c) ARINC 635.

5. Which one of the following gives the
   approximate range of audio frequencies used
   for SELCAL tones?
   (a) 256 Hz to 2048 Hz
   (b) 312 Hz to 1479 Hz
   (c) 300 Hz to 3400 Hz.

6. How many alphanumeric characters are
   transmitted in a SELCAL code?
   (a) 4
   (b) 8
   (c) 16.

7. How many bits are used in an ICAO aircraft
   address?
   (a) 16
   (b) 24
   (c) 32.

8. The typical RF output power from an aircraft
   HF transmitter is:
   (a) 25 W pep
   (b) 50 W pep
   (c) 400 W pep.

9. An HF radio is required for use on oceanic
   routes because:
   (a) VHF coverage is inadequate
   (b) higher power levels can be produced
   (c) HF radio is more reliable.

10. The function of an HF antenna coupler is to:
    (a) reduce static noise and interference
    (b) increase the transmitter output power
    (c) match the antenna to the radio.

Chapter 6 | Flight-deck audio systems

As well as systems for communication with ground stations, modern passenger aircraft require a number of facilities for local communication within the aircraft. In addition, there is a need for communications with those who work on the aircraft when it is being serviced on the ground.

Systems used for local communications need to consist of nothing more than audio signals, suitably amplified, switched and routed, and incorporating a means of alerting appropriate members of the crew and other personnel.

These flight-deck audio systems include:

- passenger address (PA) system
- service interphone system
- cabin interphone system
- ground crew call system
- flight interphone system.

The **passenger address system** provides the flight crew and cabin crew with a means of making announcements and distributing music to passengers through cabin speakers. Circuits in the system send chime signals to the cabin speakers.

The **service interphone system** provides the crew and ground staff with interior and exterior communication capability. Circuits in the system connect service interphone jacks to the flight compartment.

The **cabin interphone system** provides facilities for communication among cabin attendants, and between the flight compartment crew members and attendants. The system can be switched to the input of the passenger address system for PA announcements.

The **ground crew call system** provides a signalling capability (through the ground crew call horn) between the flight compartment and nose landing gear area.

The **flight interphone system** provides facilities for interphone communication among flight compartment crew members and provides the means for them to receive, key and transmit using the various aircraft radio systems. The flight interphone system also extends communication to ground personnel at the nose gear interphone station and allows flight compartment crew members to communicate and to make passenger address announcements. The flight interphone system also incorporates amplifiers and mixing circuits in the audio accessory unit, audio selector panels, cockpit speakers, microphone/headphone jacks and press-to-talk (PTT) switches.

In addition to the audio systems used for normal operation of the aircraft, large commercial aircraft are also required to carry a **cockpit voice recorder** (CVR). This device captures and stores information derived from a number of the aircraft's audio channels. Such information may later become invaluable in the event of a crash or malfunction.

6.1 Flight interphone system

The flight interphone system provides the essential connecting link between the aircraft's communication systems, navigation receivers and flight-deck crew members. The flight interphone system also extends communication to ground personnel at external stations (e.g. the nose gear interphone station). It also provides the means by which members of the flight crew can communicate with the cabin crew and also make passenger address announcements. The flight interphone system comprises a number of sub-systems including amplifiers and mixing circuits in the audio accessory unit, audio selector panels, cockpit speakers, microphone/headphone jacks and **press-to-talk** (PTT) switches.

The flight interphone components provided for the captain and first officer usually comprise the following components:

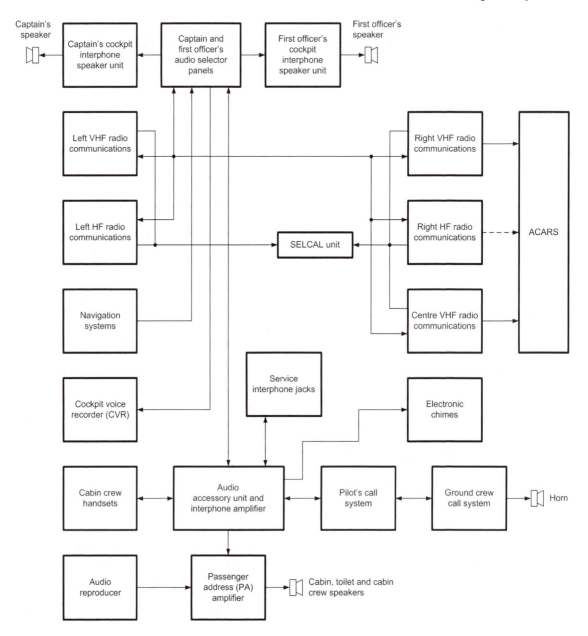

Figure 6.1 Simplified block schematic diagram of a typical flight interphone system

- audio selector panel
- headset, headphone, and hand microphone jack connectors
- audio selector panel and control wheel press-to-talk (PTT) switches
- cockpit speakers.

Note that, where a third (or fourth) seat is provided on the flight deck, a third (or fourth) set of flight interphone components will usually be available for the observer(s) to use. In common with other communication systems fitted to the aircraft, the flight interphone system normally derives its power from the aircraft's 28 V DC battery bus through circuit breakers on the overhead panel.

The simplified block schematic diagram of a typical flight interphone system is shown in Figure 6.1. Key subsystem components are the captain and first officer's audio selector panels and the **audio accessory unit** that provides a link from the flight deck audio system to the passenger address, cabin and service interphones, and ground crew call systems. It is also worth noting from Figure 6.1 that the audio signals (inputs and outputs) from the HF and VHF radio communications equipment as well as the audio signals derived from the navigation receivers (outputs only) are also routed via the audio selector panels. This arrangement provides a high degree of configuration flexibility together with a degree of redundancy sufficient to cope with failure of individual subsystem components.

Finally, it should be emphasised that the arrangement depicted in Figure 6.1 is typical and that minor variations can and do exist. For example, most modern aircraft incorporate SATCOM facilities (not shown in Figure 6.1).

The **flight interphone amplifier** is usually located in the audio accessory unit in the main avionic equipment rack). The amplifier receives low-level microphone inputs and provides audio to all flight interphone stations. The amplifier has preset internal adjustments for compression, squelch and volume.

Audio selector panels are located in the flight compartment within easy of reach of the crew members. Audio selector panels are provided for the captain and first officer as well as any observers that may be present on the flight deck. Depending on aircraft type and flight deck configuration, audio selector panels may be fitted in the central pedestal console or in the overhead panels. Typical examples of cockpit audio selector panel layouts are shown in Figures 6.2 and 6.3. Each audio selector panel contains microphone selector switches which connect microphone circuits to the interphone systems, to the radio communication systems, or to the passenger address system. The push-to-talk (PTT) switch on the audio selector panels can be used to key the flight compartment microphones. Volume control is provided by switches on each audio selector panel.

Figure 6.2 First officer's audio selector panel (top) and radio panel (bottom) fitted in the overhead panel of an A320 aircraft

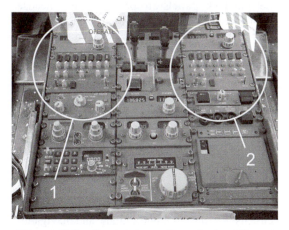

Figure 6.3 Captain's audio selector (1) and first officer's audio selector (2) fitted in the central console of a Boeing 757 aircraft

Two **cockpit speaker units** are usually fitted in the flight compartment. These are usually located in the sidewall panels adjacent to the captain's and first officer's stations. Each cockpit speaker unit contains a loudspeaker, amplifier, muting circuits, and a **volume control**. The speakers receive all audio signals provided to the audio selector panels. The speakers are muted whenever a PTT switch is pushed at the captain's or first officer's station.

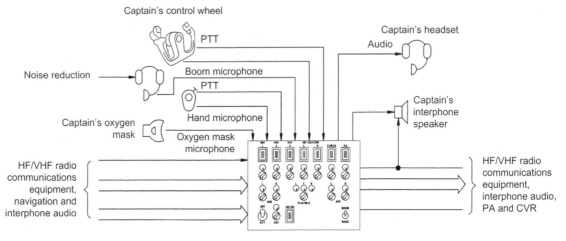

Figure 6.4 Typical arrangement for the captain's audio selector. A similar arrangement is used for the first officer's audio selector as well as any supernumerary crew members that may be present on the flight deck

Several **jack panels** are provided for a headset with integral boom microphone for the captain, first officer and observer. Hand microphones may also be used. Push-to-talk (PTT) switches are located at all flight interphone stations. The hand microphone, control wheel, and audio selector panels all have PTT switches. The switch must be pushed before messages are begun or no transmission can take place. Audio and control circuits to the audio selector panel are completed when the PTT switch is operated.

The flight interphone system provides common microphone circuits for the communications systems and common headphone and speaker circuits for the communications and navigation systems.

Figure 6.4 shows a typical arrangement for the captain's audio selector panel (note that the flight interphone components and operation are identical for both the captain and first officer). Similar (though not necessarily identical) systems are available for use by the observer and any other supernumerary crew members (one obvious difference is the absence of a control wheel push-to-talk switch and cockpit speaker). Switches are provided to select boom microphone, hand microphone (where available) as well as microphones located in the oxygen masks (for

emergency use). Outputs can be selected for use with the headset or cockpit loudspeakers.

Amplifiers, summing networks, and filters in the audio selector panel provide audio signals from the interphone and radio communication systems to the headphones and speakers. Audio signals from the navigation receivers are also monitored through the headphones and speakers. Reception of all audio signals is controlled by the volume switches. The captain's INT microphone switch is illuminated when active. Note that this switch is interlocked with the other microphone switches so that only one at a time can be pushed.

The navigation system's (ADF, VOR, ILS, etc.) audio is also controlled by switches on the audio selector panel. The left, centre, or right (L, C, R) switches control selection and volume of the desired receiver. The VOICE-BOTH-RANGE switch acts as a filter that separates voice signals and range signals. The filter switch can also combine both voice and range signals. All radio communication, interphone, and navigation outputs are received and recorded by the **cockpit voice recorder** (CVR).

A typical procedure for checking that the microphone audio is routed to the radio communication, interphone, or passenger address system is as follows:

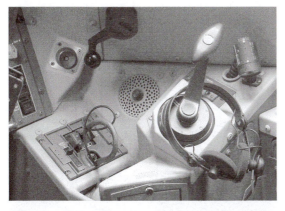

Figure 6.5 First officer's loudspeaker (centre) in a Boeing 757 aircraft (the volume control is mounted in the centre of the loudspeaker panel)

Figure 6.6 Captain's headset and boom microphone in a Boeing 757 aircraft. The press-to-talk (PTT) switch can be seen on the left-hand section of the control wheel

Figure 6.7 First officer's headset and boom microphone in an A320 aircraft

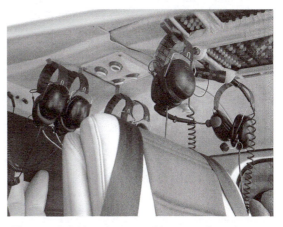

Figure 6.8 Headsets and boom microphones in a four-seat rotorcraft

1. Push the microphone select switch on the audio selector panel to select the required communication system.
2. If a handheld microphone is used, push the PTT switch on the microphone and talk.
3. If a boom microphone or oxygen mask microphone is used, select MASK or BOOM with the toggle switch on the audio selector panel and push the audio selector panel or control wheel PTT switch and talk.

The following procedure is used to listen to navigation and communication systems audio:

1. For communications systems, adjust the volume control switch on the audio selector panel and listen to the headset.
2. For navigation systems audio, select desired left-centre-right (L-C-R) and filter (VOICE-BOTH-RANGE) positions on the audio selector panel, adjust volume control switch and listen to headset.

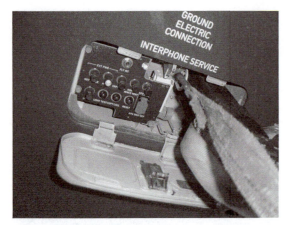

Figure 6.9 Ground staff interphone jack connector

3. The captain's and first officer's cockpit speakers (see Figure 6.5) can be used to listen to navigation as well as communication system audio. A control in the centre of the cockpit speaker (Boeing aircraft) or on an adjacent panel (Airbus) adjusts the speaker volume to the desired level.
4. External interphone panels (as appropriate to the aircraft type—see Figures 6.9 and 6.10) should be similarly tested by connecting a headset or handset (as appropriate) to each interphone jack.

Figures 6.5 to 6.10 show examples of some typical flight deck audio communications equipment used on modern passenger aircraft.

Test your understanding 6.1

Explain the differences between (a) the flight interphone and (b) the cabin interphone systems.

Test your understanding 6.2

1. Explain the function of the audio selector panels used by members of the flight crew.

2. List THREE different examples of inputs to an audio selector panel and THREE different examples of outputs from an audio selector panel.

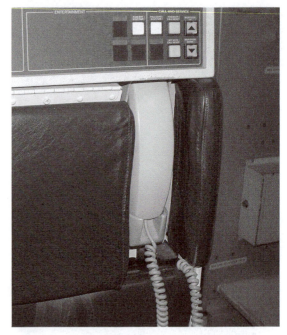

Figure 6.10 Cabin interphone/passenger address handset

6.2 Cockpit voice recorder

The cockpit voice recorder (CVR) can provide valuable information that can later be analysed in the event of an accident or serious malfunction of the aircraft or any of its systems. The voice recorder preserves a continuing record of typically between 30 and 120 minutes of the most recent flight crew communications and conversations.

The storage medium used with the CVR fitted to modern aircraft is usually based on one or more solid state memory devices whereas on older aircraft the CVR is usually based on a continuous loop of magnetic tape.

The CVR storage unit must be recoverable in the event of an accident. This means that the entire recorder unit including storage media must be mounted in an enclosure that can withstand severe mechanical and thermal shock as well as the high pressure that exists when a body is immersed at depth in water.

The CVR is usually fitted with a test switch, headphone jack, status light (green) and an externally mounted **underwater locator beacon**

(ULB) to facilitate undersea recovery. The ULB is a self-contained device (invariably attached to the front panel of the CVR) that emits an ultrasonic vibration (typically at 37.5 kHz) when the water-activated switch is activated as a result of immersion in either sea water or fresh water. A label on the ULB indicates the date by which the internal battery should be replaced. A typical specification for a ULB is shown in Table 6.1. An external view of a CVR showing its externally mounted ULB is shown in Figure 6.11.

The audio input to the CVR is derived from the captain, first officer, observer (where present) and also from an **area microphone** in the flight compartment which is usually mounted in the overhead panel and thus collects audio input from the entire flight-deck area.

In order to improve visibility and aid recovery, the external housing of the CVR is painted bright orange. The unit is thermally insulated and hermetically sealed to prevent the ingress of water. Because of the crucial nature of the data preserved by the flight, the unit should only be opened by authorised personnel following initial recovery from the aircraft.

Magnetic CVR use a multi-track tape transport mechanism. This normally comprises a tape drive, four recording heads, a single (full-width) erase head, a monitor head and a bulk erase coil. The bias generator usually operates at around 65 kHz and an internal signal (at around 600 Hz) is often provided for test purposes. Bulk erase can be performed by means of an erase switch (which is interlocked so that bulk tape erasure can only be performed when the aircraft is on the ground and the parking brake is set). The erase current source is usually derived directly from the aircraft's 115 V AC 400 Hz supply. The magnetic tape (a continuous loop) is usually 308 ft in length and ¼ in wide.

More modern solid-state recording media uses no moving parts (there is no need for a drive mechanism) and is therefore much more reliable. Erasure can be performed electronically and there is no need for a separate erase coil and AC supply. Finally, it is important to note that the CVR is usually mounted in the aft passenger cabin ceiling. This location offers the greatest amount of protection for the unit in the event of a crash.

Figure 6.11 Cockpit voice recorder fitted with an underwater locator beacon (ULB)

Table 6.1 Typical ULB specification

Parameter	Specification
Operating frequency	37.5 kHz (± 1 kHz)
Acoustic output	160 dB relative to 1 μPa at 1 m
Pulse repetition rate	0.9 pulses per sec
Pulse duration	10 ms
Activation	Immersion in either salt water or fresh water
Power source	Internal lithium battery
Battery life	6 years standby (shelf-life)
Beacon operating life	30 days
Operating depth	20,000 ft (6,096 m)
Housing material	Aluminium
Length	3.92 in (9.95 cm)
Diameter	1.3 in (3.3 cm)
Weight	6.7 oz (190 g)

Test your understanding 6.3

1. Explain the function and principle of operation of the underwater locator beacon (ULB) fitted to a cockpit voice recorder (CVR).

2. Explain why the CVR is located in the ceiling of the aft passenger cabin.

6.3 Multiple choice questions

1. Audio selector panels are located:
   (a) in the main avionic equipment bay
   (b) close to the pilot and first officer stations
   (c) in the passenger cabin for use by cabin
   crew members.

2. When are the flight-deck speaker units muted?
   (a) when a PTT switch is operated
   (b) when a headset is connected
   (c) when a navigation signal is received

3. Input to the captain's interphone speaker unit
   is derived from:
   (a) the audio selector panel
   (b) the passenger address system
   (c) the audio accessory unit and interphone
   amplifier.

4. The microphone PTT system is interlocked in
   order to prevent:
   (a) unwanted acoustic feedback
   (b) more than one switch being operated
   at any time
   (c) loss of signal due to parallel connection of
   microphones.

5. Bulk erasure of the magnetic tape media used
   in a CVR is usually carried out:
   (a) immediately after take-off
   (b) as soon as the aircraft has touched down
   (c) on the ground with the parking brake set.

6. The typical bias frequency used in a magnetic
   CVR is:
   (a) 3.4 kHz
   (b) 20 kHz
   (c) 65 kHz.

7. The typical frequency emitted by a ULB is:
   (a) 600 Hz
   (b) 3.4 kHz
   (c) 37.5 kHz.

8. Which one of the following is a suitable audio
   tone frequency for testing a CVR?
   (a) 60 Hz
   (b) 600 Hz
   (c) 6 kHz.

9. A ULB is activated:
   (a) automatically when immersed in water
   (b) manually when initiated by a crew
   member
   (c) when the unit is subjected to a high impact
   mechanical shock.

10. The CVR flight deck area microphone is
    usually mounted:
    (a) on the overhead panel
    (b) on the left-side flight deck floor
    (c) immediately behind the jump seat.

11. The typical pulse rate for a ULB is:
    (a) 0.9 pulses per sec
    (b) 10 pulses per sec
    (c) 60 pulses per sec.

12. The CVR is usually located:
    (a) on the flight deck
    (b) in the avionic equipment bay
    (c) in the ceiling of the aft passenger cabin.

13. What colour is used for the external housing
    of a CVR?
    (a) red
    (b) green
    (c) orange.

14. A ULB usually comprises:
    (a) a separate externally fitted canister
    (b) an internally fitted printed circuit module
    (c) an external module that derives its power
    from the CVR.

15. A ULB will operate:
    (a) only in salt water
    (b) only in fresh water
    (c) in either salt water or fresh water.

16. The typical shelf-life of the battery fitted to a
    ULB is:
    (a) six months
    (b) 18 months
    (c) six years.

Chapter 7

Emergency locator transmitters

The detection and location of an aircraft crash is vitally important to the search and rescue (SAR) teams and to potential survivors. Studies show that while the initial survivors of an aircraft crash have less than a 10% chance of survival if rescue is delayed beyond two days, the survival rate is increased to over 60% if the rescue can be accomplished within eight hours. For this reason, emergency locator transmitters (ELT) are required for most general aviation aircraft. ELT are designed to emit signals on the VHF and UHF bands thereby helping search crews locate aircraft and facilitating the timely rescue of survivors. This chapter provides a general introduction to the types and operating principles of ELT fitted to modern passenger aircraft.

7.1 Types of ELT

Several different types of ELT are in current use. These include the older (and simpler) units that produce a modulated RF carrier on one or both of the two spot VHF frequencies used for distress beacons (121.5 MHz and its second harmonic 243.0 MHz). Note that the former frequency is specified for civil aviation use whilst the latter is sometimes referred to as the military aviation distress frequency. Simultaneous transmission on the two frequencies (121.5 MHz and 243.0 MHz) is easily possible and only requires a frequency doubler and dual-band output stage.

Simple VHF ELT devices generate an RF carrier that is modulated by a distinctive siren-like sound. This sweeps downwards at a repetition rate of typically between 2 and 4 Hz. This signal can be readily detected by Sarsat and Cospas satellites (see later), or by any aircraft monitoring 121.5 MHz or 243.0 MHz.

More modern ELT operate on a spot UHF frequency (460.025 MHz). These devices are much more sophisticated and also operate at a significantly higher power (5 W instead of the 150 mW commonly used at VHF). Unlike the simple amplitude modulation used with their VHF counterparts, 460 MHz ELT transmit digitally encoded data which incorporates a code that is unique to the aircraft that carries them.

Provided they have been properly maintained, most ELT are capable of continuous operation for up to 50 hours. It is important to note that ELT performance (and, in particular, the operational range and period for which the signal is maintained) may become seriously impaired when the batteries are out of date. For this reason, routine maintenance checks are essential and any ELT which contains outdated batteries should be considered unserviceable.

The different types of ELT are summarised in Table 7.1. These are distinguished by application and by the means of activation. Modern passenger aircraft may carry several different types of ELT. Figure 7.1 shows a typical example of the Type-W (water activated) survival ELT carried on a modern transport aircraft.

Most ELT in general aviation aircraft are of the automatic type. Fixed automatic units contain a crash activation sensor, or G-switch, which is designed to detect the deceleration characteristics of a crash and automatically activate the transmitter.

With current Sarsat and Cospas satellites now in orbit, ELT signals will usually be detected, within 90 minutes, and the appropriate search and rescue (SAR) agencies alerted. Military aircrew monitor 121.5 MHz or 243.0 MHz and they will also notify ATS or SAR agencies of any ELT transmissions they hear.

It is worth noting that the detection ranges for Type-W and Type-S ELT can be improved if the ELT is placed upright, with the antenna vertical, on the highest nearby point with any accessible metal surface acting as a ground plane. Doubling the height will increase the range by about 40%.

Table 7.1 Types of ELT

Type	Class	Description
A or AD	Automatic ejectable or automatic deployable	This type of ELT automatically ejects from the aircraft and is set in operation by inertia sensors when the aircraft is subjected to a crash deceleration force acting through the aircraft's flight axis. This type is expensive and is seldom used in general aviation.
F or AF	Fixed (non-ejectable) or automatic fixed	This type of ELT is fixed to the aircraft and is automatically set in operation by an inertia switch when the aircraft is subjected to crash deceleration forces acting in the aircraft's flight axis. The transmitter can be manually activated or deactivated and in some cases may be remotely controlled from the cockpit. Provision may also be made for recharging the ELT's batteries from the aircraft's electrical supply. Most general aviation aircraft use this ELT type, which must have the function switch placed to the ARM position for the unit to function automatically in a crash (see Figure 7.5).
AP	Automatic portable	This type of ELT is similar to Type-F or AF except that the antenna is integral to the unit for portable operation.
P	Personnel activated	This type of ELT has no fixed mounting and does not transmit automatically. Instead, a switch must be manually operated in order to activate or deactivate the ELT's transmitter.
W or S	Water activated or Survival	This type of ELT transmits automatically when immersed in water (see Figure 7.1). It is waterproof, floats and operates on the surface of the water. It has no fixed mounting and should be tethered to survivors or life rafts by means of the supplied cord.

7.2 Maintenance and testing of ELT

ELT should be regularly inspected in accordance with the manufacturer's recommendations. The ELT should be checked to ensure that it is secure, free of external corrosion, and that antenna connections are secure. It is also important to ensure that the ELT batteries have not reached their expiry date (refer to external label) and that only approved battery types are fitted.

Air testing normally involved first listening on the beacon's output frequency (e.g. 121.5 MHz), checking first that the ELT is not transmitting before activating the unit and then checking the radiated signal. Simple air tests between an aircraft and a ground station (or between two aircraft) can sometimes be sufficient to ensure that an ELT is functional; however, it is important to follow manufacturer's instructions when

testing an ELT. Two-station air testing (in conjunction with a nearby ground station) is usually preferred because, due to the proximity of the transmitting and receiving antennae, a test carried out with the aircraft's own VHF receiver may not reveal a fault condition in which the ELT's RF output has become reduced.

To avoid unnecessary SAR missions, all accidental ELT activations should be reported to the appropriate authorities (e.g. the nearest rescue coordination centre) giving the location of the transmitter, and the time and duration of the accidental transmission. Promptly notifying the appropriate authorities of an accidental ELT transmission can be instrumental in preventing the launch of a search aircraft. Any testing of an ELT must be conducted only during the first five minutes of any UTC hour and restricted in duration to not more than five seconds.

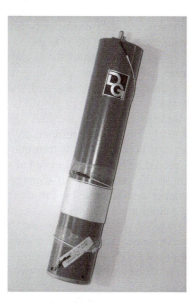

Figure 7.1 Type-W ELT with attachment cord secured by water-soluble tape (the antenna has been removed)

Figure 7.2 Interior view of the ELT shown in Figure 7.1. Note how the battery occupies approximately 50% of the internal volume

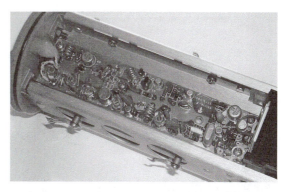

Figure 7.3 ELT transmitter and modulator printed circuit board (the crystal oscillator is located on the right with the dual-frequency output stages on the left)

Figure 7.4 ELT test switch and test light (the antenna base connector is in the centre of the unit)

7.3 ELT mounting requirements

In order to safeguard the equipment and to ensure that it is available for operation should the need arise, various considerations should be observed when placing and mounting an ELT and its associated antenna system in an aircraft. The following requirements apply to Type-F, AF, AP ELT installations in fixed wing aircraft and rotorcraft:

1. When installed in a fixed wing aircraft, ELT should be mounted with its sensitive axis pointing in the direction of flight
2. When installed in a rotorcraft ELT should be mounted with its sensitive axis pointing

Figure 7.5 Type-AF ELT control panel (note the three switch positions marked ON, ARMED and TEST/RESET)

Test your understanding 7.1

Distinguish between the following types of ELT: (a) Type-F, (b) Type-AF, and (c) Type-W.

Table 7.2 Typical Type-AF ELT specification

Parameter	Specification
Operating frequencies	121.5 MHz, 243 MHz and 406.025 MHz
Frequency tolerance	±0.005% (121.5 MHz and 243 MHz); ±2 kHz (406.025 MHz)
RF output power	250 mW typical (121.5 MHz and 243 MHz); 5 W ±2 dB (406.025 MHz)
Pulse duration	10 ms
Activation	G-switch
Power source	Internal lithium battery
Battery life	5 years (including effects of monthly operational checks)
Beacon operating life	50 hours
Digital message repetition period (406.025 MHz only)	Every 50 s
Modulation	AM (121.5 MHz and 243 MHz); phase modulation (406.025 MHz)
Housing material	Aluminium alloy

approximately 45° downward from the normal forward direction of flight

3. ELT should be installed to withstand ultimate inertia forces of 10 g upward, 22.5 g downward, 45 g forward and 7.5 g sideways
4. The location chosen for the ELT should be sufficiently free from vibration to prevent involuntary activation of the transmitter
5. ELT should be located and mounted so as to minimise the probability of damage to the transmitter and antenna by fire or crushing as a result of a crash impact
6. ELT should be accessible for manual activation and deactivation. If it is equipped with an antenna for portable operation, the ELT should be easily detachable from inside the aircraft
7. The external surface of the aircraft should be marked to indicate the location of the ELT
8. Where an ELT has provision for remote operation it is important to ensure that appropriate notices are displayed.

The antenna used by a fixed type of ELT should conform to the following:

1. ELT should not use the antenna of another avionics system
2. ELT antenna should be mounted as far away as possible from other very high frequency (VHF) antennas
3. The distance between the transmitter and antenna should be in accordance with the ELT manufacturer's installation instructions or other approved data
4. The position of the antenna should be such as to ensure essentially omnidirectional radiation characteristics when the aircraft is in its normal ground or water attitude
5. The antenna should be mounted as far aft as possible
6. ELT antenna should not foul or make contact with any other antennas in flight.

The following considerations apply to Type-W and Type-S ELT:

1. ELT should be installed as specified for Type-F but with a means of quick release, and located as near to an exit as practicable without being an obstruction or hazard to aircraft occupants

2. Where the appropriate regulations require the carriage of a single ELT of Type-W or Type-S, the ELT should be readily accessible to passengers and crew
3. Where the appropriate regulations require the carriage of a second Type-W or Type-S ELT, that ELT should be either located near a life raft pack, or attached to a life raft in such a way that it will be available or retrievable when the raft is inflated
4. An ELT fitted with a lithium or magnesium battery must not be packed inside a life raft in an aircraft.

7.4 Typical ELT

Figures 7.1 to 7.4 show the external and internal construction of a basic Type-W ELT. The unit is hermetically sealed at each end in order to prevent the ingress of water. The procedure for disassembling the ELT usually involves withdrawing the unit from one end of the cylindrical enclosure. When reassembling an ELT care must be taken to reinstate the hermetic seals at each end of the enclosure.

The specification for a modern Type-AF ELT is shown in Table 7.2. This unit provides outputs on all three ELT beacon frequencies; 121.5 MHz,

243 MHz, and 406.025 MHz. The ELT uses amplitude modulation (AM) on the two VHF frequencies (121.5 MHz and 243 MHz) and phase modulation (PM) on the UHF frequency (406.025 MHz). The AM modulating signal consists of an audio tone that sweeps downwards from 1.5 kHz to 500 Hz with three sweeps every second. The modulation depth is greater than 85%.

The block schematic diagram for a simple Type-W ELT is shown in Figure 7.6. The supply is connected by means of a water switch (not shown in Figure 7.6). The unit shown in Figure 7.6 only provides outputs at VHF (121.5 MHz and 243 MHz). These two frequencies are harmonically related which makes it possible to generate the 243 MHz signal using a frequency doubler stage.

Test your understanding 7.2

1. State THREE requirements that must be observed when an ELT is mounted in an aircraft.

2. Describe two methods of activating an ELT.

3. What precautions must be taken when an ELT is tested?

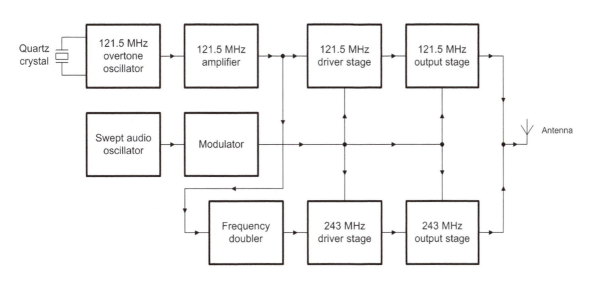

Figure 7.6 Block schematic diagram for a Type-W ELT

7.5 Cospas–Sarsat satellites

Cospas–Sarsat is a satellite system designed to supply alert and location information to assist search and rescue operations. The Russian Cospas stands for 'space system for the search of vessels in distress' whilst Sarsat stands for 'search and rescue satellite-aided tracking'.

The system uses satellites and ground stations to detect and locate signals from ELT operating at frequencies of 121.5 MHz, 243 MHz and/or 406 MHz. The system provides worldwide support to organizations responsible for air, sea or ground SAR operations.

The basic configuration of the Cospas–Sarsat system features:

- ELT that transmit VHF and/or UHF signals in case of emergency
- Instruments on board geostationary and low-orbiting satellites detecting signals transmitted by the ELT
- **Local user terminals** (LUT), which receive and process signals transmitted via the satellite downlink to generate distress alerts
- **Mission control centres** (MCC) which receive alerts from LUTs and send them to a **Rescue coordination centre** (RCC)
- **Search and rescue** (SAR) units.

There are two Cospas–Sarsat systems. One operates at 121.5 MHz (VHF) whilst the other operates at 406 MHz (UHF). The Cospas–Sarsat 121.5 MHz system uses low earth orbit (LEO) polar-orbiting satellites together with associated ground receiving stations. The basic system is shown in Figure 7.7.

The signals produced by ELT beacons are received and relayed by Cospas–Sarsat LEO-SAR satellites to Cospas–Sarsat LUTs that process the signals to determine the location of the ELT. The computed position of the ELT transmitter is relayed via an MCC to the appropriate RCC or **search and rescue point of contact** (SPOC).

The Cospas–Sarsat system uses Doppler location techniques (using the relative motion between the satellite and the distress beacon) to accurately locate the ELT. The carrier frequency transmitted by the ELT is reasonably stable during the period of mutual beacon-satellite visibility. Doppler performance is enhanced due to the low-altitude near-polar orbit used by the Cospas–Sarsat satellites. However, despite this it is important to note that the location accuracy of the 121.5 MHz system is not as good as the accuracy that can be achieved with the 406 MHz system. The low altitude orbit also makes it possible for the system to operate with very low uplink power levels.

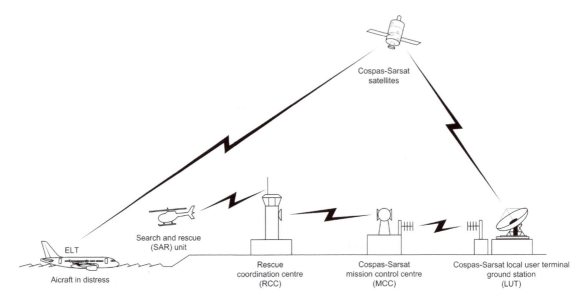

Figure 7.7 The Cospas–Sarsat system in operation

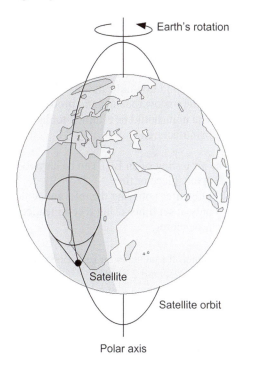

Figure 7.8   Polar orbit for a low altitude earth orbit (LEO) search and rescue (SAR) satellite

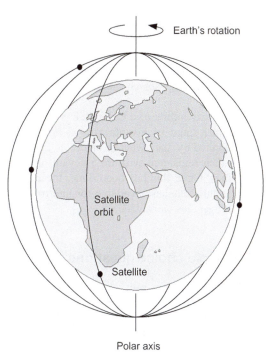

Figure 7.9   The constellation of four LEO–SAR satellites

A near polar orbit could provide full global coverage but 121.5 MHz can only be produced if the uplink signals from the ELT are actually received by an LUT. This constraint of the 121.5 MHz system limits the useful coverage to a geographic area of about 3,000 km radius around each LUT. In this region, the satellite can 'see' both the ELT and the LUT.

Figure 7.8 shows the polar orbit of a single satellite. The path (or 'orbital plane') of the satellite remains fixed, while the earth rotates underneath. At most, it takes only one half rotation of the earth (i.e. 12 hours) for any location to pass under the orbital plane. With a second satellite, having an orbital plane at right angles to the first, only one quarter of a rotation is required, or six hours maximum. Similarly, as more satellites orbit the earth in different planes, the waiting time is further reduced.

The complete Cospas–Sarsat system uses four satellites as shown in Figure 7.9. The system provides a typical waiting time of less than one hour at mid-latitudes. However, users of the 121.5 MHz system have to wait for a satellite

pass which provides for a minimum of four minutes, simultaneous visibility of an ELT and an LUT. This additional constraint may increase the waiting time to several hours if the transmitting beacon is at the edge of the LUT coverage area. The Doppler location provides two positions for each beacon: the true position and its mirror image relative to the satellite ground track. In the case of 121.5 MHz beacons, a second pass is usually required to resolve the ambiguity.

Sarsat satellites are also equipped with 243 MHz repeaters which allow the detection and location of 243 MHz distress beacons. The operation of the 243 MHz system is identical to the 121.5 MHz system except for the smaller number of satellites available.

The Cospas–Sarsat 406 MHz System is much more sophisticated and involves both orbiting and geostationary satellites. The use of 406 MHz beacons with digitally encoded data allows unique beacon identification.

In order to provide positive aircraft identification, it is essential that 406 MHz ELT are registered in a recognised ELT database accessible to search and rescue authorities. The information held in the database includes data on the ELT, its owner, and the aircraft on which the ELT is mounted. This information can be invaluable in a search and rescue (SAR) operation.

The unique coding of a UHF ELT is imbedded in the final stage of manufacture using aircraft data supplied by the owner or operator. The ELT data is then registered with the relevant national authorities. Once this has been done, the data is entered into a database available for interrogation by SAR agencies worldwide.

7.6 Multiple choice questions

1. ELT transmissions use:
    (a) Morse code and high-power RF at HF
    (b) pulses of acoustic waves at 37.5 kHz
    (c) low-power RF at VHF or UHF.

2. A Type-P ELT derives its power from:
    (a) aircraft batteries
    (b) internal batteries
    (c) a small hand-operated generator.

3. Transmission from an ELT is usually initially detected by:
    (a) low-flying aircraft
    (b) one or more ground stations
    (c) a satellite.

4. The operational state of an ELT is tested using:
    (a) a test switch and indicator lamp
    (b) immersion in a water tank for a short period
    (c) checking battery voltage and charging current.

5. A Type-W ELT needs checking. What is the first stage in the procedure?
    (a) Inspect and perform a load test on the battery
    (b) Open the outer case and inspect the hermetic seal
    (c) Read the label on the ELT in order to determine the unit's expiry date.

6. If bubbles appear when an ELT is immersed in a tank of water, which one of the following statements is correct?
    (a) This is normal and can be ignored
    (b) This condition indicates that the internal battery is overheating and producing gas
    (c) The unit should be returned to the manufacturer.

7. The air testing of an ELT can be carried out:
    (a) at any place or time
    (b) only after notifying the relevant authorities
    (c) only at set times using recommended procedures.

8. On which frequencies do ELT operate?
    (a) 125 MHz and 250 MHz
    (b) 122.5 MHz and 406.5 MHz
    (c) 121.5 MHz and 406.025 MHz

9. A Type-W ELT is activated by:
    (a) a member of the crew
    (b) immersion in water
    (c) a high G-force caused by deceleration.

10. The location accuracy of a satellite-based beacon locator system is:
    (a) better on 121.5 MHz than on 406 MHz
    (b) better on 406 MHz than on 121.5 MHz
    (c) the same on 121.5 MHz as on 406 MHz.

11. An ELT fitted with a lithium battery is:
    (a) safe for packing in a life raft
    (b) unsafe for packing in a life raft
    (c) not suitable for use with a Type-F ELT.

12. A Type-W or Type-S ELT will work better when the antenna is:
    (a) held upright
    (b) slanted downwards slightly
    (c) carefully aligned with the horizontal.

13. The satellites used by the Cospas–Sarsat 121.5 MHz system are:
    (a) in high earth orbit
    (b) in low earth orbit
    (c) geostationary.

Aircraft navigation

Navigation is the science of conducting journeys over land and/or sea. Whether the journey is to be made across deserts or oceans, we need to know the ultimate destination and how the journey's progress will be checked along the way. Finding a position on the earth's surface and deciding on the direction of travel can be simply made by observations or by mathematical calculations. Aircraft navigation is no different, except that the speed of travel is much faster! Navigation systems for aircraft have evolved with the nature and role of the aircraft itself. Starting with visual references and the basic compass, leading onto radio ground aids and self-contained systems, many techniques and methods are employed.

Although the basic requirement of a navigation system is to guide the crew from point A to point B, increased traffic density and airline economics means that more than one aircraft is planning a specific route. Flight planning takes into account such things as favourable winds, popular destinations and schedules. Aircraft navigation is therefore also concerned with the management of traffic and safe separation of aircraft. This chapter reviews some basic features of the earth's geometry as it relates to navigation, and introduces some basic aircraft navigation terminology. The chapter concludes by reviewing a range of navigation systems used on modern transport and military aircraft (a full description of these systems follows in subsequent chapters).

8.1 The earth and navigation

Before looking at the technical aspects of navigation systems, we need to review some basic features of the earth and examine how these features are employed for aircraft navigation purposes. Although we might consider the earth to be a perfect sphere, this is not the case. There is a flattening at both the poles such that the earth

is shaped more like an orange. For short distances, this is not significant; however, for long-range (i.e. global) navigation we need to know some accurate facts about the earth. The mathematical definition of a sphere is where the distance (radius) from the centre to the surface is equidistant. This is not the case for the earth, where the actual shape is referred to as an oblate spheroid.

8.1.1 Position

To define a unique two-dimensional position on the earth's surface, a coordinate system using imaginary lines of **latitude** and **longitude** are drawn over the globe, see Figure 8.1. Lines of longitude join the poles in **great circles** or **meridians**. A great circle is defined as the intersection of a sphere by a plane passing through the centre of the sphere; this has a radius measured from the centre to the surface of the earth. These north–south lines are spaced around the globe and measured in angular distance from the **zero (or prime) meridian**, located in Greenwich, London. Longitude referenced to the prime-meridian extends east or west up to 180 **degrees**. Note that the distance between lines of longitude converge at the poles. Latitude is the angular distance north or south of the equator; the poles are at latitude 90 degrees.

For accurate navigation, the degree (symbol ° after the value, e.g. 90° north) is divided by 60 giving the unit of **minutes** (using the symbol ' after numbers), e.g. one half of a degree will be 30'. This can be further refined into smaller units by dividing again by 60 giving the unit of **seconds** (using the symbol " after numbers), e.g. one half of a minute will be 30". A second of latitude (or longitude at the equator) is approximately 31 metres, just over 100 feet. Defining a unique position on the earth's surface, e.g. Land's End in Cornwall, UK, using latitude

Longitude

Meridians (described by the angle they make, measured east or west from the Greenwich, or 'prime', meridian)

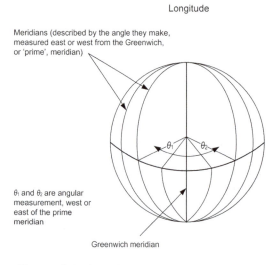

θ_1 and θ_2 are angular measurement, west or east of the prime meridian

Greenwich meridian

Latitude

Parallels of latitude (described by the angle they make, measured north or south of the equator)

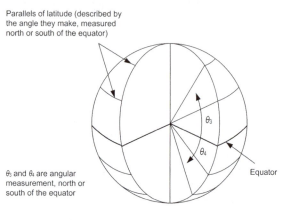

θ_3 and θ_4 are angular measurement, north or south of the equator

Equator

Figure 8.1 Longitude and latitude

and longitude is written as:

Latitude N 50° 04' 13" Longitude W 5° 42' 42"

8.1.2 Direction

Direction to an observed point (**bearing**) can be referenced to a known point on the earth's surface, e.g. **magnetic north**. Bearing is defined as the angle between the vertical plane of the reference point through to the vertical plane of the observed point. Basic navigational information is expressed in terms of **compass points** from zero referenced to north through 360° in a clockwise direction, see Figure 8.2. For practical navigation purposes, north has been taken from the natural feature of the earth's magnetic field; however; magnetic north is not at 90° latitude; the latter defines the position of **true north**. The location of magnetic north is in the Canadian Arctic, approximately 83° latitude and 115° longitude west of the prime meridian, see Figure 8.3. Magnetic north is a natural feature of the earth's geology; it is slowly drifting across the Canadian Arctic at approximately 40 km northwest per year. Over a long period of time, magnetic north describes an elliptical path. The Geological Survey of Canada keeps track of this motion by periodically carrying out magnetic surveys to re-determine the pole's location. In addition to this

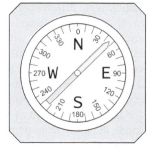

(a) Typical compass display

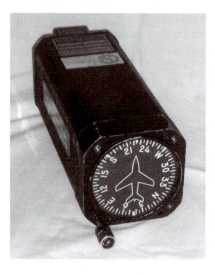

(b) Compass indicator

Figure 8.2 Compass indications

long-term change, the earth's magnetic field is also affected on a random basis by the weather, i.e. electrical storms.

Navigation charts based on magnetic north have to be periodically updated to consider this gradual drift. Compass-based systems are referenced to magnetic north; since this is not at 90° latitude there is an angular difference between magnetic and true north. This difference will be zero if the aircraft's position happens to be on the same longitude as magnetic north, and maximum at longitudes ±90° either side of this longitude. The angular difference between magnetic north and true north is called **magnetic variation**. It is vital that when bearings or headings are used, we are clear on what these are referenced to.

The imaginary lines of latitude and longitude described above are curved when superimposed on the earth's surface; they also appear as straight lines when viewed from above. The shortest distance between points A and B on a given route is a straight line. When this route is examined, the projection of the path (the **track**) flown by the aircraft over the earth's surface is described by a great circle.

Flying in a straight line implies that we are maintaining a constant heading, but this is not the case. Since the lines of longitude converge, travelling at a constant angle at each meridian yields a track that actually curves as illustrated in Figure 8.4. A track that intersects the lines of longitude at a constant angle is referred to as a **rhumb line**. Flying a rhumb line is readily achieved by reference to a fixed point, e.g. magnetic north. The great circle route; however, requires that the direction flown (with respect to the meridians) changes at any given time, a role more suited to a navigation computer.

8.1.3 Distance and speed

The standard unit of measurement for distance used by most countries around the world (the exceptions being the UK and USA) is the kilometre (km). This quantity is linked directly to the earth's geometry; the distance between the poles and equator is 10,000 km. The equatorial radius of the earth is 6378 km; the polar radius is 6359 km.

For aircraft navigation purposes, the quantity

Figure 8.3 Location of magnetic north

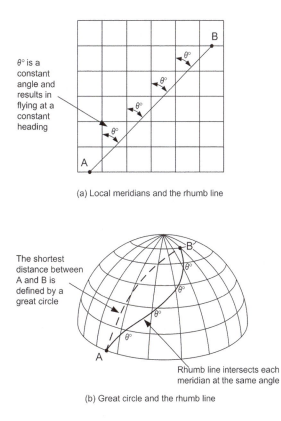

$\theta°$ is a constant angle and results in flying at a constant heading

(a) Local meridians and the rhumb line

The shortest distance between A and B is defined by a great circle

Rhumb line intersects each meridian at the same angle

(b) Great circle and the rhumb line

Figure 8.4 Flying a constant heading

of distance used is the **nautical mile (nm)**. This quantity is defined by distance represented by one minute of arc of a great circle (assuming the earth to be a perfect sphere). The nautical mile (unlike the statute mile) is therefore directly linked to the geometry of the earth. Aircraft speed, i.e. the rate of change of distance with respect to time, is given by the quantity '**knots**'; nautical miles per hour.

Calculating the great circle distance between two positions defined by an angle is illustrated in Figure 8.5. The distance between two positions defined by their respective latitudes and longitudes, (*lat1*, *lon1*) and (*lat2*, *lon2*), can be calculated from the formula:

$$d = \cos^{-1}(\sin(lat1) \times \sin(lat2) + \cos(lat1) \times \cos(lat2) \times \cos(lon1-lon2))$$

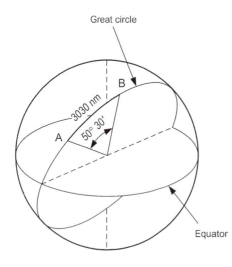

$$50° \ 30' = (50 \times 60) + 30 = 3030' = 3030 \text{ nm}$$

Figure 8.5 Calculation of great circle distances

Key point

Although we might consider the earth to be a perfect sphere, this is not the case. The actual shape of the earth is referred to as an oblate spheroid.

Key point

Longitude referenced to the prime-meridian extends east or west up to 180 degrees. Latitude is the angular distance north or south of the equator; the poles are at a latitude of 90 degrees.

Key point

The nautical mile (unlike the statute mile) is directly linked to the geometry of the earth. This quantity is defined by distance represented by one minute of arc of a great circle (assuming the earth to be a perfect sphere).

Key point

Both latitude and longitude are angular quantities measured in degrees. For accurate navigation, degrees can be divided by 60 giving the unit of 'minutes'; this can be further divided by 60 giving the unit of 'seconds'.

Test your understanding 8.1

Explain each of the following terms:

1. Latitude
2. Longitude
3. Great circle
4. Rhumb line.

8.2 Dead reckoning

Estimating a position by extrapolating from a known position and then keeping note of the direction, speed and elapsed time is known as **dead reckoning**. An aircraft passing over a given point on a heading of 90° at a speed of 300 knots will be five miles due east of the given point after one minute. If the aircraft is flying in zero wind conditions, this simple calculation holds true. In realistic terms, the aircraft will almost certainly be exposed to wind at some point during the flight and this will affect the navigation calculation. With our aircraft flying on a heading of 90° at a speed of 300 knots, let's assume that the wind is blowing from the south at 10 knots, see Figure 8.6. In a one hour time period, the air that the aircraft is flying in will have moved north by ten nautical miles. This means that the aircraft's path (referred to as its **track**) over the earth's surface is not due east. In other words, the aircraft track is not the same as the direction in which the aircraft is heading. This leads to a horizontal displacement (**drift**) of the aircraft from the track it would have followed in zero wind conditions.

The angular difference between the heading and track is referred to as the **drift angle** (quoted as being to port/left or starboard/right of the heading). If the wind direction were in the same direction as the aircraft heading, i.e. a tail wind, the aircraft speed of 300 knots through the air would equate to a ground speed of 310 knots. Likewise, if the wind were from the east (a headwind) the ground speed would be 290 knots.

Knowledge of the wind direction and speed allows the crew to steer the aircraft into the wind such that the wind actually moves the aircraft onto the desired track. For dead reckoning purposes, we can resolve these figures in mathematical terms and determine a position by triangulation as illustrated in Figure 8.7. Although the calculation is straightforward, the accuracy of navigation by dead reckoning will depend on up to date knowledge of wind speed and direction. Furthermore, we need accurate measurements of speed and direction. Depending on the accuracy of measuring these parameters, positional error will build up over time when navigating by dead reckoning. We therefore need a means of checking our calculated position on a periodic basis; this is the process of **position fixing**.

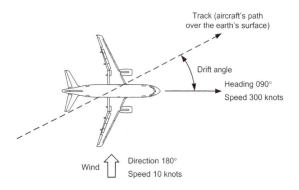

Figure 8.6   Effect of crosswind

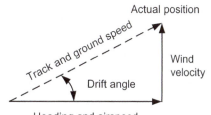

Figure 8.7   Resolving actual position

Key point

The angular difference between the heading and track is referred to as the drift angle.

8.3 Position fixing

When travelling short distances over land, natural terrestrial features such rivers, valleys, hills etc. can be used as direct observations to keep a check on (**pinpointing**) the journey's progress. If the journey is by sea, then we can use the coastline and specific features such as lighthouses to confirm our position. If the journey is now made at night or out of sight of the coast, we need other means of fixing our position.

The early navigators used the sun, stars and planets very effectively for navigation purposes; if the position of these celestial objects is known,

---

Key point

Dead reckoning is used to estimate a position by extrapolating from a known position and then keeping note of the direction and distance travelled.

---

then the navigator can confirm a position anywhere on the earth's surface. **Celestial navigation** (or **astronavigation**) was used very effectively in the early days of long distance aircraft navigation. Indeed, it has a number of distinct advantages when used by the military: the aircraft does not radiate any signals; navigation is independent of ground equipment; the references cannot be jammed; navigation references are available over the entire globe.

The disadvantage of celestial navigation for aircraft is that the skies are not always clear and it requires a great deal of skill and knowledge to be able to fix a position whilst travelling at high speed. Although automated celestial navigation systems were developed for use by the military, they are expensive; modern avionic equipment has now phased out the use of celestial navigation for commercial aircraft.

The earliest ground-based references (**navigation aids**) developed for aircraft navigation are based on radio beacons. These beacons can provide angular and/or distance information; when using this information to calculate a position fix, the terms are referred to mathematically as theta (θ) and rho (ρ). By utilising the directional properties of radio waves, the intersection of signals from two or more navigation aids can be used to fix a position (**theta–theta**), see Figure 8.8. Alternatively, if we know the distance and direction (bearing) to a navigation aid, the aircraft position can be confirmed (**rho–theta**). Finally, we can establish our position if we know the aircraft's distance (**rho–rho**) from any two navigation aids, i.e. without knowledge of the bearing.

8.4 Maps and charts

Maps provide the navigator with a representative diagram of an area that contains a variety of physical features, e.g. cities, roads and topographical information. Charts contain lines of latitude and longitude, together with essential data such as the location of navigation aids. Creating charts and maps requires that we transfer distances and geographic features from the earth's spherical surface onto a flat piece of paper. This is not possible without some kind of compromise in geographical shape, surface area, distance or direction. Many methods of producing charts have been developed over the centuries; the choice of projection depends on the intended purpose.

In the sixteenth century Gerhardus Mercator, the Flemish mathematician, geographer and cartographer, developed what was to become the standard chart format for nautical navigation: the **Mercator projection**. This is a cylindrical map projection where the lines of latitude and longitude are projected from the earth's centre, see Figure 8.9. Imagine a cylinder of paper wrapped around the globe and a light inside the globe; this projects the lines of latitude and longitude onto the paper. When the cylinder is unwrapped, the lines of latitude appear incorrectly as having equal length. Directions and the shape of geographic features remain true; however, distances and sizes become distorted. The advantage of using this type of chart is that the navigator sets a constant heading to reach the destination. The meridians of the Mercator projected chart are crossed at the same angle; the track followed is referred to as a **rhumb line** (see Figure 8.4).

For aircraft navigation the Mercator projection might be satisfactory; however, if we want to navigate by great circle routes then we need true directions. An alternative projection used for aircraft navigation, and most popular maps and charts, is the **Lambert** azimuthul equal-area projection. This projection was developed by Johann Heinrich Lambert (1728–77) and is particularly useful in high latitudes. The projection is developed from the centre point of the geographic feature to be surveyed and charted. Using true north as an example, Figure 8.10 illustrates the Lambert projection.

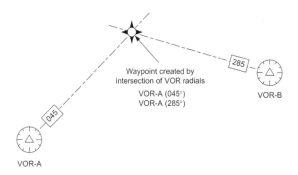

Figure 8.8 Position fixing

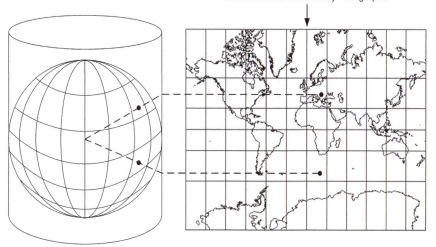

Figure 8.9 Mercator projection

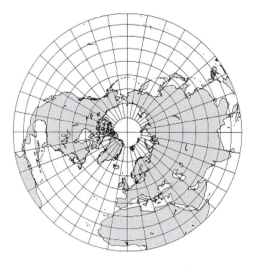

Figure 8.10   Lambert projection (viewed from true north)

8.5 Navigation terminology

The terms shown in Table 8.1 are used with numerous navigation systems including INS and RNAV; computed values are displayed on a control display unit (CDU) and/or primary flight instruments. These terms are illustrated in Figure 8.11, all terms are referenced to true north.

8.6 Navigation systems development

This section provides a brief overview of the development of increasingly sophisticated navigation systems used on aircraft.

8.6.1 Gyro-magnetic compass

The early aviators used visual aids to guide them along their route; these visual aids would have included rivers, roads, rail tracks, coastlines etc. This type of navigation is not possible at high altitudes or in low visibility and so the earth's magnetic field was used as a reference leading to the use of simple **magnetic compasses** in aircraft. We have seen that magnetic variation has to be taken into account for navigation; there are additional considerations to be addressed for compasses in aircraft. The earth's magnetic field around the aircraft will be affected by:

- the aircraft's own 'local' magnetic fields, e.g. those caused by electrical equipment
- sections of the aircraft with high permeability causing the field to be distorted.

Magnetic compasses are also unreliable in the short-term, i.e. during turning manoeuvres. **Directional gyroscopes** are reliable for azimuth guidance in the short term, but drift over longer time periods. A combined magnetic compass

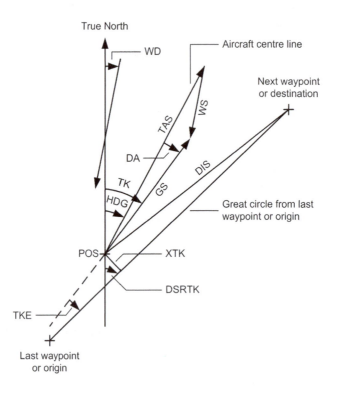

Figure 8.11   Navigation terminology

Table 8.1   Navigation terminology

Term	Abbreviation	Description
Cross track distance	XTK	Shortest distance between the present position and desired track
Desired track angle	DSRTK	Angle between north and the intended flight path of the aircraft
Distance	DIS	Great circle distance to the next waypoint or destination
Drift angle	DA	Angle between the aircraft's heading and ground track
Ground track angle	TK	Angle between north and the flight path of the aircraft
Heading	HDG	Horizontal angle measured clockwise between the aircraft's centreline (longitudinal axis) and a specified reference
Present position	POS	Latitude and longitude of the aircraft's position
Track angle error	TKE	Angle between the actual track and desired track (equates to the desired track angle minus the ground track angle)
Wind direction	WD	Angle between north and the wind vector
True airspeed	TAS	True airspeed measured in knots
Wind speed	WS	Measured in knots
Ground speed	GS	Measured in knots

stabilised by a directional gyroscope (referred to as a **gyro-magnetic compass**) can overcome these deficiencies. The gyro-magnetic compass (see Figure 8.12), together with an airspeed indicator, allowed the crew to navigate by dead reckoning, i.e. estimating their position by extrapolating from a known position and then keeping note of the direction and distance travelled.

In addition to directional references, aircraft also need an attitude reference for navigation, typically from a vertical gyroscope. Advances in sensor technology and digital electronics have led to combined attitude and heading reference systems (AHRS) based on laser gyros and micro-electromechanical sensors (see Chapter 17).

Instrumentation errors inevitably lead to deviations between the aircraft's actual and calculated positions; these deviations accumulate over time. Crews therefore need to be able to confirm and update their position by means of a fixed ground-based reference, e.g. a radio navigation aid.

8.6.2 Radio navigation

Early airborne navigation systems using ground-based navigation aids consisted of a loop antenna in the aircraft tuned to amplitude modulated (AM) commercial radio broadcast stations transmitting in the low-/medium-frequency (LF/MF) bands. Referring to Figure 8.13, pilots would know the location of the radio station (indeed, it would invariably have been located close to or even in the town/city that the crew wanted to fly to) and this provided a means of fixing a position. Although technology has moved on, these **automatic direction finder** (ADF) systems are still in use today.

Operational problems are encountered using low-frequency (LF) and medium-frequency (MF) transmissions. During the mid to late 1940s, it was evident to the aviation world that an accurate and reliable short-range navigation system was needed. Since radio communication systems based on very high frequency (VHF) were being successfully deployed, a decision was made to develop a radio navigation system based on VHF. This system became the **VHF omnidirectional range** (VOR) system, see Figure 8.14; a system that is in widespread use throughout the world

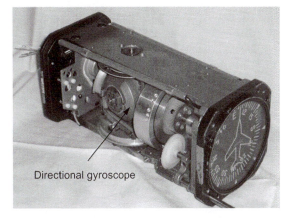

Figure 8.12   Gyro-magnetic compass

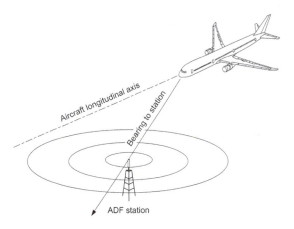

Figure 8.13   ADF radio navigation

today. VOR is the basis of the current network of 'airways' that are used in navigation charts.

The advent of radar in the 1940s led to the development of a number of navigation aids including **distance measuring equipment** (DME). This is a short-/medium-range navigation system, often used in conjunction with the VOR system to provide accurate navigation fixes. The system is based on secondary radar principles, see Figure 8.15.

Navigation aids such as automatic direction finder (ADF), VHF omnidirectional range (VOR) and distance measuring equipment (DME) are used to define airways for en route navigation, see Figure 8.16 . They are also installed at airfields to assist with approaches to those airfields. These navigation aids cannot, however,

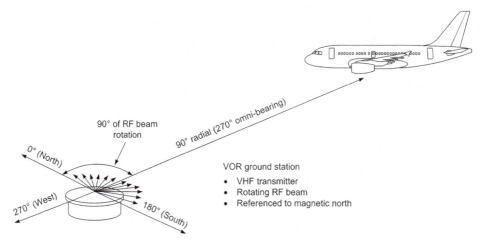

Figure 8.14 VOR radio navigation

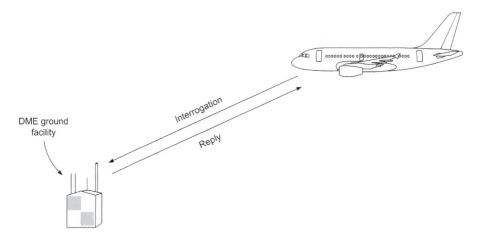

Figure 8.15 Distance measuring equipment (DME)

be used for precision approaches and landings. The standard approach and landing system installed at airfields around the world is the **instrument landing system** (ILS), see Figure 8.17. The ILS uses a combination of VHF and UHF radio waves and has been in operation since 1946. There are a number of shortcomings with ILS; in 1978 the **microwave landing system** (MLS) was adopted as the long-term replacement. The system is based on the principle of time referenced scanning beams and provides precision navigation guidance for approach and landing. MLS provides three-dimensional approach guidance, i.e. azimuth, elevation and range. The system provides multiple approach angles for both azimuth and elevation guidance. Despite the advantages of MLS, it has not yet been introduced on a worldwide basis for commercial aircraft. Military operators of MLS often use mobile equipment that can be deployed within hours.

The aforementioned radio navigation aids have one disadvantage in that they are land based and only extend out beyond coastal regions. Long-range radio navigation systems based on **hyperbolic navigation** were introduced in the

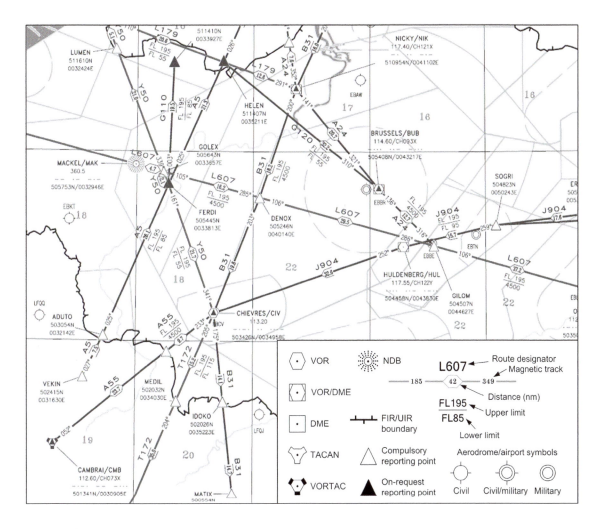

Figure 8.16 Airways defined by navigation aids

1940s to provide for en route operations over oceans and unpopulated areas. Several hyperbolic systems have been developed since, including Decca, Omega and Loran. The operational use of Omega and Decca navigation systems ceased in 1997 and 2000 respectively. Loran systems are still very much available today as stand-alone systems; they are also being proposed as a complementary navigation aid for global navigation satellite systems. The **Loran-C** system is based on a master station and a number of secondary stations; the use of VLF radio provides an increased area of coverage, see Figure 8.18.

The advent of computers, in particular the increasing capabilities of integrated circuits using digital techniques, has led to a number of advances in aircraft navigation. One example of this is the **area navigation system** (RNAV); this is a means of combining, or filtering, inputs from one or more navigation sensors and defining positions that are not necessarily co-located with ground-based navigation aids. Typical navigation sensor inputs to an RNAV system can be from external ground-based navigation aids such as VHF omnirange (VOR) and distance measuring equipment (DME), see Figure 8.19.

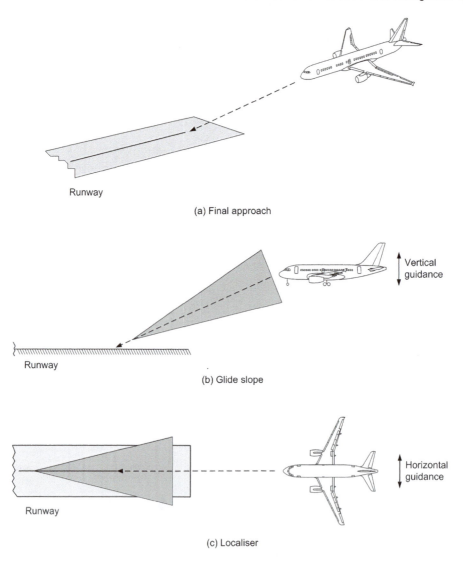

Runway

(a) Final approach

(b) Glide slope

(c) Localiser

Figure 8.17   Instrument landing system

8.6.3 Dead reckoning systems

Dead reckoning systems require no external inputs or references from ground stations. Doppler navigation systems were developed in the mid-1940s and introduced in the mid-1950s as a primary navigation system. Ground speed and drift can be determined using a fundamental scientific principle called **Doppler shift**. Being self-contained, the system can be used for long distance navigation over oceans and undeveloped areas of the globe.

A major advance in aircraft navigation came with the introduction of the **inertial navigation system** (INS). This is an autonomous dead reckoning system, i.e. it requires no external inputs or references from ground stations. The system was developed in the 1950s for use by the US military and subsequently the space programmes. Inertial navigation systems (INS) were introduced into commercial aircraft service during the early 1970s. The system is able to compute navigation data such as present position, distance to waypoint, heading, ground speed,

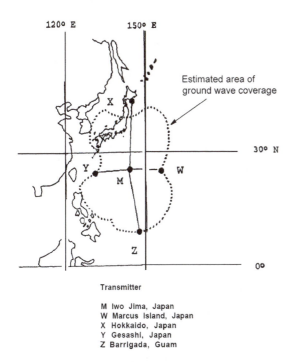

Transmitter

M Iwo Jima, Japan
W Marcus Island, Japan
X Hokkaido, Japan
Y Gesashi, Japan
Z Barrigada, Guam

Figure 8.18 Loran-C oceanic coverage using VLF transmissions

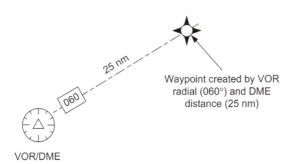

VOR/DME

Figure 8.19 Area navigation

wind speed, wind direction etc. It does not need radio navigation inputs and it does not transmit radio frequencies. Being self-contained, the system can be used for long distance navigation over oceans and undeveloped areas of the globe.

8.6.4 Satellite navigation

Navigation by reference to the stars and planets has been employed since ancient times;

commercial aircraft used to have periscopes to take celestial fixes for long distance navigation. An artificial constellation of navigation aids was initiated in 1973 and referred to as **Navstar** (navigation system with timing and ranging). The global positioning system (GPS) was developed for use by the US military; the first satellite was launched in 1978 and the full constellation was in place and operating by 1994. GPS is now widely available for use by many applications including aircraft navigation; the system calculates the aircraft position by triangulating the distances from a number of satellites, see Figure 8.20.

8.6.5 Radar navigation

The planned journey from A to B could be affected by adverse weather conditions. Radar was introduced onto passenger aircraft during the 1950s to allow pilots to identify weather conditions (see Figure 8.21) and subsequently re-route around these conditions for the safety and comfort of passengers. A secondary use of weather radar is a terrain-mapping mode that allows the pilot to identify features on the ground, e.g. rivers, coastlines and mountains.

8.6.6 Air traffic control

Increasing traffic density, in particular around airports, means that we need a method of **air traffic control** (ATC) to manage the flow of traffic and maintain safe separation of aircraft. The ATC system is based on secondary surveillance radar (SSR) facilities located at strategic sites, at or near airfields. Ground controllers use the SSR system to identify individual aircraft on their screens, see Figure 8.22.

With ever increasing air traffic congestion, and the subsequent demands on air traffic control (ATC) resources, the risk of a mid-air collision increases. The need for improved traffic flow led to the introduction of the **traffic alert and collision avoidance system** (TCAS). This is an automatic surveillance system that helps aircrews and ATC to maintain safe separation of aircraft. It is an airborne system (see Figure 8.23) based on secondary radar that interrogates and replies directly with aircraft via a high-integrity data link. The system is functionally independent of ground stations, and alerts the crew if another aircraft

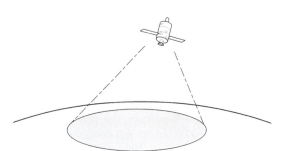

(a) Single satellite describes a circle on the earth's surface

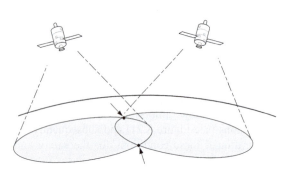

(b) Two satellites define two unique positions.
A third satellite defines a unique position

Figure 8.20   Satellite navigation

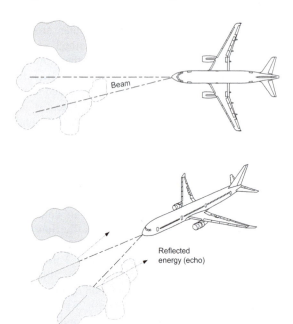

Beam

Figure 8.21   Weather radar

comes within a predetermined time to a potential collision. It is important to note that TCAS is a back-up system, i.e. it provides warnings when other navigation systems (including ATC) have failed to maintain safe separation of aircraft that could lead to a collision.

8.7 Navigation systems summary

Navigation systems for aircraft have evolved with the nature and role of the aircraft itself. These individual systems are described in detail in the following chapters. Each system has been developed to meet specific requirements within the available technology and cost boundaries. Whatever the requirement, all navigation systems are concerned with several key factors:

- **Accuracy**: conformance between calculated and actual position of the aircraft
- **Integrity**: ability of a system to provide timely warnings of system degradation
- **Availability**: ability of a system to provide the required function and performance
- **Continuity**: probability that the system will be available to the user
- **Coverage**: geographic area where each of the above are satisfied at the same time.

Test your understanding 8.2

The nautical mile is directly linked to the geometry of the earth; how is a nautical mile defined?

Test your understanding 8.3

Explain the difference between dead reckoning and position fixing.

Test your understanding 8.4

For a given airspeed, explain how tailwinds and headwinds affect groundspeed

Reflected
energy (echo)

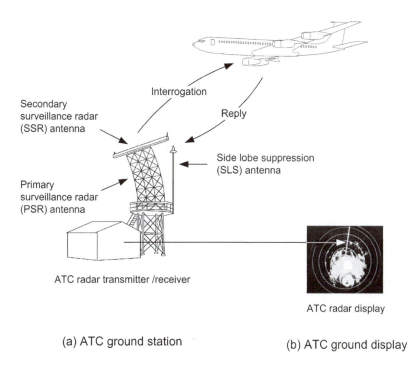

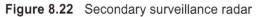

(a) ATC ground station   (b) ATC ground display

Figure 8.22   Secondary surveillance radar

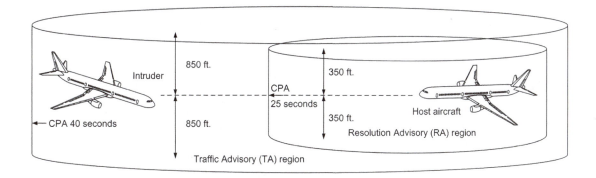

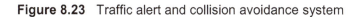

Figure 8.23   Traffic alert and collision avoidance system

Test your understanding 8.5

Explain the following terms: accuracy, integrity, availability.

Test your understanding 8.6

Describe three ways that bearings and ranges can be used for position fixing.

Test your understanding 8.7

1. Explain the difference between Mercator and Lambert projections.

2. Where on the earth's surface is the difference between a rhumb line and great circle route the greatest?

8.8 Multiple choice questions

1. Longitude referenced to the prime meridian extends:
   (a) north or south up to 180°
   (b) east or west up to 180°
   (c) east or west up to 90°.

2. Latitude is the angular distance:
   (a) north or south of the equator
   (b) east or west of the prime meridian
   (c) north or south of the prime meridian.

3. The distance between lines of longitude converge at the:
   (a) poles
   (b) equator
   (c) great circle.

4. Lines of latitude are always:
   (a) converging
   (b) parallel
   (c) the same length

5. Degrees of latitude can be divided by 60 giving the unit of:
   (a) longitude
   (b) minutes
   (c) seconds.

6. The location of magnetic north is approximately:
   (a) 80° latitude and 110° longitude, east of the prime meridian
   (b) 80° longitude and 110° latitude, west of the prime meridian
   (c) 80° latitude and 110° longitude, west of the prime meridian.

7. One minute of arc of a great circle defines a:
   (a) nautical mile
   (b) kilometre
   (c) knot.

8. The angular difference between magnetic north and true north is called the:
   (a) magnetic variation
   (b) great circle
   (c) prime meridian.

9. Mercator projections produce parallel lines of:
   (a) the earth's magnetic field
   (b) longitude
   (c) great circle routes.

10. With respect to the polar radius, the equatorial radius of the earth is:
    (a) equal
    (b) larger
    (c) smaller.

11. Dead reckoning is the process of:
    (a) fixing the aircraft's position
    (b) correcting the aircraft's position
    (c) estimating the aircraft's position.

12. The angle between the aircraft's heading and ground track is known as the:
    (a) drift angle
    (b) cross track distance
    (c) wind vector.

13. Magnetic compasses are unreliable in the:
    (a) long-term, flying a constant heading
    (b) short-term, during turning manoeuvres
    (c) equatorial regions.

14. The angle between north and the flight path of the aircraft is the:
    (a) ground track angle
    (b) drift angle
    (c) heading.

15. When turning into a 25 knot head wind at constant indicated airspeed, the ground speed will:
    (a) increase by 25 knots
    (b) remain the same
    (c) decrease by 25 knots.

Automatic direction finder

Radio waves have directional characteristics as we have seen from earlier chapters. This is the basis of the automatic direction finder (ADF); one of earliest forms of radio navigation that is still in use today. ADF is a short-/medium-range (200 nm) navigation system providing directional information; it operates within the frequency range 190–1750 kHz, i.e. low and medium frequency bands. The term 'automatic' is somewhat misleading in today's terms; this refers to the introduction of electromechanical equipment to replace manually operated devices. In this chapter we will look at the historical background to radio navigation, review some typical ADF hardware that is fitted to modern commercial transport aircraft, and conclude with some practical aspects associated with the operational use of ADF.

9.1 Introducing ADF

The early aviators used visual aids to guide them along their route; these visual aids would have included rivers, roads, rail tracks, coastlines etc. This type of navigation is not possible in low visibility and so magnetic compasses were introduced. Magnetic compasses are somewhat unreliable in the short term, i.e. during turning manoeuvres. Directional gyroscopes are reliable in the short term, but drift over longer time periods. A combined magnetic compass stabilised by a directional gyroscope (referred to as a **gyro-magnetic compass**) can overcome these deficiencies. The gyro-magnetic compass, together with an airspeed indicator, allowed the crew to navigate by dead reckoning, i.e. estimating their position by extrapolating from a known position and then keeping note of the direction and distance travelled. Instrumentation errors inevitably lead to deviations between the aircraft's actual and calculated positions; these

deviations accumulate over time. Crews therefore need to be able to confirm and update their position by means of a fixed ground-based reference.

The early airborne navigation systems using ground-based navigation aids consisted of a fixed-loop antenna in the aircraft tuned to an amplitude modulated (AM) commercial radio broadcast station. Pilots would know the location of the radio station (indeed, it would invariably have been located close to or even in the town/city that the crew wanted to fly to). The fixed-**loop antenna** was aligned with the longitudinal axis of the aircraft, with the pilot turning the aircraft until he received the minimum signal strength (null reading). By maintaining a null reading, the pilot could be sure that he was flying towards the station. This constant turning was inefficient in terms of fuel consumption and caused inherent navigation problems in keeping note of the aircraft's position during these manoeuvres! The effects of crosswind complicated this process since the aircraft's heading is not aligned with its track.

9.2 ADF principles

The introduction of an 'automatic' direction finder (ADF) system addresses this problem. A loop antenna that the pilot could rotate by hand solves some of these problems; however, this still requires close attention from the crew. Later developments of the equipment used an electrical motor to rotate the loop antenna. The received signal strength is a function of the angular position of the loop with respect to the aircraft heading and bearing to the station, see Figure 9.1 (a) and (b). If a plot is made of loop angle and signal strength, the result is a sine wave as shown in Figure 9.1(c). The **null point** is easier to determine than the maximum signal strength

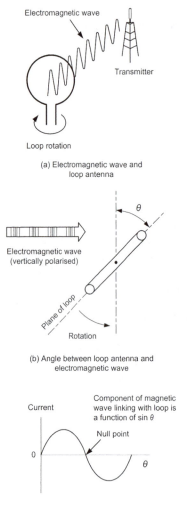

(a) Electromagnetic wave and loop antenna

(b) Angle between loop antenna and electromagnetic wave

(c) Loop angle and signal strength

Figure 9.1 Loop antenna output

since the rate of change is highest. Rotating the antenna (rather than turning the aircraft) to determine the null reading from the radio station was a major advantage of the system. The pilot read the angular difference between the aircraft's heading and the direction of the radio station, see Figure 9.2(a), from a graduated scale and a bearing to the station could then be determined. The industry drive towards solid-state components, i.e. with no moving parts, has led to the equipment described in Section 9.3.

Navigation based on ADF (using AM commercial radio stations broadcasting in the

frequency range 540–1620 kHz) became an established method of travelling across country. With the growth of air travel, dedicated radio navigation aids were installed along popular air transport routes. These radio stations, known as **non-directional beacons** (NDB), gradually supplemented the commercial radio stations and a network of NDBs sprang up in the nations developing their aeronautical infrastructure. These NDBs broadcast in the low-frequency (LF) range 190–415 kHz and medium-frequency (MF) range 510–535 kHz. As the quantity of NDBs increased, air navigation charts were produced and the NDBs were identified by a two or three letter alpha code linked to the location and frequency. In Figure 2(c), the NDB located at Mackel in Belgium transmits on 360.5 kHz and is identified as MAK; note the Morse code, latitude and longitude details on the chart. Beacons are deployed with varying power outputs classified as high (2 kW), medium (50W to 2 kW) and low (less than 50 W).

Table 9.1 provides a list of typical NDBs associated with airports and cities in a typical European country (note that these are provided for illustration purposes only). Beacons marked with an asterisk in this table are referred to as locator beacons; they are part of the final approach procedures for an airfield (see Chapter 12).

9.3 ADF equipment

9.3.1 Antenna

The rotating loop antenna was eventually replaced with a fixed antenna consisting of two loops combined into a single item; one aligned with the centreline of the fuselage, the other at right angles as shown in Figure 9.3. This orthogonal antenna is still referred to as the 'loop' antenna. Measuring the signal strength from each of the loops, and deriving an angular position in a dedicated ADF receiver, determines the direction to the selected beacon (or commercial radio station). The loop antenna resolves the directional signal; however, this can have two possible solutions 180 degrees apart. A second **'sense' antenna** is therefore required to detect non-directional radio waves from the beacon; this

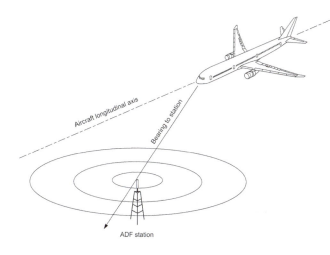

(a) Using an ADF system for navigation

(b) ADF non-directional beacon (NDB)
(photo courtesy of T. Diamond)

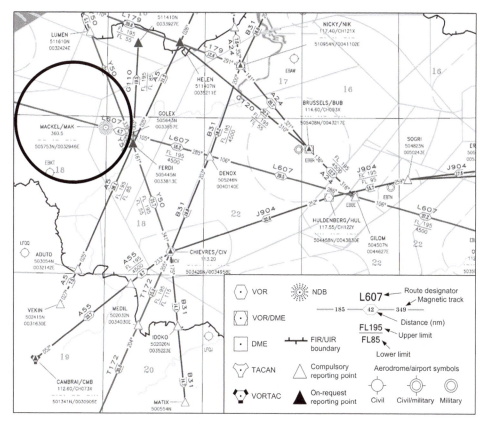

(c) MACKEL NDB shown on a navigation chart

Figure 9.2 Navigation by non-directional beacons NDB

Table 9.1 Examples of NDB codes and frequencies

Name	Identification code	Frequency (kHz)
Eelde	SO	330.00
Eindhoven	EHN	397.00
Eindhoven	PH	316.00
Gull	GUL	383.50
Maastricht	NW	373.00
Maastricht *	ZL	339.00
Rotterdam	ROT	350.50
Rotterdam *	PS	369.00
Rotterdam *	RR	404.50
Schiphol	CH	388.50
Schiphol *	NV	332.00
Stad	STD	386.00
Stadskanaal	STK	315.00
Thorn	THN	434.00
Twenthe	TWN	335.50

* Locator beacons

signal is combined with the directional signals from the loop antenna to produce a single directional solution. The polar diagram for a loop and sense antenna is shown in Figure 9.4; when the two patterns are combined, it forms a **cardioid**. Most commercial transport aircraft are fitted with two independent ADF systems typically identified as left and right systems; the antenna locations for a typical transport aircraft are shown in Figure 9.5.

9.3.2 Receiver

ADF receivers are located in the avionic equipment bay. The signal received at the antenna is coupled to the receiver in three ways:

- The sense signal
- A loop signal proportional in amplitude to the cosine of the relative angle of the aircraft centreline and received signal
- A loop signal proportional in amplitude to the sine of the relative angle of the aircraft centreline and received signal.

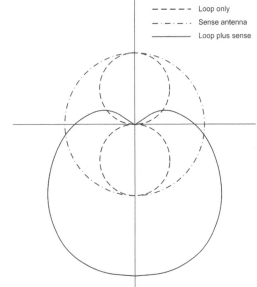

Figure 9.3 ADF antenna

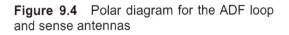

Figure 9.4 Polar diagram for the ADF loop and sense antennas

The sense antenna signal is processed in the receiver via a superhet receiver (see page 46) which allows weak signals to be identified, together with discrimination of adjacent frequencies. The output of the superhet receiver is then integrated into the aircraft's audio system. Loop antenna signals are summed with the sense antenna signal; this forms a phase-modulated (PM) carrier signal. The superhet intermediate frequency (IF) is coupled with the PM signal into a coherent demodulator stage that senses the

Figure 9.5   Location of left and right ADF antennas on a typical transport aircraft

presence of a sense antenna signal from the IF stage. The PM component of the signal is recovered from the voltage controlled oscillator (VCO) phase lock circuit (see page 53). This recovered signal contains the bearing information received by the antenna and is compared to a reference modulation control signal.

Receivers based on analogue technology send bearing data to the flight deck displays using synchro systems. Digital receivers transmit bearing data to the displays using a data bus system, typically ARINC 429. The ADF receiver is often incorporated into a multi-mode receiver along with other radio navigation systems.

9.3.3 Control panel

Aircraft with analogue (electromechanical) avionics have a dedicated ADF control panel, located on the centre pedestal, see Figure 9.6(a). An alternative panel shown in Figure 9.6(b) enables the crew to select a range of functions including: frequency selectors/displays and the beat frequency oscillator (BFO). This function is used when they want to create an audio frequency for carrier wave transmissions through their audio panel.

NDB carrier waves that are not modulated with an audio component use the beat frequency

oscillator (BFO) circuit in the ADF receiver. To produce an audio output, the receiver heterodynes (beats) the carrier wave signal with a separate signal derived from an oscillator in the receiver.

Some ADF panels have an ADF/ANT switch where 'ADF' selects normal operation, i.e. combined sense and loop antennas; and 'ANT' selects the sense antenna by itself so that the crew can confirm that a station is broadcasting, i.e. without seeking a null. General aviation products combine the control panel and receiver into a single item, see Figure 9.6(c). A changeover switch is used to select the active and standby frequencies.

9.3.4 ADF bearing display

The output from the ADF receiver is transmitted to a display that provides the pilot with both magnetic heading and direction to the tuned NDB, this can either be a dedicated ADF instrument as shown in Figure 9.7(a), or a **radio magnetic indicator** (RMI), see Figure 9.7(b).

In the RMI, two bearing pointers (coloured red and green) are associated with the two ADF systems and allow the crew to tune into two different NDBs at the same time. RMIs can have a dual purpose; pilots use a switch on the RMI to select either ADF and/or VHF omnidirectional

(a) Location of ADF control panel

(b) Typical ADF control panel

(c) ADF panel/receiver for general aviation

Figure 9.6 ADF control panels

range (VOR) bearings (see Chapter 10 for the latter). Referring to Figure 9.7(c), some aircraft have a bearing source indicator (located adjacent to the RMI) that confirms ADF or VOR selection.

The evolution of digital electronics together with integration of other systems has led to the introduction of the flight management system (FMS: see Chapter 19) control display unit (CDU) which is used to manage the ADF system. Aircraft fitted with electronic flight instrument systems (EFIS) have green NDB icons displayed on the electronic horizontal situation indicator (EHSI) as shown in Figure 9.7(e).

Key point

ADF is a short/medium-range (200 nm) navigation system operating within the frequency range 190 to 1750 kHz, i.e. low and medium frequency bands. The ADF system uses an orthogonal antenna consisting of two loops; one aligned with the centreline of the fuselage, the other at right angles.

Test your understanding 9.1

Why does the ADF system seek a null rather than the maximum signal strength from a transmitting station?

Test your understanding 9.2

Explain the function of the ADF/ANT switch that is present on some ADF panels.

9.4 Operational aspects of ADF

ADF radio waves are propagated as ground waves and/or sky waves. Problems associated with ADF are inherent in the frequency range that the system uses. ADF transmissions are susceptible to errors from:

(a) ADF bearing indicator

(b) RMI with two bearing indications

(c) Location of RMI and source indicator

(d) RMI and source indicator

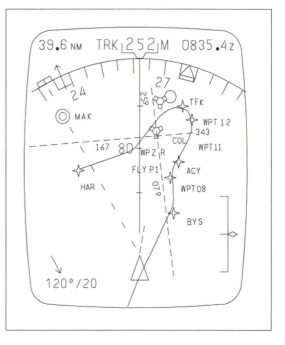

(e) EHSI with an NDB icon (shown as MAK on the upper left of the display)

Figure 9.7 ADF displays

Test your understanding 9.3

Explain the purpose of a beat frequency oscillator (BFO) and why it is needed in an ADF receiver.

- **Atmospheric conditions:** the height and depth of the ionosphere will vary depending on solar activity. The sky waves (see Figure 9.8) will be affected accordingly since their associated skip distances will vary due to refraction in the ionosphere. This is particularly noticeable at sunrise and sunset.
- **Physical aspects of terrain:** mountains and valleys will reflect the radio waves causing multipath reception.
- **Coastal refraction:** low-frequency waves that are propagated across the surface of the earth as ground waves will exhibit different characteristics when travelling over land versus water. This is due to the attenuation of the ground wave being different over land and water. The direction of a radio wave across land will change (see Figure 9.9) when it reaches the coast and then travels over water. This effect depends on the angle between the radio wave and the coast.
- **Quadrantal error (QE):** many parts of the aircraft structure, e.g. the fuselage and wings, are closely matched in physical size to the wavelength of the ADF radio transmissions. Radiated energy is absorbed in the airframe and re-radiated causing interference; this depends on the relative angle between the direction of travel, the physical aspects of the aircraft and location of the ADF transmitter. Corrections can be made for QE in the receiver.
- **Interference:** this can arise from electrical storms, other radio transmissions, static build-up/discharges and other electrical equipment on the aircraft.

The accuracy of an ADF navigation system is in the order of ±5 degrees for locator beacons and ±10 degrees for en route beacons. Any of the above conditions will lead to errors in the bearing information displayed on the RMI. If these conditions occur in combination then the navigation errors will be significant. Pilots cannot use ADF for precision navigation due to these limitations.

The increased need for more accuracy and reliability of navigation systems led to a new generation of en route radio navigation aids; this is covered in the next chapter. In the meantime,

ADF transmitters remain installed throughout the world and the system is used as a secondary radio navigation aid. The equipment remains installed on modern aircraft, albeit integrated with other radio navigation systems.

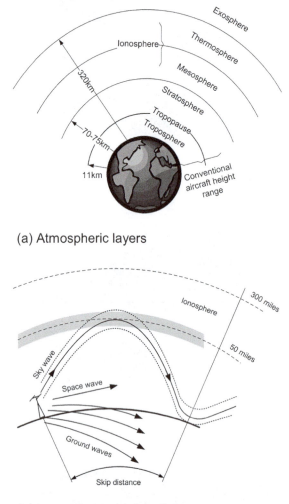

(a) Atmospheric layers

(b) Ionosphere and skip distance

Figure 9.8   Sky waves and the ionosphere

Key point

ADF radio waves are propagated as ground waves and/or sky waves. Problems associated with ADF are inherent in the frequency range that the system uses.

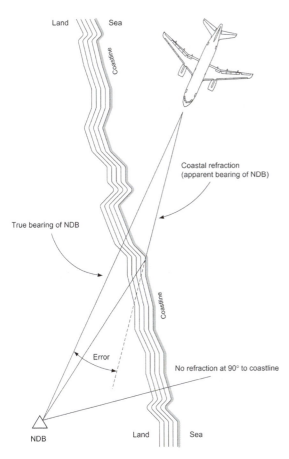

Figure 9.9 Effect of coastal refraction

Key point

ADF cannot be used for precision navigation due to inherent performance limitations; it remains as a backup to other navigation systems.

Test your understanding 9.4

Why do ADF antennas need a sense loop?

Test your understanding 9.5

How are NDBs identified on navigation charts?

Test your understanding 9.6

Where would locator beacons be found?

Test your understanding 9.7

Why are there two pointers on the RMI?

Test your understanding 9.8

Describe how ground and sky waves are affected by:

    (a) local terrain
    (b) the ionosphere
    (c) attenuation over land and water
    (d) electrical storms.

Test your understanding 9.9

Explain, in relation to an ADF system, what is meant by quadrantal error. What steps can be taken to reduce this error?

9.5 Multiple choice questions

1. ADF antennas are used to determine what aspect of the transmitted signal?
   (a) Wavelength
   (b) Null signal strength
   (c) Maximum signal strength.

2. The ADF antennas include:
   (a) one sense loop and two directional loops
   (b) two sense loops and two directional loops
   (c) two sense loops and one directional loop.

3. ADF operates in the following frequency range:
   (a) MF to VHF
   (b) LF to MF
   (c) VLF.

4. Bearing to the tuned ADF station is displayed on the:
   (a) RMI
   (b) NDB
   (c) HSI.

5. The purpose of an ADF sense antenna is to:
   (a) provide directional information to the receiver
   (b) discriminate between NDBs and commercial broadcast stations
   (c) combine with the loop antenna to determine a station bearing.

6. The RMI has two pointers coloured red and green; these are used to indicate:
   (a) two separately tuned ADF stations
   (b) AM broadcast stations (red) and NDBs (green)
   (c) heading (red) and ADF bearing (green).

7. The bearing source indicator adjacent the RMI confirms:
   (a) ADF or VOR selection
   (b) the NDB frequency
   (c) the NDB bearing.

8. NDBs on navigation charts can be identified by:
   (a) five letter codes
   (b) two/three letter codes
   (c) triangles.

9. Morse code is used to confirm the NDB's:
   (a) frequency
   (b) name
   (c) bearing.

10. During sunrise and sunset, ADF transmissions are affected by:
    (a) coastal refraction
    (b) static build-up in the airframe
    (c) variations in the ionosphere.

11. NDBs associated with the final approach to an airfield are called:
    (a) locator beacons
    (b) reporting points
    (c) en route navigation aids.

12. Quadrantal error (QE) is associated with the:
    (a) ionosphere
    (b) physical aspects of terrain
    (c) physical aspects of the aircraft structure.

13. ADF ground waves are affected by:
    (a) the ionosphere
    (b) coastal refraction and terrain
    (c) solar activity.

14. ADF sky waves are affected by:
    (a) the ionosphere
    (b) coastal refraction
    (c) the local terrain.

15. A BFO can be used to establish:
    (a) the non-directional output of an NDB
    (b) which loop antenna is receiving a null
    (c) an audio tone for an NDB.

16. Referring to Figure 9.10, which icon is for the NDB?
    (a) HAR
    (b) MAK
    (c) WPT08.

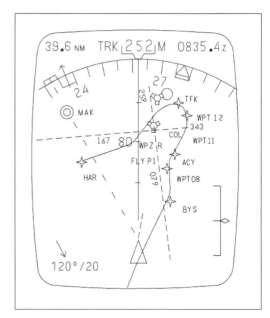

Figure 9.10 See Question 16

Chapter 10 VHF omnidirectional range

We have seen from earlier chapters that radio waves have directional characteristics. In Chapter 9, we looked at the early use of radio navigation, and some of the operational problems in using low-frequency (LF) and medium-frequency (MF) transmissions. During the mid- to late 1940s, it was evident to the aviation world that an accurate and reliable short-range navigation system was needed. Since radio communication systems based on very high frequency (VHF) were being successfully deployed, a decision was made to develop a radio navigation system based on VHF. This system became the VHF omnidirectional range (VOR) system; a system that is in widespread use throughout the world today and is the basis for the current network of 'airways' that are used for navigation.

10.1 VOR principles

10.1.1 Overview

VOR is a short/medium-range navigation system operating in the 108–117.95 MHz range of frequencies. This means that the radio waves are now propagated as space waves. The problems that were encountered with ground and sky waves in the LF and MF ranges are no longer present with a VHF system. VOR navigation aids are identified by unique three-letter codes (derived from their name, e.g. London VOR is called LON, Dover VOR is called DVR etc.). The code is modulated onto the carrier wave as a 1020 Hz tone that the crew can listen to as a Morse code signal. Some VOR navigation aids have an automatic voice identification announcement that provides the name of the station; this alternates with the Morse code signal. The location of the VOR navigation aids (specified by latitude and longitude) together with their carrier wave frequencies is provided on navigation charts as with ADF.

VOR operates in the same frequency range as the instrument landing system (ILS), described in Chapter 12. Although the two systems are completely independent and work on totally different principles, they often share the same receiver. The two systems are differentiated by their frequency allocations within this range. ILS frequencies are allocated to the odd tenths of each 0.5 MHz increment, e.g. 109.10 MHz, 109.15 MHz, 109.30 MHz etc. VOR frequencies are allocated even tenths of each 0.5 MHz increment, e.g. 109.20 MHz, 109.40 MHz, 109.60 MHz etc. Table 10.1 provides an illustration of how these frequencies are allocated within the 109 MHz range. This pattern is applied from 108 to 111.95 MHz.

Table 10.1 Allocation of ILS and VOR frequencies

ILS frequency (MHz)	VOR frequency (MHz)
	109.00
109.10	
109.15	
	109.20
109.30	
109.35	
	109.40
109.50	
109.55	
	109.60
109.70	
109.75	
	109.80
109.90	
109.95	

10.1.2 Overview

In addition to the inherently improved system performance and navigation reliability, VOR has another feature that makes it extremely useful for air navigation. The VOR system has the ability to transmit specific bearing information, referred to as a '**radial**', see Figure 10.1(a). The pilot can select any radial from a given VOR navigation aid and fly to or from that aid. Since the system is 'line of sight', i.e. receiving signals as space waves, the altitude of the aircraft will have a direct relationship with the range within which the system can be used, see Figure 10.1(b).

Using VHF navigation aids imposes a limit on the theoretical working range that can be obtained. The maximum theoretical line-of-sight (LOS) distance between an aircraft and the ground station is given by the relationship:

$$d = 1.1\sqrt{h}$$

where d is the distance in nautical miles, and h is the altitude in feet above ground level (assumed to be flat terrain). The theoretical LOS range for altitudes up to 20,000 feet is given in Table 10.2.

At higher altitudes, it is possible to receive VOR signals at greater distances but with reduced signal integrity. Although the actual range also depends on transmitter power and receiver sensitivity, the above relationship provides a good approximation. In practice, navigation aids have a designated standard service volume (SSV); this defines the reception limits within an altitude envelope as shown in Table 10.3.

Key point

VOR radials are referenced to magnetic north; they are the basis of airways for en route navigation.

Key point

VOR transmissions are 'line of sight' therefore range increases with increased altitude.

Table 10.2 Theoretical LOS range

Altitude (feet)	Range (nm)
100	10
1,000	32
5,000	70
10,000	100
20,000	141

Table 10.3 Navigation aid classifications

Classification	Altitude (feet)	Range (nm)
Terminal	1,000–12,000	25
Low altitude	1,000–18,000	40
High altitude	18,000–45,000	130

10.1.3 Conventional VOR (CVOR)

There are two types of VOR ground navigation transmitter: conventional and Doppler. The conventional VOR (CVOR) station radiates two signals: omnidirectional and directional on a continuous basis. The omnidirectional (or reference) signal is the carrier wave frequency of the station, amplitude modulated to 30 Hz. The directional signal is radiated as a cardioid pattern rotating at 30 revolutions per second. The sub-carrier frequency is 9960 Hz, frequency modulated in the range 9960 ±480 Hz at 30 Hz.

The directional signal is arranged to be in phase with the reference signal when the aircraft is due north (magnetic) of the VOR station. As the cardioid pattern rotates around the station, the two signals become out of phase on a progressive basis, see Figure 10.2. The bearing between any given angle and magnetic north is determined by the receiver as the phase angle difference between the reference and directional signals. This difference in phase angle is resolved in the aircraft receiver and displayed to the crew as a radial from the VOR station, see Figure 10.3.

Locations of conventional VOR (CVOR) ground stations have to be carefully planned to take into account local terrain and obstacles.

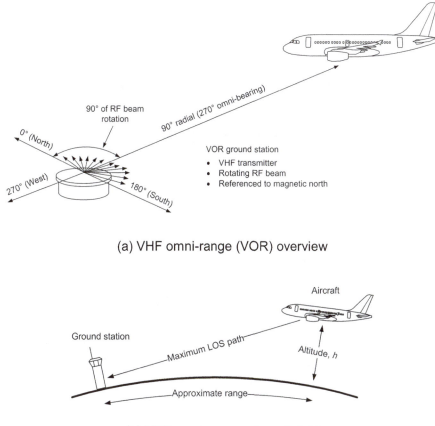

(a) VHF omni-range (VOR) overview

(b) VHF omni-range—line of sight

Figure 10.1   VOR overview

Mountains and trees can cause multipath reflections resulting in distortion (known as siting errors) of the radiated signal. These errors can be overcome with an enhanced second-generation system known as Doppler VOR (DVOR).

10.1.4 Doppler VOR (DVOR)

Doppler is usually associated with self-contained navigation systems, and this subject is described in a separate chapter. The Doppler effect is also applied to the second-generation version of VOR ground transmitters. The Doppler effect can be summarised here as: '…the frequency of a wave apparently changes as its source moves closer to, or farther away from an observer'.

The DVOR ground station has an omnidirectional transmitter in the centre, amplitude modulated at 30 Hz; this is the reference phase. The directional signal is derived from a 44 feet diameter circular array comprising up to 52 individual antennas, see Figure 10.4(a) and (b). Each antenna transmits in turn to simulate a rotating antenna.

Consider two aircraft using the DVOR station as illustrated in Figure 10.4(c). The effect of the rotating 9960 Hz signal is to produce a Doppler shift; aircraft A will detect a decreased frequency, aircraft B will detect an increased frequency. Doppler shift creates a frequency modulated (FM) signal in the aircraft receiver over the range 9960 Hz ±480 Hz varying at 30 Hz in a sine wave. Note that the perceived frequency will be 9960

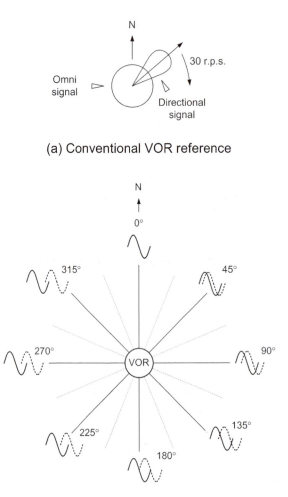

(a) Conventional VOR reference

(b) Variable signal phase relationship

(c) Conventional VOR navigation aid

Figure 10.2   Conventional VOR (CVOR)

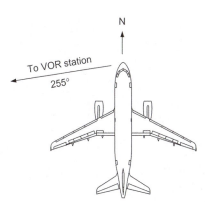

(a) VOR bearing

255° bearing to VOR station
(tuned by left VOR receiver)

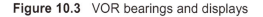

(b) RMI

Figure 10.3   VOR bearings and displays

10.2 Airborne VOR equipment

Modern transport aircraft have two VOR systems often designated left and right; note that airborne equipment is the same for conventional and Doppler VOR. Radio frequency (RF) signals from the antenna are processed in the receiver as determined by frequency and course selections from the control panel; outputs are sent to various displays.

10.2.1 Antennas

The VOR antenna is a horizontally polarised, omnidirectional half-wave dipole, i.e. a single conductor with a physical length equal to half the wavelength of the VOR signals being received. Two such antennas are formed into a single package and usually located in the aircraft fin as indicated in Figure 10.5. They are packaged within a composite fairing for aerodynamic streamline purposes. The antennas are connected to the receivers via coaxial cables.

10.2.2 Receivers

VOR receivers are often combined with other radio navigation functions, e.g. the instrument landing system; receivers are located in the avionic equipment bay, see Figure 10.6.

VOR receivers are based on the super-heterodyne principle with tuning from the control panel. The received radio frequency signal is passed through an amplitude modulation filter to separate out the:

- 30 Hz tone from the rotating pattern
- voice identification (if provided from the navigation aid)
- Morse code tone; 9960 Hz signal FM by ±480 Hz at 30 Hz reference tone.

Voice and Morse code tones are integrated with the audio system. A comparison of the phase angles of the variable and reference 30 Hz signals produces the VOR radial signal. Receivers based on analogue technology interface with the flight deck displays using synchro systems. Digital receivers interface with other systems using a data bus system, typically ARINC 429. Receivers usually combine both VOR and instrument landing system functions (see Chapter 12).

Hz when the aircraft are in the positions shown in Figure 10.4(c). The phase difference measured in the airborne equipment depends on the bearing of the aircraft relative to the station. Since the FM variable signal is less prone to interference, DVOR performance is superior to CVOR.

DVOR actually uses two rotating patterns as shown in Figure 10.4(d). These patterns (diagonally opposite each other) rotate at 30 revolutions per second; one pattern is 9960 Hz above the reference, the other is 9960 Hz below the reference frequency. The diameter of the array, together with the speed of pattern rotation creates a Doppler shift of 480 Hz (at VOR frequencies).

(a) Doppler VOR (DVOR) navigation aid
(photo courtesy of T. Diamond)

(b) Doppler VOR (DVOR) navigation aid
(photo courtesy of T. Diamond)

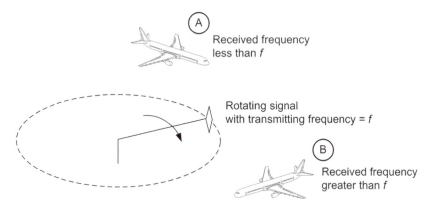

(c) Rotating transmitting signal

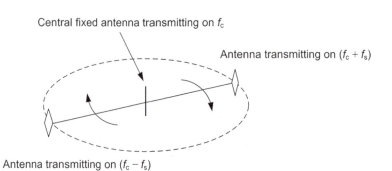

(d) Double sideband Doppler VOR

Figure 10.4   Doppler VOR (DVOR)

Figure 10.5 Location of VOR antennas

10.2.3 Control panels

Control panels identified as 'VHF NAV' can be located on the glare shield (as shown in Figure 10.7) or centre pedestal. This panel is used to select the desired course and VOR frequencies. General aviation products have a combined VHF navigation and communications panel—see Figure 10.6(b)—this can be integrated with the GPS navigation panel.

10.2.4 VOR displays

The bearing to a VOR navigation aid (an output from the receiver) can be displayed on the radio magnetic indicator (RMI); this is often shared with the ADF system as discussed in the previous chapter. The RMI, Figure 10.8(a), provides the pilot with both magnetic heading and direction to the tuned VOR navigation aid. The two bearing pointers (coloured red and green) are associated with the two VOR systems and allow the crew to tune into two different VOR navigation aids at the same time. On some instruments, a switch on the RMI is used to select either ADF or VOR bearing information, see Figure 10.8(b). RMIs therefore have a dual purpose; pilots use a switch

on the RMI to select either ADF and/or VOR bearings (see Chapter 9 for ADF). Some aircraft have a bearing source indicator adjacent to the RMI to confirm ADF or VOR selection, see Figure 10.8(c).

In order to fly along an airway, first it has to be intercepted. This is achieved by flying towards the desired radial on a specified heading. The method of displaying the VOR radial depends on the type of avionic fit. Electromechanical instruments include the omni-bearing selector (OBS) and course deviation indicator (CDI), see Figure 10.9. The omni-bearing selector (OBS) indicator has a number of features; the selector is used to manually rotate the course card. This card is calibrated from 0 to 360° and indicates the VOR bearing selected to fly TO or FROM.

In Figure 10.9(a), a VOR radial of 345° has been selected. The deviation pointer moves to the left or right to guide the pilot in the required direction to maintain the selected course. Each dot on the scale represents a 2° deviation from the selected course. The back-course (BC) indicator indicates when flying FROM the VOR navigation aid. On some instruments, the BC indicator is replaced by a TO/FROM display in the form of

Figure 10.7 Typical VOR control panel

an arrow. A red OFF flag indicates when the:

- VOR navigation aid is beyond reception range
- pilot has not selected a VOR frequency
- VOR system is turned off, or is inoperative.

An updated version of this instrument is the CDI. This has a compass display and course selector as shown in Figure 10.9(b). The course selector (lower right-hand side of instrument) is set to the desired VOR radial; a deviation pointer moves left or right of the aircraft symbol to indicate if the aircraft is to the right or left of the selected radial.

For aircraft with electronic flight instruments, the desired radial is displayed on the electronic horizontal situation indicator (EHSI). This EHSI display can either be selected to show a conventional compass card (Figure 10.10(a)) or expanded display (Figure 10.10(b)). As the radial is approached the deviation bar gradually aligns with the selected course.

If flying manually, the crew turn the aircraft onto the selected course whilst monitoring the deviation bar; when it is centred, the radial has been intercepted and the EHSI will display 'TO' confirming that the inbound radial is being followed. The flight continues until the VOR navigation is reached, and the radial to the next navigation aid is selected. If the EHSI were still selected to the original inbound radial, the EHSI would display 'FROM'. The lateral deviation bar therefore shows if the aircraft is flying on the selected radial, or if it is to the left or right of the radial.

(a) VOR receiver (remotely located in the aircraft's avionic equipment bay)

(b) Navigation and communications panel/ receiver used in general aviation

Figure 10.6 VOR receivers

(a) RMI with two bearing pointers

Right omni-bearing pointer   Left omni-bearing pointer

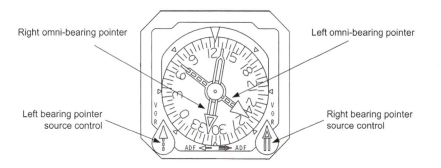

Left bearing pointer   Right bearing pointer
source control   source control

(b) RMI source control (VOR/ADF)

(c) RMI and bearing source annunciator

Figure 10.8   VOR displays and indicators

(a) Omnibearing selector

(b) Course deviation indicator (CDI)

Figure 10.9 Omni-bearing selector and course deviation indicator

10.3 Operational aspects of VOR

Radials from any given VOR navigation aid are the basis of **airways**; system accuracy is typically within one degree. These are the standard routes flown by aircraft when flying on instruments. When two VOR radials intersect, they provide a unique navigation position fix based (theta–theta). The accuracy of this fix is greatest when the radials intersect at right angles. Figure 10.11 illustrates how navigation charts are built up on a network of VOR radials; the accuracy of VOR radials is generally very good (±1 degree). In this illustration, the navigation aid located near Brussels (abbreviated BUB) transmits on 114.6 MHz. Three radials can be seen projected from this navigation aid on 136°, 310° and 321°. These radials are used to define airways A24 and G120. Note the Morse code output and latitude/ longitude for the navigation aid.

The intersecting radials from navigation aids are used to define reporting points for en route navigation. These reporting points are given five-letter identification codes associated with their geographic location. For example, the reporting point HELEN is defined by airways G5 and A120. The flexibility of VOR is greatly increased when co-located distance measuring equipment (DME) is used, thereby providing rho–theta fixes from a single navigation aid. There are examples of VOR-only navigation aids, e.g. Perth in Scotland (identification code PTH, frequency 110.40 MHz). The majority of VOR navigation aids are paired with DME; this system is described in the next chapter.

Key point

Two intersecting VOR radials can be used to define unique locations known as reporting points; these are used for air traffic control purposes.

Key point

Doppler VOR was introduced to overcome siting problems found with conventional VOR. The two systems operate on different principles; however, the airborne equipment is the same.

Test your understanding 10.1

Why are VOR transmissions 'line of sight' only?

Test your understanding 10.2

Calculate (a) the line of sight range for an aircraft at an altitude at 7,500 feet and (b) the altitude of an aircraft that would yield a line of sight range of 200 nautical miles.

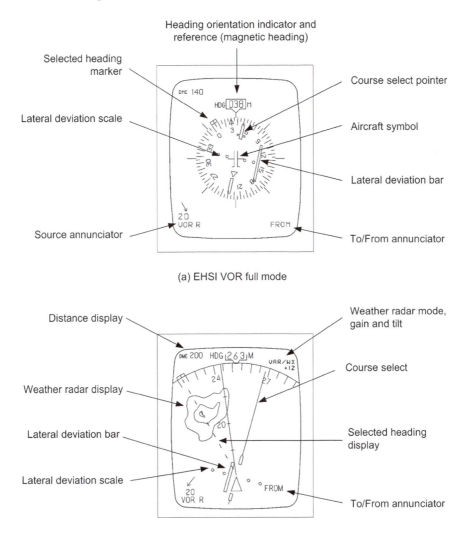

(a) EHSI VOR full mode

Heading orientation indicator and reference (magnetic heading)

Selected heading marker

Lateral deviation scale

Source annunciator

Course select pointer

Aircraft symbol

Lateral deviation bar

To/From annunciator

Distance display

Weather radar display

Lateral deviation bar

Lateral deviation scale

Weather radar mode, gain and tilt

Course select

Selected heading display

To/From annunciator

(b) EHSI VOR expanded mode

Figure 10.10 VOR electronic displays

Test your understanding 10.3

How can the crew identify a specific VOR navigation aid?

Test your understanding 10.4

Where can a VOR radial be displayed?

Test your understanding 10.5

Explain how a VOR radial is captured.

Test your understanding 10.6

Why does an RMI have two VOR pointers?

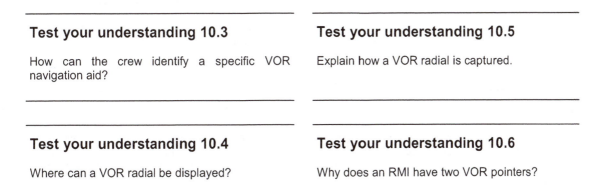

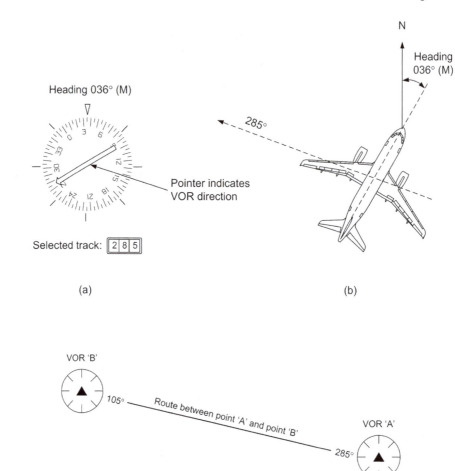

(a)   (b)

(c) Airways using VOR radials

Figure 10.11   VOR radials and airways

Test your understanding 10.7

What is the difference in aircraft equipment between conventional and Doppler VOR?

Key point

Navigation charts are built up on a network of VOR radials. We shall see in later chapters how these charts are supplemented by area navigation waypoints.

Test your understanding 10.8

What is the Morse code output used for in a VOR transmission?

Key point

VOR operates in the frequency band extending from 108 MHz to 117.95 MHz. Three alpha characters are used to identify specific VOR navigation aids.

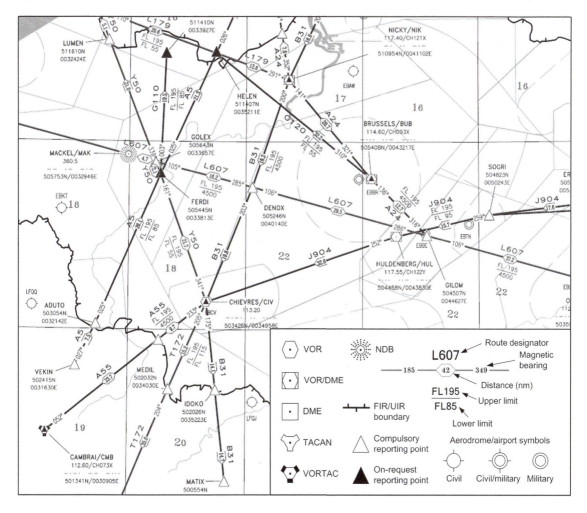

(c) Airway network over Belgium

Figure 10.11 (continued)

10.4 Multiple choice questions

1. VOR operates in which frequency range?
   (a) LF
   (b) MF
   (c) VHF.

2. VOR signals are transmitted as what type of wave?
   (a) Sky wave
   (b) Ground wave
   (c) Space wave.

3. Where is the deviation from a selected VOR radial displayed?
   (a) RMI
   (b) HSI
   (c) NDB.

4. At which radial will the directional wave be out of phase by 90 degrees with the non-directional wave?
   (a) 090 degrees
   (b) 000 degrees
   (c) 180 degrees.

5. At which radial will the directional signal be in phase with the non-directional signal?
    (a) 090 degrees
    (b) 000 degrees
    (c) 180 degrees.

6. VOR navigation aids are identified by how many alpha characters?
    (a) Two
    (b) Three
    (c) Five.

7. VOR radials are referenced to:
    (a) non-directional signals from the navigation aid
    (b) magnetic north
    (c) true north.

8. The RMI has two pointers coloured red and green; these are used to indicate:
    (a) the bearing of two separately tuned VOR stations
    (b) directional (red) and non-directional transmissions (green)
    (c) the radials of two separately tuned VOR stations.

9. Morse code tones are used to specify the VOR:
    (a) identification
    (b) frequency
    (c) radial.

10. The intersection of two VOR radials provides what type of position fix?
    (a) Rho–rho
    (b) Theta–theta
    (c) Rho–theta.

11. An aircraft is flying on a heading of 090 degrees to intercept the selected VOR radial of 180 degrees; the HSI will display that the aircraft is:
    (a) right of the selected course
    (b) left of the selected course
    (c) on the selected course.

12. The DVOR navigation aid has an omnidirectional transmitter located in the:
    (a) centre
    (b) outer antenna array
    (c) direction of magnetic north.

13. When flying overhead a VOR navigation aid, the reliability of directional signals:
    (a) decreases
    (b) increases
    (c) stays the same.

14. Reporting points using VOR navigation aids are defined by the:
    (a) identification codes
    (b) intersection of two radials
    (c) navigation aid frequencies.

15. With increasing altitude, the range of a VOR transmission will be:
    (a) increased
    (b) decreased
    (c) the same.

16. Referring to Figure 10.12, the instrument shown is called the:
    (a) omni-bearing selector (OBS)
    (b) radio magnetic indicator (RMI)
    (c) course deviation indicator (CDI).

Figure 10.12 See Question 16

Chapter 11   Distance measuring equipment

The previous two chapters have been concerned with obtaining directional information for the purposes of airborne navigation. In this chapter we will look at a system that provides the crew with the distance to a navigation aid. Distance measuring equipment (DME) is a short/medium-range navigation system, often used in conjunction with the VOR system to provide accurate navigation fixes. The system is based on secondary radar principles, and operates in the L-band of radar. Before looking at what the system does and how it operates in detail, we need to take at look at some basic radar principles.

11.1 Radar principles

The word **radar** is derived from <u>ra</u>dio <u>d</u>etection <u>a</u>nd <u>r</u>anging; the initial use of radar was to locate aircraft and display their range and bearing on a monitor (either ground based or in another aircraft). This type of radar is termed **primary radar**: Energy is radiated via a rotating radar antenna to illuminate a 'target'; this target could be an aircraft, the ground or a cloud. Some of this energy is reflected back from the target and is collected in the same antenna, see Figure 11.1. The strength of the returned energy is measured and used to determine the range of the target. A rotating antenna provides the directional information such that the target can be displayed on a screen.

Primary radar has its disadvantages; one of which is that the amount of energy being transmitted is very large compared with the amount of energy reflected from the target. An alternative method is **secondary radar** that transmits a specific low energy signal (the interrogation) to a known target. This signal is analysed and a new (or secondary) reply signal, i.e. not a reflected signal, is sent back to the origin, see Figure 11.2(a). Secondary radar was

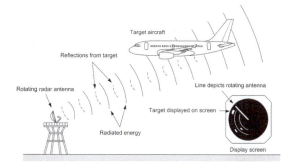

Figure 11.1   Primary radar

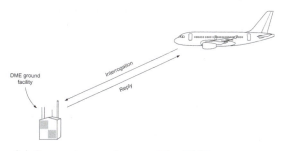

(a) Secondary radar used for DME

(b) DME transponder (right of photo)
    (photo courtesy of T. Diamond)

Figure 11.2   DME overview

developed during the Second World War to differentiate between friendly aircraft and ships: Identification Friend or Foe (IFF). The principles of secondary radar now have a number of applications including distance measuring equipment (DME).

11.2 DME overview

The DME navigation aid contains a transponder (receiver and transmitter) contained within a single navigation aid, Figure 11.2(b). The aircraft equipment radiates energy pulses to the DME navigation aid; secondary signals are then transmitted back to the aircraft. An on-board **interrogator** measures the time taken for the signals to be transmitted and received at the aircraft. Since we know the speed of radio wave propagation, the interrogator can calculate the distance to the DME navigation aid. DME navigation aids can either be self-contained ground stations, or co-located with a VOR navigation aid, Figure 11.2(c).

Since the system is 'line of sight', the altitude of the aircraft will have a direct relationship with the range that the system can be used, see Figure 11.3(a). Using DME navigation aids imposes a limit on the working range that can be obtained. The maximum line-of-sight (LOS) distance between an aircraft and the ground station is given by the relationship:

$$d = 1.1\sqrt{h}$$

where d is the distance in nautical miles, and h is the altitude in feet above ground level (assumed to be flat terrain). The theoretical LOS range for altitudes up to 20,000 feet is given in Table 11.1.

Table 11.1 Theoretical LOS range

Altitude (feet)	Range (nm)
100	10
1,000	32
5,000	70
10,000	100
20,000	141

Referring to Figure 11.3(b) it can be seen that the actual distance being measured by the interrogator is the '**slant**' **range**, i.e. not the true distance (horizontal range) over the ground. The effects of slant range in relation to the horizontal range are greatest at high altitudes and/or when the aircraft is close to the navigation aid. Taking this to the limit, when the aircraft is flying over a DME navigation aid, it would actually be measuring the aircraft's altitude!

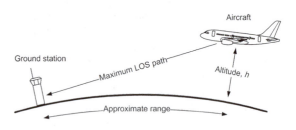

(a) Line of sight versus altitude

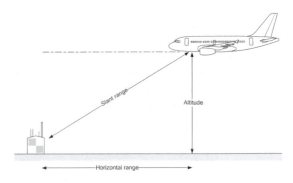

(b) DME slant range

Figure 11.3 DME range terminology

Test your understanding 11.1

What is the difference between primary and secondary radar?

Test your understanding 11.2

Distinguish between slant range and horizontal range.

11.3 DME operation

The signals transmitted by the interrogator are a pair of pulses, each of 3.5 ms duration and 12 ms apart modulated on the DME navigation aid frequency. The interrogator generates a pulse-pair repetition rate between 5 and 150 pulse-pairs per second. At the DME navigation aid, the transponder receives these pulses and, after a 50 ms time delay, transmits a new pair of pulses at a frequency 63 MHz above or below the interrogator's frequency. The aircraft's interrogator receives the pulses and matches the time interval between the transmitted pair of pulses. This ensures that other aircraft interrogating the same DME navigation aid at the same time only process their own pulses.

By measuring the elapsed time between transmitting and receiving (and taking into account the 50 ms time delay) the interrogator calculates the distance to the navigation aid. DME is a line of sight system with a maximum range of approximately 200 nm; this equates to approximately 2400 ms elapsed time taken for a pair of pulses to be transmitted and received, taking into account the 50 ms time delay in the ground station. System accuracy is typically ± 0.5 nm, or 3% of the calculated distance, whichever is the greater.

Test your understanding 11.3

What is the typical accuracy and maximum range of a DME system?

Key point

The varying interval between pulse-pairs ensures that the DME interrogator recognises its own signals and rejects other signals.

Key point

DME is based on secondary radar; it operates in the L-band between 962 MHz and 1215 MHz (UHF) with channel spacing at 1MHz.

11.4 Equipment overview

Commercial transport aircraft are usually fitted with two independent DME systems, comprising antennas and interrogators.

The DME antennas are L-band blades, located on the underside of the aircraft fuselage, see Figure 11.4(a); note that the antenna is dual purpose in that it is used for both transmitting and receiving.

The interrogators are located in the equipment bays (Figure 11.4(b)) and provide three main functions: transmitting, receiving and calculation of distance to the selected navigation aid. Transmission is in the range 1025 to 1150 MHz; receiving is in the range 962 to 1215 MHz; channel spacing is 1 MHz. The interrogator operates in several modes:

- Standby
- Search
- Track
- Scan
- Memory
- Fault
- Self-test.

When the system is first powered up, it enters the standby mode; transmissions are inhibited, the receiver and audio are operative; the DME display is four dashes to indicate **no computed data** (NCD). The receiver monitors pulse-pairs received from any local ground stations. If sufficient pulse-pairs are counted, the interrogator enters the search mode. The transmitter now transmits pulse-pairs and monitors any returns; synchronous pulse-pairs are converted from time into distance and the system enters the track mode. Distance to the navigation aid will now be displayed on the DME indicator (see Figure 11.5). The scan mode has two submodes: directed scanning for multiple navigation aid tuning; up to five stations can be scanned in accordance with a predetermined area navigation auto-tuning programme (described in more detail in Chapter 16). Alternatively, free scanning occurs for any DME navigation aids within range. If pulse-pairs from any navigation aids are not received after a short period of time (two seconds typical), the interrogator goes into memory mode whereby distance is calculated from the most recently

(a) Location of DME antennas

(b) Location of DME transceiver

Figure 11.4   DME equipment

received pulse-pairs. Memory mode expires after a short period of time, typically ten seconds, or until pulse-pairs are received again. If the system detects any fault conditions, the distance display is blanked out. Self-test causes the system to run through a predetermined sequence causing the indicators to read: blank, dashes (NCD) and 0.0 nm.

DME outputs can be displayed in a variety of ways, see Figure 11.5. These displays include dedicated readouts, electronic flight instrument systems (EFIS), combined panels/transceivers (for general aviation) and radio distance magnetic indicators (RDMI). When selecting a co-located VOR-DME navigation aid, the crew only needs to tune into the VOR frequency; the DME frequency is automatically selected.

Key point

When no computed data (NCD) is available this condition is displayed as four dashes.

Test your understanding 11.4

List and describe four modes in which a DME interrogator can operate.

Key point

VOR and DME systems operate on different frequencies. When they are co-located, the DME frequency is automatically selected when the pilot tunes into the VOR frequency.

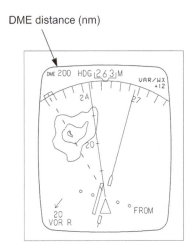

(d) Electronic instrument—DME display

(a) Self-contained DME displays

Figure 11.5   Various types of DME display

11.5 En route navigation using radio navigation aids

Basic en route navigation guidance for commercial aircraft can be readily accomplished using co-located VOR and DME systems, thereby providing rho–theta fixes from a single navigation aid. The DME frequency is paired with the VOR frequency; this means that only the VOR frequency needs to be tuned, the DME frequency is automatically tuned as a result. Alternatively, rho–rho fixes can be established from a pair of DME navigation aids. Note that this produces an ambiguous fix unless another DME is used, see Figure 11.6. An example of DME transponder locations and co-located VOR-DME navigation aids in Switzerland is provided in Table 11.2.

(b) DME panel/transceiver for general aviation

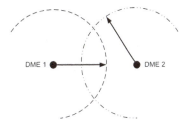

(c) Radio distance magnetic indicator (RDMI)

Figure 11.6   Ambiguous DME–DME position

Table 11.2 Locations of VOR and DME navigation aids in Switzerland

Name	Identification code	Type
Corvatsch	CVA	DME
Fribourg	FRI	VOR-DME
Geneva Cointrin	GVA	VOR-DME
Grenchen	GRE	VOR-DME
Hochwald	HOC	VOR-DME
Kloten	KLO	VOR-DME
La Praz	LAP	DME
Montana	MOT	VOR-DME
Passeiry	PAS	VOR-DME
Sion	SIO	VOR-DME
St. Prex	SPR	VOR-DME
Trasadingen	TRA	VOR-DME
Willisau	WIL	VOR-DME
Zurich East	ZUE	VOR-DME

Figure 11.7 TACAN navigation aid

In the US, a combined rho–theta system was introduced for military aircraft known as **TACAN** (<u>ta</u>ctical <u>a</u>ir <u>n</u>avigation). This system is a short-range bearing and distance navigation aid operating in the 962–1215 MHz band. TACAN navigation aids (see Figure 11.7) are often co-located with VOR navigation aids; these are identified on navigation charts as '**VORTAC**'.

The TACAN navigation aid is essentially a DME transponder (using the same pulse pair and frequency principles as the standard DME) to which directional information has been added; both operate in the same UHF band. An important feature of TACAN is that both distance and bearing are transmitted on the same frequency; this offers the potential for equipment economies. Furthermore, because the system operates at a higher frequency than VOR, the antennas and associated hardware can be made smaller. This has the advantage for military use since the TACAN equipment can be readily transported and operated from ships or other mobile platforms.

When co-located with a VOR navigation aid, military and commercial aircraft can share the VORTAC facility. Referring to Figure 11.8, military aircraft obtain their distance and bearing information from the TACAN part of the VORTAC; commercial aircraft obtain their distance information from the TACAN, and bearing information from the VOR part of the TACAN. Reporting points (shown as triangles) based on DME navigation aids, e.g. the VORTAC navigation aid located at Cambrai (CMB), northern France, are illustrated in Figure 11.9. The intersecting radials from navigation aids are used to define reporting points for en route navigation. These **reporting points** are given five-letter identification codes associated with their geographic location. For example, the reporting point 'HELEN' (at the top of the chart) is defined by a distance and bearing from the Brussels VOR/DME navigation aid.

TACAN frequencies are specified as channels that are allocated to specific frequencies, e.g. Raleigh–Durham VORTAC in North Carolina,

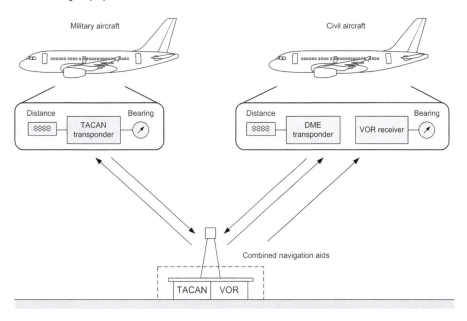

Figure 11.8 VORTAC navigation aid and associated aircraft functions

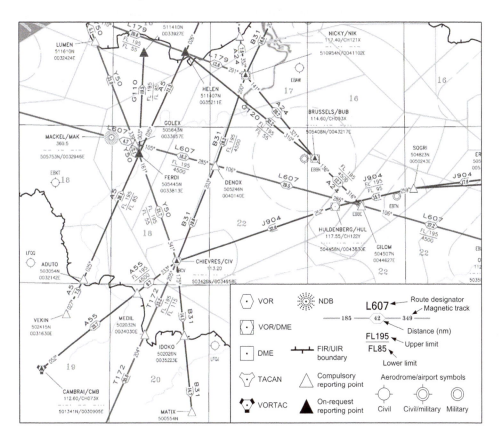

Figure 11.9 Reporting points defined by VOR–DME

USA, operates on channel 119X. This corresponds to a:

- VOR frequency of 117.2 MHz
- DME interrogation frequency of 1143 MHz
- DME reply frequency of 1206 MHz
- Pulse code of 12 ms.

Note that since DME, VOR and VORTAC navigation aids have to be located on land, the airways' network does not provide a great deal of coverage beyond coastal regions.

Referring to Figure 11.10, a combination of VOR, DME and VORTAC stations (see Figure 11.11) located in a number of European countries provides a certain amount of navigation guidance in the North Atlantic, Norwegian Sea and North Sea. This diagram assumes a line-of-sight range of approximately 200 nm. The gaps in this radio navigation network can be overcome by the use of alternative navigation systems including: inertial navigation (INS), Doppler, satellite navigation and Loran-C, these are all described elsewhere in this book.

Figure 11.11 Typical VOR–DME navigation aid (photo courtesy T. Diamond)

Test your understanding 11.5

Explain what is meant by frequency pairing.

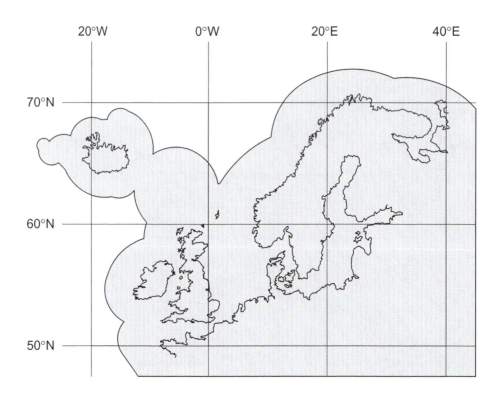

Figure 11.10 Approximate max. line of sight navigation coverage for northern Europe

Test your understanding 11.6

Describe two ways in which DME distance information is displayed.

Test your understanding 11.7

DME ground stations could be responding to numerous aircraft; how does the airborne DME system recognise its own signals and reject signals intended for other aircraft?

Test your understanding 11.8

What information does an RDMI provide the crew?

Test your understanding 11.9

What type of information does a VORTAC provide?

11.6 Multiple choice questions

1. DME is based on what type of radar?
   (a) Primary
   (b) Secondary
   (c) VHF.

2. DME provides the following information to the crew:
   (a) bearing to a navigation aid
   (b) deviation from a selected course
   (c) distance to a navigation aid.

3. When tuned into a VORTAC, commercial aircraft obtain their distance and bearing information from the:
   (a) TACAN and VOR
   (b) DME and VOR
   (c) DME and TACAN.

4. DME signals are transmitted:
   (a) by line of sight
   (b) as ground waves
   (c) as sky waves.

5. An RDMI provides the following information:
   (a) distance and bearing to a navigation aid
   (b) deviation from a selected course
   (c) the frequency of the selected navigation aid.

6. Slant range errors are greatest when the aircraft is flying at:
   (a) high altitudes and close to the navigation aid
   (b) high altitudes and far from the navigation aid
   (c) low altitudes and far from the navigation aid.

7. To select a co-located VOR-DME navigation aid, the crew tunes into the:
   (a) DME frequency
   (b) VOR frequency
   (c) NDB frequency.

8. The DME interrogator is part of the:
   (a) airborne equipment
   (b) DME navigation aid
   (c) VORTAC.

9. The varying interval between pulse-pairs ensures that the interrogator:
   (a) recognises its own pulse-pairs and rejects other signals
   (b) recognises other pulse-pairs and rejects its own signal
   (c) tunes into a VOR station and DME navigation aid.

10. When a DME indicator is receiving no computed data, it will display:
    (a) dashes
    (b) zeros
    (c) eights.

11. Using a collocated VOR–DME navigation aid produces what type of position fix?
    (a) Rho–rho
    (b) Rho–theta
    (c) Theta–theta.

12. Distance and bearing signals from a TACAN
navigation aid are transmitted on:
(a) HF
(b) UHF
(c) VHF.

13. Using two DME navigation aids provides how
many calculated positions?
(a) two
(b) one
(c) three.

14. DME operates in which frequency band?
(a) UHF
(b) VHF
(c) LF/MF.

15. The instrument shown in Figure 11.12 is
called the:
(a) RMI
(b) RDMI
(c) CDI.

16. Referring to Figure 11.13, the display is
providing:
(a) maximum distance
(b) minimum distance
(c) no computed data.

17. Referring to Figure 11.14, the installation on
the right is a DME:
(a) transponder
(b) transmitter
(c) receiver.

Figure 11.13 See Question 16

Figure 11.12 See Question 15

Figure 11.14 See Question 17

Instrument landing system

Navigation aids such as automatic direction finder (ADF), VHF omnidirectional range (VOR) and distance measuring equipment (DME) are used to define airways for en route navigation. They are also installed at airfields to assist with approaches to those airfields. These navigation aids cannot, however, be used for precision approaches and landings. The standard approach and landing system installed at airfields around the world is the instrument landing system (ILS). The ILS uses a combination of VHF and UHF radio waves and has been in operation since 1946. In this chapter we will look at ILS principles and hardware in detail, concluding with how the ILS combines with the automatic flight control system (AFCS) to provide fully automatic approach and landing.

12.1 ILS overview

The instrument landing system is used for the final approach and is based on directional beams propagated from two transmitters at the airfield, see Figure 12.1. One transmitter (the **glide slope**) provides guidance in the vertical plane and has a range of approximately 10 nm. The second transmitter (the **localizer**) guides the aircraft in the horizontal plane. In addition to the directional beams, two or three marker beacons are located at key points on the extended runway centreline defined by the localizer, see Figure 12.4.

12.2 ILS ground equipment

12.2.1 Localizer transmitter

The localizer transmits in the VHF frequency range, 108–112 MHz in 0.5 MHz increments. Note that this is the same frequency range as used by the VOR system (see Chapter 10). Although

the two systems are completely independent and work on totally different principles, they often share the same receiver. The two systems are differentiated by their frequency allocations within this range. ILS frequencies are allocated to the odd tenths of each 0.5 MHz increment, e.g. 109.10 MHz, 109.15 MHz, 109.30 MHz etc. VOR frequencies are allocated even tenths of each 0.5 MHz increment, e.g. 109.20 MHz, 109.40 MHz, 109 60 MHz etc. Table 12.1 provides an illustration of how these frequencies are allocated within the 109 MHz range. This pattern applies from 108 to 111.95 MHz.

Table 12.1 Allocation of ILS and VOR frequencies

ILS frequency (MHz)	VOR frequency (MHz)
	109.00
109.10	
109.15	
	109.20
109.30	
109.35	
	109.40
109.50	
109.55	
	109.60
109.70	
109.75	
	109.80
109.90	
109.95	

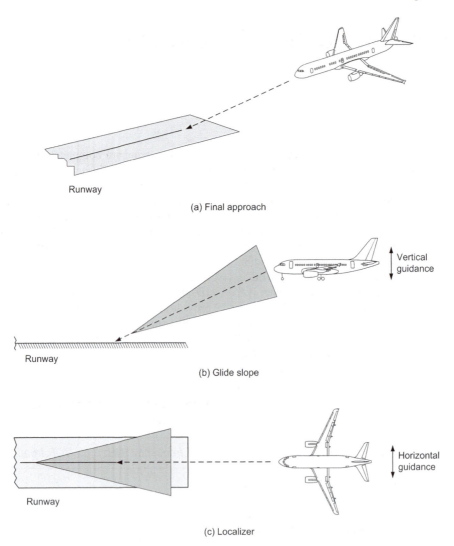

(a) Final approach

(b) Glide slope

(c) Localizer

Figure 12.1 ILS overview

The localizer antenna is located at the far end of the runway, and transmits two lobes to the left and right of the runway centreline modulated at 90 Hz and 150 Hz respectively. On the extended runway centreline, see Figure 12.2, the combined depth of modulation is equal. Either side of the centreline will produce a **difference in depth of modulation** (DDM); this difference is directly proportional to the deviation either side of the extended centreline of the runway. The localizer also transmits a two or three letter Morse code identifier that the crew can hear on their audio panels.

Key point

The instrument landing system is based on directional beams propagated from two transmitters at the airfield: localizer and glide slope.

Test your understanding 12.1

What frequency bands do the localizer and glide slope use?

(a) Localizer beams (plan view)

(b) Localizer antenna (viewed across the runway end)

(c) Localizer antenna (viewed down the runway)

Figure 12.2 Localizer beams and antenna

12.2.2 Glide slope antenna

The glide slope antenna transmits in the UHF frequency band, 328.6 to 335 MHz at 150 kHz spacing. Upper and lower lobes are modulated at 90 HZ and 150 Hz respectively. When viewed from the side, see Figure 12.3, the two lobes overlap and produce an approach path inclined at a fixed angle between 2.5 and 3.5 degrees. Glide slope frequency is automatically selected when the crew tunes the localizer frequency.

12.2.3 Marker beacons

Two or three beacons are sited on the extended runway centreline at precise distances; these are specified in the approach charts for specific runways. These beacons operate at 75 MHz and radiate approximately 3–4 W of power. The beacons provide visual and audible cues to the crew to confirm their progress on the ILS, see Figure 12.4. The **outer marker** is located between four and seven miles from the runway

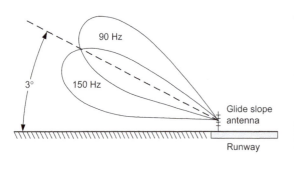

(a) Glide slope beams

(b) Glide slope antenna

Figure 12.3 Glide slope beams and antenna

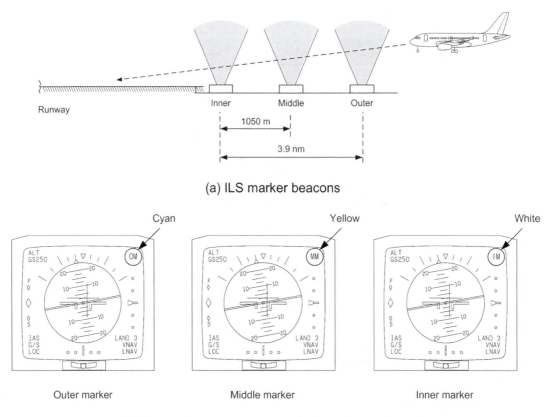

(a) ILS marker beacons

(b) Marker beacon indications (primary flying display)

Figure 12.4 ILS marker beacon system

threshold; it transmits Morse code dashes at a tone frequency of 400 Hz and illuminates a blue light (or cyan 'OM' icon for electronic displays) when the aircraft passes over the beacon.

The outer marker provides the approximate point at which an aircraft on the localizer will intercept the glide slope. Some airfields use non-directional beacons (NDBs) in conjunction with (or in place) of the outer marker. These are referred to as **locator beacons** (compass locator in the USA).

The middle marker is located approximately 3500 feet from the runway threshold. When passing over the middle marker, the crew receive an alternating Morse code of dots/dashes modulated at 1300 Hz and a corresponding amber light (or yellow 'MM' icon for electronic displays) is illuminated. The middle marker coincides with the aircraft being 200 feet above the runway touchdown point.

Runways that are used for low visibility approach and landings (see later in this chapter) have a third **inner marker**. When passing over the inner marker, the crew receive Morse code dots modulated at 3000 Hz on the audio system and a corresponding white light (or 'IM' icon for

electronic displays) is illuminated. The marker beacon system is currently being phased out with the introduction of DME and GPS approaches.

12.3 ILS airborne equipment

The airborne equipment comprises localizer and glide slope antennas, ILS receiver, marker receiver, and flight deck controls and displays. Most aircraft are fitted with two or three independent ILS systems (typically named left, centre and right). The localizer and glide slope frequencies are in different wave bands, the crew tunes the localizer frequency (via the 'Nav' control panel) and this automatically tunes a paired glide slope frequency for a particular runway.

12.3.1 Antennas

The typical ILS antenna installation on a transport aircraft is illustrated in Figure 12.5. In this installation, two dual channel antennas are used for localizer and two dual channel antennas for the glide slope. One channel from each of the

Figure 12.5 ILS antennas—glide slope and localizer

antennas is not used; the received signals are fed to the corresponding ILS receiver.

12.3.2 Receivers

ILS receivers are often combined with other radio navigation functions, e.g. VHF omni-range (VOR); these are located in the avionic equipment bay, see Figure 12.6. ILS receivers are based on the super-heterodyne principle with remote tuning from the control panel. The signal received from the localizer antenna is modulated with 90 and 150 Hz tones for left/right deviation; a 1020 Hz tone contains the navigation aid identification in Morse code. Filters in the ILS receiver separate out the 90 and 150 Hz tones for both localizer and glide slope. The identification signal is integrated with the audio system.

The marker beacon function is often incorporated with other radio navigation receivers, e.g. a combined VOR and marker beacon unit as illustrated in Figure 12.6(b). The marker beacon receiver filters out the 75 MHz tone and sends the signal to an RF amplifier. Three bandpass filters are then employed at 400 Hz, 1300 Hz and 3000 Hz to identify the specific marker beacon. The resulting signals are sent to an audio amplifier and then integrated into the audio system. Discrete outputs drive the visual warning lights (or PFD icons).

12.3.3 Controls and displays

A control panel typically located on the centre pedestal, see Figure 12.7(a), is used to select the runway heading and ILS frequency. Alternatively, it can be a combined navigation/ controller and display as shown in Figure 12.7(b). Outputs from the marker receiver are sent to three indicator lights (or PFD icons) and the crew's audio system as described in the previous section.

Typical electromechanical displays are provided by an omni-bearing indicator or course deviation indicator (CDI), see Figure 12.8(a). The omni-bearing selector is used to rotate the course card. This card is calibrated from 0 to 360° and indicates the selected runway heading. In Figure 12.8(a), a runway heading of 182° has been selected. Each dot on the scale represents a 2° deviation from the selected runway heading. A

(a) ILS receivers

(b) Combined VOR/marker beacon receiver

Figure 12.6 VHF/navigation receivers

(a) ILS control panel (centre pedestal)

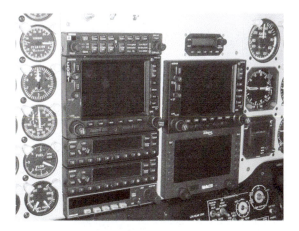

(b) Navigation and communications unit

Figure 12.7 ILS control panels

second pointer displays glide slope deviation. Two flags are used to indicate when the:

- localizer and/or glide slope signals are beyond reception range
- pilot has not selected an ILS frequency
- ILS system is turned off, or is inoperative.

Note that this type of indicator can also be used with the VOR navigation system; refer to Chapter 10 for a detailed description of this feature. An updated version of this instrument is the CDI; this has a compass display and course selector as

shown in Figure 12.8(b). The course selector (lower right-hand side of instrument) is set to the desired runway heading. Figure 12.8(c) illustrates ILS information on the electronic horizontal situation indicator (EHSI). The two/three letter Morse code identifier is sent to the audio system to allow the crew to confirm their selected ILS frequency.

A pointer moves left/right over a deviation scale to display lateral guidance information. The glide slope deviation pointer moves up/down over a scale to indicate vertical deviation. The strength of the 90 Hz and 150 Hz tones is summed to confirm the presence of the localizer and glide slope transmissions; this summed output is displayed in the form of a 'flag'. If either of the two transmissions is not present, the warning flag is displayed.

Key point

ILS frequencies are selected by tuning the localizer, which automatically selects the glide slope.

Key point

The marker beacon system is being phased out and replaced by GPS/DME approaches.

Key point

The ILS glide slope is inclined at a fixed angle between 2.5 and 3.5 degrees from the ground.

Test your understanding 12.2

Where are the localizer and glide slope antennas located?

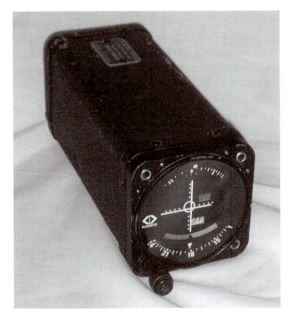

(a) Omni-bearing indicator   (b) Course deviation indicator

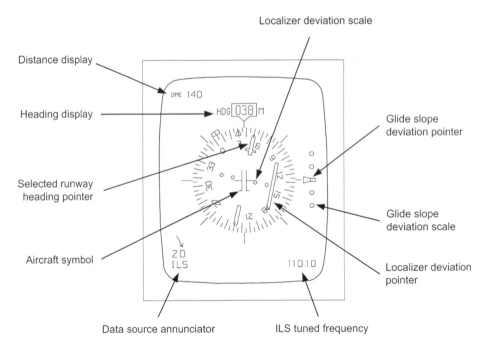

(c) Electronic display of ILS

Figure 12.8   ILS displays

12.4 Low range radio altimeter

The low range radio altimeter (LRRA) is a self-contained vertically directed primary radar system operating in the 4.2 to 4.4 GHz band. Airborne equipment comprises a transmitting/receiving antenna, LRRA transmitter/receiver and a flight deck indicator. Most aircraft are fitted with two independent systems. Radar energy is directed via a transmitting antenna to the ground; some of this energy is reflected back from the ground and is collected in the receiving antenna, see Figure 12.9.

Two types of LRRA methods are used to determine the aircraft's radio altitude. The pulse modulation method measures the elapsed time taken for the signal to be transmitted and received; this time delay is directly proportional to altitude. The frequency modulated, continuous wave (FM/CW) method uses a changeable FM signal where the rate of change is fixed. A proportion of the transmitted signal is mixed with the received signal; the resulting beat signal frequency is proportional to altitude.

Radio altitude is either displayed on a dedicated instrument, or incorporated into an electronic display, see Figure 12.9. Note that radio altitude used for approach and landing is only indicated from 2,500 feet. The decision height is selected during ILS approaches and

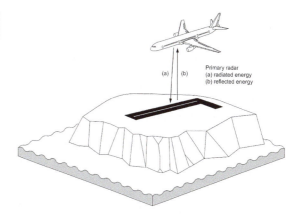

(a) Low range radio altimeter

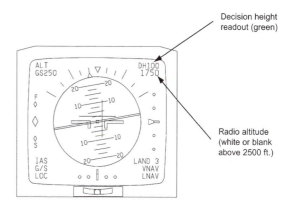

(b) Radio altimeter display (electronic instrument)

Test your understanding 12.3

What type of radar system is used for the low range radio altimeter (LRRA)?

Test your understanding 12.4

(a) When flying overhead ILS marker beacons, what indications are provided to the crew?

(b) What is the preferred sequence to capture the localizer and glide slope?

(c) What are the decision heights for Category 1, 2 and 3 landings?

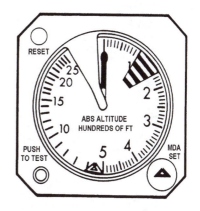

(c) Radio altimeter display

Figure 12.9 LRRA system

12.5 ILS approach

The normal procedure is to capture the localizer first and then the glide slope. The crew select the ILS frequency on the navigation control panel as described above. Runway heading also needs to be sent to the ILS receiver; the way of achieving this depends on the avionic fit of the aircraft. Desired runway heading can either be selected on a CDI, or via a remote selector located on a separate control panel. Deviation from the localizer and glide slope is monitored throughout the approach, together with confirmation of position from the marker beacons. The ILS can be used to guide the crew on the approach using instruments when flying in good visibility. In the event that visibility is not good, then the approach is flown using the automatic flight control system (AFCS). The crew select localizer and glide slope as the respective roll and pitch modes on the AFCS mode control panel (MCP), see Figure 12.10. With approved ground and airborne equipment, qualified crew can continue the approach through to an automatic landing. To complete an automatic landing (autoland) the pitch and roll modes need precise measurement of altitude above the ground; this is provided by the low range radio altimeter (LRRA).

12.6 Autoland

The development of airborne and ground equipment, together with crew training led to trials being carried out on the effectiveness and reliability of fully automatic landings using the ILS. In 1947, the Blind Landing Experimental Unit (BLEU) was established within the UK's Royal Aircraft Establishment. The world's first fully automatic landing was achieved in 1950. Equipment and procedures were further developed leading to the world's first automatic landing in a passenger carrying aircraft (the HS121 Trident) in July 1965.

Automatic approach and landings are categorised by the certifying authorities as a function of ground equipment, airborne equipment and crew training. The categories are quoted in terms of decision height (DH) and runway visual range (RVR). These categories are summarised in Table 12.1; JAR OPS provides further details and notes. Category 3 figures depend on aircraft type and airfield equipment, e.g. quality of ILS signals and runway lighting (centreline, edges, taxi ways etc.).

An operator has to have approval from the regulatory authorities before being permitted to operate their aircraft with automatic Category 2

Figure 12.10   AFCS mode control panel (LOC/GS modes)

Table 12.2 Automatic approach and landing categories

Category	DH	RVR (min)	RVR (max)
1	200'	550 m	1000 m
2	100'	300 m	–
3A	<100'	200 m	–
3B	<50'	75 m	–
3C	None	<75 m	–

and 3 approach and landings. This applies in particular to Category 3 decision heights.

Automatic approaches are usually made by first capturing the localizer (LOC) and then capturing the glide slope (GS), see Figure 12.11. The localizer is intercepted from a heading hold mode on the automatic flight control system (AFCS), with LOC armed on the system. The active pitch mode at this point will be altitude hold, with the GS mode armed.

Once established on the localizer, the glide slope is captured and becomes the active pitch mode. The approach continues with deviations from the centreline and glide slope being sensed by the ILS receiver; these deviations are sent to roll and pitch channels of the AFCS, with sensitivity of pitch and roll modes being modified by radio altitude. The auto throttle controls desired airspeed. Depending on aircraft type, two or three AFCS channels will be engaged for fully automatic landings thus providing levels of redundancy in the event of channel disconnects.

Although the glide slope antenna is located adjacent to the touchdown point on the runway, it departs from the straight-line guidance path below 100 feet. The approach continues with radio altitude/descent rate being the predominant control input into the pitch channel. At approximately 50 feet, the throttles are retarded and the aircraft descent rate and airspeed are reduced by the '**flare**' mode, i.e. a gradual nose-up attitude that is maintained until touchdown. The final pitch manoeuvre is to put the nose of the aircraft onto the runway. Lateral guidance is still provided by the localizer at this point until such time as the crew take control of the aircraft.

12.7 Operational aspects of ILS

ILS remains installed throughout the world and is the basis of automatic approach and landing for many aircraft types. Limitations of ILS are the single approach paths from the glide slope and localizer; this can be a problem for airfields located in mountainous regions. Furthermore, any vehicle or aircraft approaching or crossing the runway can cause a disturbance to the localizer beam, which could be interpreted by the airborne equipment as an unreliable signal. This often causes an AFCS channel to disconnect with the possibility of a missed approach. The local terrain can also have an effect on ILS performance, e.g. multipath errors can be caused by reflections of the localizer; the three-degree glide slope angle may not be possible in mountainous regions or in cities with tall buildings. These limitations led to the development of the microwave landing system (MLS); see Chapter 13.

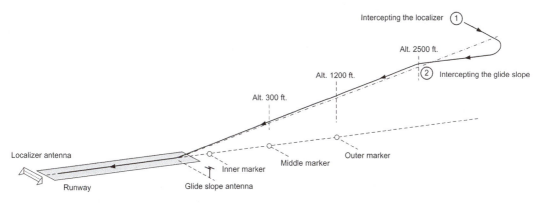

Figure 12.11 Automatic approach and landing

Key point

When the aircraft has touched down with an automatic landing, the ILS continues to provide lateral control via the localizer.

12.8 Multiple choice questions

1. Frequency bands for ILS are:
   (a) localizer (UHF) and glide slope (VHF)
   (b) localizer (VHF) and glide slope (VHF)
   (c) localizer (VHF) and glide slope (UHF).

2. Localizer transmitters are located:
   (a) at the threshold of the runway, adjacent to the touchdown point
   (b) at the stop end of the runway, on the centreline
   (c) at three locations on the extended centreline of the runway.

3. The LRRA provides:
   (a) deviation from the runway centreline
   (b) deviation from the glide path angle
   (c) altitude in feet above the ground.

4. Some airfields use NDBs in conjunction with (or in place of) the:
   (a) localizer
   (b) glide slope
   (c) outer marker.

5. When viewed from the antenna, the localizer is characterised by two lobes modulated:
   (a) 90 Hz to the right, 150 Hz to the left of the centreline
   (b) 150 Hz to the right, 90 Hz to the left of the centreline
   (c) equally either side of the centreline.

6. ILS frequencies are selected by tuning:
   (a) the glide slope which automatically selects the localizer
   (b) the localizer which automatically selects the glide slope
   (c) the localizer and glide slope frequencies independently.

7. The ILS glide slope is inclined at a fixed angle between:
   (a) 2.5 and 3.5 degrees
   (b) zero and 2.5 degrees
   (c) 2.5 degrees and above.

8. The glide slope is characterised by two lobes modulated:
   (a) 90 Hz above, 150 Hz below the glide slope angle
   (b) 150 Hz above, 90 Hz below the glide slope angle
   (c) equally either side of the glide slope angle.

9. Marker beacons transmit on which frequency?
   (a) 75 MHz
   (b) 1300 Hz
   (c) 400 Hz.

10. With three marker beacons installed in an ILS system, they will be encountered along the approach as:
    (a) outer, middle, inner
    (b) inner, middle, outer
    (c) outer, inner, middle.

11. Marker beacon outputs are given by:
    (a) coloured lights and Morse code tones
    (b) deviations from the runway centreline
    (c) deviations from the glide slope.

12. The decision height and runway visual range for a Category 2 automatic approach are:
    (a) 100 ft. and 300 m respectively
    (b) 200 ft. and 550 m respectively
    (c) less than 100 ft. and 200 m respectively.

13. The outer marker is displayed on the primary flying display as a coloured icon that is:
    (a) yellow
    (b) white
    (c) cyan.

14. When the aircraft has touched down with an automatic landing, the ILS continues to provide:
    (a) lateral control via the localizer
    (b) lateral control via the glide slope
    (c) vertical control via the LRRA.

Chapter 13

Microwave landing system

The microwave landing system (MLS) was adopted in 1978 as the long-term replacement for instrument landing systems (ILS). The system is based on the principle of time referenced scanning beams and provides precision navigation guidance for approach and landing. MLS provides three-dimensional approach guidance, i.e. azimuth, elevation and range. The system provides multiple approach angles for both azimuth and elevation guidance. Despite the advantages of MLS, it has not yet been introduced on a worldwide basis for commercial aircraft. Military operators of MLS often use mobile equipment that can be deployed within hours. In this chapter, we will review MLS principles and discuss its advantages over the ILS.

13.1 MLS overview

MLS was introduced to overcome a number of problems and limitations associated with ILS. The principle of MLS allows curved, or segmented approaches in azimuth together with selectable glide slope angles. All of these features are beneficial in mountainous regions, or for environmental reasons, e.g. over residential areas of a town or city. MLS installations are not affected by ground vehicles or taxiing aircraft passing through the beam as with the localizer. Aircraft making an approach using ILS in low visibility have to maintain sufficient separation to preserve the integrity of the localizer beams; with MLS, this separation is not required. The combination of all these features allows for increased air traffic control flexibility and higher take-off and landing rates for a given airfield. Two ground transmitters provide azimuth and elevation guidance; these **scanning beams** extend the coverage for an approach compared with the ILS, see Figure 13.1.

13.2 MLS principles

The system is based on the principle of time referenced scanning beams and operates in the C-band at 5 GHz. Two directional fan-shaped beams are used for azimuth and elevation guidance. The **azimuth** approach transmitter is located at the stop end of the runway; the **elevation** transmitter is located near the threshold of the runway. Azimuth scanning is through ±40° either side of the runway centreline with a range of 20 nm, see Figure 13.2(a). An expansion capability can extend azimuth coverage to ±60°, but with a reduced range of 14 nm. Elevation scanning sweeps over an angle of 15 degrees (with 20 degrees as an option) providing coverage up to 20,000 feet, see Figure 13.2(b).

At the aircraft receiver, a pulse is detected each time the respective beams sweep past the aircraft. Consider an aircraft on the approach as illustrated in Figure 13.3. The (azimuth) time referenced scanning beam sweeps from left to right ('TO'), and then returns from right to left ('FRO'). If the aircraft is in position A, it is to the left of the centreline and will receive a pulse at time interval t_1 as the beam sweeps 'TO', and then at time t_2 when the beam sweeps 'FRO'. The two pulses are therefore close together with the aircraft to the left of centreline. If the aircraft were in position B, i.e. to the right of the centreline, it would receive pulses at t_3 and t_4 due to the relative position of the aircraft.

The aircraft receiver in a given aircraft will interpret the timing of each pulse, in terms of when they occurred and the time difference between each pulse. These **pulse timings** provide a precise position fix for the aircraft with respect to the runway centreline. Elevation guidance is calculated in the same way as in azimuth, except that the beam is scanning up and down. Timing signals are referenced to a selected elevation approach angle.

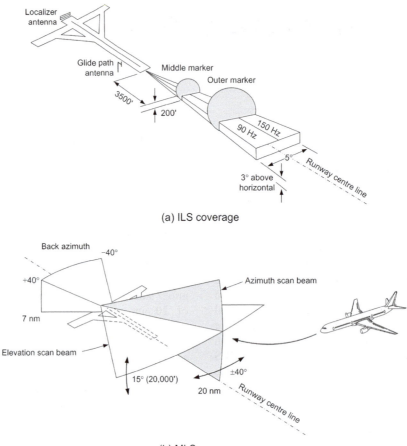

(a) ILS coverage

(b) MLS coverage

Figure 13.1   Comparison of ILS/MLS coverage

Key point

MLS was introduced to overcome a number of problems and limitations associated with ILS. The scanning principle of MLS allows curved, or segmented approaches in azimuth together with selectable glide slope angles.

Key point

MLS is based on the principle of time referenced scanning beams; two ground transmitters provide azimuth and elevation guidance. MLS operates at around 5 GHz in the C-band.

Key point

MLS installations are not affected by ground vehicles or taxiing aircraft passing through the beam as with the localizer.

Key point

Locations of the MLS ground equipment are not as critical as with ILS; this is particularly useful in mountainous regions.

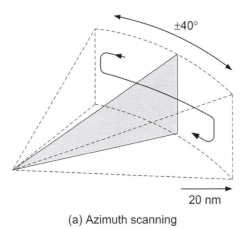

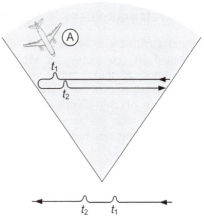

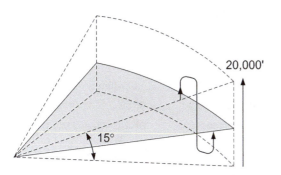

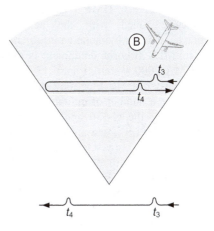

(a) Azimuth scanning

(a) The pulses are close together on this side of the beam

(b) Elevation scanning

(b) The pulses are further apart on this side of the beam

Figure 13.2 MLS azimuth and elevation scanning

Figure 13.3 Time referenced MLS scanning beam

Key point

MLS installations are not affected by ground vehicles or taxiing aircraft passing through the beam as with the localizer.

Test your understanding 13.2

In what frequency band does MLS operate?

Test your understanding 13.1

Explain the principle of operation of the time referenced azimuth scanning beam used in the MLS.

Test your understanding 13.3

What range and altitude does MLS cover?

13.3 Aircraft equipment

The aircraft is fitted with two antennas located on the nose and aft centrelines. An MLS receiver (often incorporated into a multi-mode receiver with ILS, marker beacon and VOR capability) is tuned into one of 200 channels and calculates azimuth and elevation guidance as described.

The receiver operates in the frequency range 5031 MHz to 5091 MHz with 300 kHz spacing. Referring to Figure 13.4 and Table 13.1, the pulse timing is used to determine the three aircraft positions.

An integral part of the MLS is a distance measuring equipment (DME) system to provide range; this can either be a conventional DME system as described in Chapter 11, or a dedicated system operating in the 962 MHz to 1105 MHz frequency range. DME frequencies are automatically tuned with the azimuth and elevation beams to provide range information.

Typical MLS airborne equipment is illustrated by the CMA-2000 system, Figure 13.5 (data and image courtesy of CMC Electronics). This system is installed on a number of military aircraft in the USA including the C-130 and Air Force One. Control of the MLS is via a control display unit (CDU), where the crew selects the desired MLS channel, together with azimuth and glide path approach angles. The system meets the requirements of ARINC 727 and provides three-dimensional positional data within a large airspace volume.

Azimuth and glide path guidance outputs are either displayed on a conventional course deviation indicator (CDI) or incorporated into multipurpose electronic displays. A summary of the CMA-2000 microwave landing system leading particulars is given in Table 13.2.

Table 13.2   CMA-2000 microwave landing system leading particulars

Feature	Specification
Range/channels	200 channels in C-band (5031 to 5090.7 MHz)
Control unit weight	6.7 kg
CDU weight	1.4 kg
Power supply	115 V AC, 400 Hz, 60 VA nominal
Control unit microprocessor	8086, 128 kbyte EPROM, 64 kbyte RAM
Range	Up to 40 nm
Azimuth range	0° through 360°
Elevation range	2° to 29.5° (in increments of 0.1°)
Resolution	0.005°
Sensitivity	−106 dBm
Dynamic range	95 dB
Digital interfaces	ARINC 429 and MIL-STD-1553B
Analogue interfaces	Synchro, DC voltages
Navigation aids	DME tuning, frequency tuning

Test your understanding 13.4

How many MLS channels are available?

Test your understanding 13.5

What frequency range does MLS use?

Table 13.1   Azimuth angle relationship

Aircraft position	TO scan	FRO scan	Difference	Angle (+ is left)
A	6.6 ms	11.5 ms	4.9 ms	+20°
B	5.7 ms	12.2 ms	6.3 ms	0°
C	3.6 ms	14.2 ms	10.6 ms	−40°

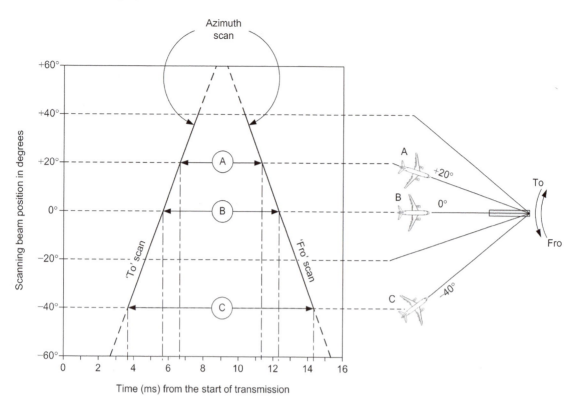

Figure 13.4 Relationship between transmissions and position of aircraft

Figure 13.5 MLS airborne equipment (courtesy of CMC electronics)

13.4 Ground equipment

The items of ground equipment needed for MLS are the azimuth and elevation transmitters and a DME navigation aid. This basic system can be expanded to provide lateral guidance for missed approaches. Both azimuth and elevation transmissions are radiated on the same frequency with a time-sharing arrangement.

In addition to guidance, the MLS also transmits data to system users. Basic data includes runway identification (four-letter Morse code), together with locations and performance levels of the azimuth, elevation transmitters and the DME transponder. The expanded data transmission provides runway conditions and meteorological data, e.g. visibility, cloud base, barometric pressures, wind speed/direction and any wind shear conditions.

Locations of the ground equipment are not as critical as with ILS; this is particularly useful in mountainous regions. Military users of MLS take advantage of this by have mobile systems that can be deployed within hours. The azimuth transmitter has an accuracy of ±4 metres at the runway threshold. The elevation transmitter has an accuracy of ±0.6 metres. The dedicated DME navigation aid has a range accuracy of 100 feet. A variety of approach patterns is possible with MLS as illustrated in Figure 13.6.

13.5 MLS summary

Despite the advantages of MLS, it has not yet been introduced on a worldwide basis for commercial aircraft. The advent and development of global navigation satellite systems (Chapter 19) has led to the reality of precision approaches and automatic landings being made under the guidance of satellite navigation systems during low visibility; however, this is not likely to be available for some time. Since MLS technology is already available, a number of European airlines have been lobbying for MLS; ground equipment has been installed at a number of airports including London Heathrow and Toulouse Blagnac for development purposes. The reader is encouraged to monitor the industry press for developments of this subject.

Key point

Despite the advantages of MLS, it has not yet been introduced on a worldwide basis for commercial aircraft. The military use mobile equipment that can be deployed within hours.

Test your understanding 13.6

Explain why MLS can be advantageous for use in mountainous areas or in areas of high population.

Test your understanding 13.7

How does the MLS provide range to the runway?

Test your understanding 13.8

Why does MLS provide more air traffic control flexibility?

13.6 Multiple choice questions

1. MLS azimuth and elevation transmitters operate in which frequency band?
   (a) 5 GHz
   (b) 962 MHz to 1105 MHz
   (c) 108 MHz to 112 MHz.

2. What are the angular extremes for azimuth guidance either side of the runway centreline in a basic MLS installation?
   (a) ±60°
   (b) ±40°
   (c) +15° to +20°.

3. What are the elevation guidance limits for an MLS installation?
   (a) ±60°
   (b) +15° to +20°
   (c) ±40°.

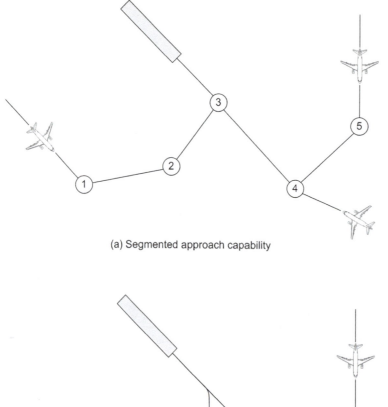

(a) Segmented approach capability

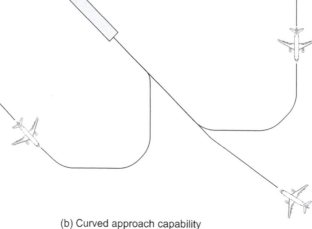

(b) Curved approach capability

Figure 13.6 MLS approach patterns

4. MLS range information is provided by the:
   (a) azimuth transmitter
   (b) DME navigation aid
   (c) elevation transmitter.

5. Time referenced scanning beams are used in the MLS to provide:
   (a) range to the airfield
   (b) azimuth and elevation guidance
   (c) altitude above the terrain.

6. How many MLS channels are available:
   (a) 40
   (b) 300
   (c) 200.

7. During an MLS approach, deviation in azimuth and elevation is displayed on the:
   (a) HSI
   (b) RMI
   (c) CDU.

8. The elevation approach angle for an approach
   is selected by:
   (a) air traffic control using the ground
       equipment
   (b) flight crew using the CDU
   (c) flight crew using the HSI.

9. With increasing elevation approach angles,
   slant range to the airfield will:
   (a) increase
   (b) decrease
   (c) stay the same.

10. MLS ground equipment identification codes
    are provided by:
    (a) two Morse code characters
    (b) three Morse code characters
    (c) four Morse code characters.

11. Referring to Figure 13.7, pulses t_1 and t_2 are
    providing:
    (a) range to the runway
    (b) elevation guidance
    (c) azimuth guidance.

12. Referring to Figure 13.8, the scanning is
    providing guidance in:
    (a) range
    (b) azimuth
    (c) elevation.

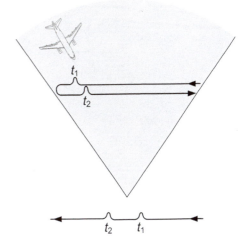

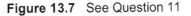

Figure 13.7   See Question 11

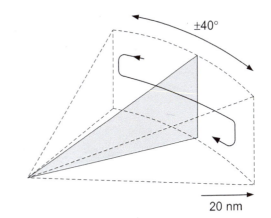

Figure 13.8   See Question 12

Chapter 14   Hyperbolic radio navigation

Hyperbolic radio navigation systems provide medium to long-range position fix capabilities and can be used for en route operations over oceans and unpopulated areas. Several hyperbolic systems have been developed since the 1940s, including Decca, Omega and Loran. The operational use of Omega and Decca navigation systems ceased in 1997 and 2000 respectively. Loran-C systems are still very much available today as stand-alone en route navigation systems; they are also being proposed as a complementary navigation aid for global navigation satellite systems. The principles of hyperbolic radio navigation are described in this chapter together with specific details for Loran-C. The development of enhanced Loran (**eLoran**) is discussed at the end of this chapter.

14.1 Hyperbolic position fixing

The principles of hyperbolic position fixing can be illustrated in Figure 14.1. Two radio stations A and B are located at a known distance apart; the imaginary line joining them is referred to as the **baseline**, Figure 14.1(a). Station A is the master and station B is the secondary. The **master station** transmits pulses at regular intervals; these pulses, represented by concentric circles in Figure 14.1(b), reach the secondary station after a fixed period of time (determined by the propagation speed of the radio wave). When the **secondary station** receives the master station's first pulse, the secondary station transmits its own pulse after a fixed time delay, as shown in Figure 14.1(c). This is a continuous process, with pulses transmitted by the master station at fixed intervals, and the secondary station replying after a fixed delay period.

The radiated pulses begin to overlap as the waves radiate away from their respective stations as illustrated in Figure 14.2. In this illustration, a

(a) Baseline between master and secondary station

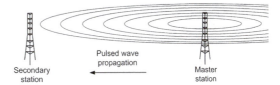

(b) Pulses transmitted from the master station form a concentric pattern around the transmitter

(c) When the first wave is received at the secondary station a pulse is transmitted from the secondary station after a fixed time delay

Figure 14.1 Hyperbolic navigation principles

series of pulses (represented by the solid lines) is radiating from the master station A at a rate of one thousand pulses per second, i.e. at intervals of 1 ms. The first pulse reaches the secondary station B depending on the distance to the station, e.g. after 7 ms. The secondary station transmits its response after a predetermined delay, e.g. 1 ms. This is represented by the dashed circle number 8, i.e. it is transmitted after the 7 ms travel time and 1 ms fixed delay. The radiated pulses from both stations form a pattern of intersecting pulses. Examine the timing differences between the intersecting circles on lines X, Y and Z. It can be seen that the time difference between the secondary and master pulses occurs at:

- 2 ms anywhere on line X
- 4 ms anywhere on line Y
- 6 ms anywhere on line Z.

The intersection of two pulses with the same time delay anywhere on this pattern can be used to determine a **line of position** (LOP). These points can be plotted to form unique curves known as hyperbolae. The foci of the hyperbolae are at each of the transmitters. Each hyperbola provides a LOP related to the time delay between receiving master and secondary pulses. Since there are two positions on any given hyperbola, a third (or fourth) secondary station will provide a unique position fix as illustrated in Figure 14.3. In this case, the three hyperbolae generated by stations A, B and C only intersect in one place.

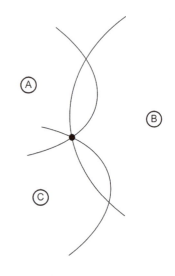

Figure 14.3 Using three stations to define a unique position fix

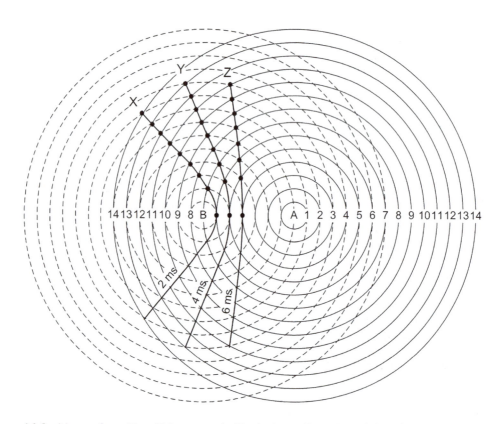

Figure 14.2 Lines of position (this example illustrates a 7 ms travel time from A to B, with a 1 ms time delay at transmitter B)

14.2 Loran overview

Loran is an acronym for long range navigation, a system based on hyperbolic radio navigation. The system was developed during the 1940s as Loran-A and has undergone many developments; the current version is Loran-C. Operating in the LF frequency range of 90–110 kHz, the system comprises ground transmitters and monitoring stations. The Loran-C system has a typical range of up to 1000 nm and an accuracy of better than 0.25 nm (460 metres) in the defined coverage areas. Transmitters are grouped together in '**chains**' thus providing a two-dimensional position fixing capability. The patterns are formed in various ways by **master** and **secondary** stations as illustrated in Figure 14.4.

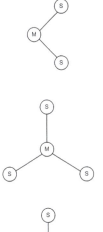

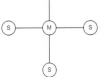

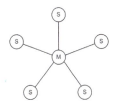

Figure 14.4 Loran-C master/secondary stations forming chains

14.3 Loran-C operation

Loran-C chains are organised in a master and secondary configuration. Each master has at least two associated secondary stations; in some cases there are five secondary stations in the chain. The elapsed time between receiving pulses from the master station and two or more secondary stations is used to determine a unique position. Pulses are formed as **variable amplitude sine waves** with a fixed frequency; the pulse duration is 270 ms representing 27 cycles of the 100 kHz carrier wave as illustrated in Figure 14.5. This unique pulse provides a recognisable signal and ensures that the majority of the pulse's bandwidth is confined to the frequency range of 90–110 kHz.

The intention for a Loran-C system is to only use ground waves for navigation purposes; sky waves are filtered out with pulse timing techniques. The approximate time taken for a transmitted wave to reflect off the ionosphere is 30 ms; since the pulse duration is 270 ms some of the transmitted pulse can be expected to be reflected from the ionosphere. To avoid this, a specific peak within the pulse is selected as the indexing pulse. This is the **third peak** within the pulse, and represents approximately 50% of the maximum amplitude.

Signals are transmitted from the master station as a group of nine pulses; secondary stations transmit eight pulses, see Figure 14.6. Groups of pulses from each of the chains are transmitted within the range of 10–25 groups per second. Each pulse is spaced at 1 ms intervals, the ninth pulse from the master station occurs after a 2 ms delay. The specific timing interval of the group of pulses (starting and finishing with the master pulses) is referred to as the **group repetition interval**, or **GRI**. This time interval is used as the

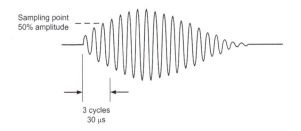

Figure 14.5 Loran-C pulse format

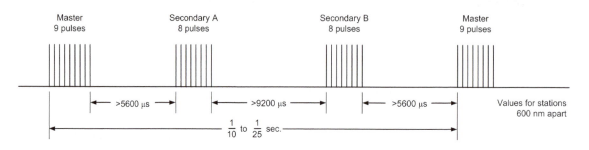

Figure 14.6   Loran-C pulse transmission format

basis of identifying the chain, e.g. a chain with GRI of 99,600 microseconds is identified as '9960'.

The first group of nine pulses from the master station is received at different times by each of the secondary stations due to the varying baseline distances between respective stations. The secondary stations transmit their pulse groups after predetermined time delays, referred to as the **coding delay**. The total time for the pulse to travel over the baseline together with the secondary station's coding delay is called the **emission delay**.

Operational aspects associated with Loran-C include:

- Electromagnetic interference affecting the signal, e.g. from power lines
- Loss of one station affects the area of coverage
- Local weather conditions (particularly electrical storms) affecting the signal.

In addition to master and secondary stations, monitoring stations are deployed to sample the chain's signal strength, timing and pulse shape. In the event that any of these are outside a specified limit an alert signal, known as a **blink**, is coded into the pulse groupings.

Key point

Loran is an acronym for long range navigation, a system based on hyperbolic radio navigation.

Key point

The Loran-C system uses ground waves at low frequencies. It has a typical range of up to 1000 nm with an accuracy of 0.25 nm. Transmitters are grouped together in 'chains' thus providing a two-dimensional position fixing capability.

Key point

Loran-C chains all transmit at 100 kHz, i.e. there is no need to tune the receiver to a specific chain.

Key point

The elapsed time between receiving pulses from the master station and two or more secondary stations is used to determine a unique position.

Key point

The operational use of Omega and Decca hyperbolic navigation systems ceased in 1997 and 2000 respectively.

14.4 Loran-C ground equipment

Master and secondary transmitting stations are located at strategic places to provide the required geometry for obtaining navigation information. Transmitter towers are typically 700-1300 feet high and radiate between 400 and 1600 W of power. The master and secondary stations are formed in groups known as **chains** as discussed earlier. Baseline distances vary from chain to chain since many stations are located on islands to provide oceanic coverage; distances of between 175 and 1000 nm are typical. The majority of these chains are in the USA and Canada; other chains are located in Russia, the northern Pacific, Europe, Asia and the Middle East. The master stations are identified as 'M' and the secondary stations are identified from the series 'W, X, Y and Z'. The US Coast Guard (USCG) provides full details of each chain, together with an on-line handbook containing very useful data and information relating to Loran; details can be found on their website www.navcen.usg.gov. The USCG introduced Loran-C into Europe, the system was transferred to the host nations in 1995.

Table 14.1 provides a list of currently available Loran-C chains, together with a summary of how many secondary stations are associated with the master. Table 14.2 provides details for the Northwest Pacific chain; this comprises stations on the Japanese mainland and a number of islands in the Pacific. Figure 14.7 gives an illustration of the area of coverage for this chain.

In previous chapters, radio navigation systems including VOR, DME and VORTAC have been described. Note that since VOR, DME and VORTAC navigation aids have to be located on land, the airways' network does not provide a great deal of coverage beyond coastal regions. Referring to Figure 14.9(a), a combination of VOR, DME and TACAN stations located in a number of European countries provides a certain amount of navigation guidance in the North Atlantic, Norwegian Sea and North Sea. This diagram assumes a line-of-sight range of approximately 200 nm. The gaps in this radio navigation network can be largely overcome by the use of Loran-C, see Figure 14.9(b). This is the Norwegian Sea chain, with the master station

Table 14.1 Loran-C chains (source USCG)

Chain	Master location; number of secondary stations
Canadian East Coast	Caribou, Maine; three secondary stations
Canadian West Coast	Williams Lake; three secondary stations
Great Lakes USA	Dana, Indiana; four secondary stations
Gulf of Alaska	Tok, Alaska; three secondary stations
Icelandic Sea	Sandur, Iceland; two secondary stations
Labrador Sea	Fox Harbor, Canada; two secondary stations
Mediterranean Sea	Sellia Marina, Italy; three secondary stations
North Central USA	Havre, Montana; three secondary stations
North Pacific	St Paul, Alaska; three secondary stations
Northeast USA	Seneca, New York; four secondary stations
Northwest Pacific	Iwo Jima, Japan; four secondary stations
Norwegian Sea	Ejde, Denmark; four secondary stations
South Central USA	Boise City, Oklahoma; five secondary stations
Southeast USA	Malone, Florida; four secondary stations
West Coast USA	Fallon, Nevada; three secondary stations

Table 14.2 Details of the Northwest Pacific chain (source USCG)

Master station (M)	Iwo Jima, Japan
Secondary station (W)	Marcus Island, Japan
Secondary station (X)	Hokkaido, Japan
Secondary station (Y)	Gesashi, Japan
Secondary station (Z)	Barrigada, Japan

located at Ejde (Denmark); four secondary stations (X, W, Y and Z) located in Bo (Norway), Sylt (Germany), Sandur (Iceland) and Jan Mayen (Norway) respectively. Note that this illustrates the estimated ground coverage, actual coverage will vary.

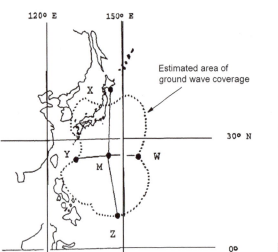

Figure 14.7 Northwest Pacific chain (courtesy USCG)

Transmitter

M Iwo Jima, Japan
W Marcus Island, Japan
X Hokkaido, Japan
Y Gesashi, Japan
Z Barrigada, Guam

14.5 Loran-C airborne equipment

Airborne equipment comprises the antenna, receiver and control display unit. The antenna is often shared with the ADF sense loop. Loran-C chains all transmit at 100 kHz, i.e. there is no need to tune the receiver to a specific chain.

The receiver searches for master stations and tracks secondary signals; this is achieved with a phase locked loop process. Since all chains transmit at 100 kHz, an aircraft in range of more than one chain will receive pulses from many stations; the receiver has to be able to identify specific chains by their emission delays. Once identified, the receiver determines which chain is providing the strongest signals, and which is providing the best navigation solution. Accurate timing signals are used to recognise the unique Loran-C pulse shape. Once acquired, the receiver needs to identify the third peak in the pulse; this peak has the highest rate of change with respect to the eighth pulse. Identification of the third peak is determined by measuring the zero crossings and amplitude growth within the pulse. In addition to this, the receiver also has to be able to reject a large amount of interference and atmospheric noise.

A navigation computing function can provide enhanced operation for the system. Chain details such as latitude and longitude of stations, GRI and secondary delay times are all stored in a database. Corrections can be applied for known propagation differences over sea, land, and ice. If the receiver is receiving pulses from more than one chain, it is possible to calculate an average position. A typical control display unit used for hyperbolic navigation is shown in Figure 14.8

Figure 14.8 Typical control display unit

Test your understanding 14.1

What frequency range does Loran-C use?

Test your understanding 14.2

What does GRI mean, and how does this define a Loran-C chain?

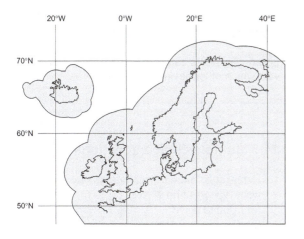

(a) VOR–DME coverage

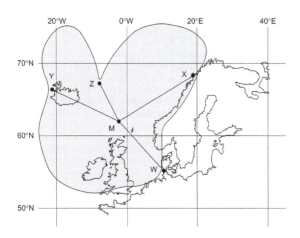

(b) Loran-C coverage

Figure 14.9 Comparison of VOR–DME and Loran-C coverage in a coastal area

14.6 Enhanced Loran (eLoran)

Loran-C has several advantages over the two other (now obsolete) hyperbolic navigation systems, Decca and Omega; these advantages include the use of ground waves at low radio frequencies and pulse techniques to discriminate against sky wave inference.

The introduction of global navigation satellite systems (GNSS) will, in theory, make the use of Loran-C unattractive, and eventually become obsolete. There were plans to decommission the system due to the emerging use and attractions of GNSS. In reality however, this situation is being reversed.

Referring to Chapter 18, it is clear that any GNSS is vulnerable to disruption; this can be either a deliberate attempt to interfere with the transmissions, satellite failure or because of adverse atmospheric conditions. With increased dependence on GNSS for aviation, marine, vehicle and location-based services, the impact of any disruption is significant. The solution to this is to have an alternative navigation system working alongside GNSS as a backup, e.g. VOR, DME, inertial navigation (described elsewhere in the book) or Loran.

The next development from Loran-C is **enhanced Loran** (eLoran) which will take advantage of new and emerging technology. Enhanced Loran introduces an additional data channel via the Loran transmission; this data includes up to sixteen message types including (but not limited to) station identity, coordinated universal time (UTC), corrections, warnings, and signal integrity information. This data channel is achieved via pulse-position modulation. The new pulse is added to the Loran transmission one millisecond after the eighth pulse on a secondary transmitting station, and between the current eighth and ninth pulses on a master transmitting station. Testing of the Loran data channel (LDC) by the FAA and US Coast Guard began in July of 2005.

The eLoran system comprises the transmitting station, monitoring sites, and control monitor station; this is a self-correcting system as illustrated in Figure 14.10.

Using a technique called **time of transmission control**, timing is held constant at each transmitting station rather than in the monitoring sites. The eLoran receiver acquires, tracks and manages stations as if they were satellites, thereby providing reliable timing measurements leading to accurate position calculations. This concept increases coverage since multiple stations from any chain can be selected by the receiver, provided that they are within range. This feature (known as **all-in-view**) treats each Loran

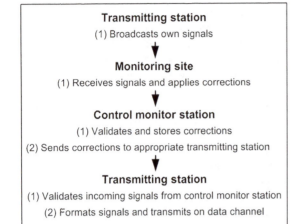

Figure 14.10 Self-correcting system used in eLoran

transmitter as an individual, i.e. it does not relate that station to a specific chain.

A combined GNSS/eLoran receiver offers a powerful solution to the problem of GNSS vulnerability. The use of eLoran will complement global navigation satellite systems (GNSS), it will also provide a backup with integrity maintained via eLoran's independence and dissimilar method of navigation.

The expected accuracy of eLoran is better than 10 metres compared to a Loran-C accuracy of 460 metres (0.25 nm). The reader is encouraged to read the industry press and monitor developments of this subject.

Key point

In addition to master and secondary stations, monitoring stations are deployed to sample the chain's signal strength, timing and pulse shape.

Test your understanding 14.3

How many unique lateral geographical positions can two hyperbolic navigation stations define?

Test your understanding 14.4

Loran-C systems can share their aircraft antennas with which other navigation system?

14.7 Multiple choice questions

1. Long-range radio navigation systems rely on what type of radio wave?
   (a) Ground wave
   (b) Sky wave
   (c) Space wave.

2. How many transmitting stations are required in a hyperbolic navigation system to provide a unique position?
   (a) One
   (b) Two
   (c) Three or more.

3. How many unique locations are defined on a hyperbolic line of position?
   (a) One
   (b) Two
   (c) None.

4. The foci of hyperbolae are located at:
   (a) each of the transmitters
   (b) the intersection of lines of position
   (c) the intersection of concentric circles.

5. The intersection of two Loran-C pulses with same time delay can be used to determine a:
   (a) line of position
   (b) baseline
   (c) unique position.

6. Loran-C operates in which frequency band?
   (a) 190–1750 kHz
   (b) 90–110 kHz
   (c) 108–112 MHz.

7. How many pulses does the master station in a Loran-C chain transmit?
   (a) 27
   (b) 8
   (c) 9.

Chapter 15 Doppler navigation

Doppler navigation is a self-contained dead reckoning system, i.e. it requires no external inputs or references from ground stations. Ground speed and drift can be determined using a fundamental scientific principle called Doppler shift. Doppler navigation systems were developed in the mid-1940s and introduced in the mid-1950s as a primary navigation system with many features including continuous calculations of ground speed and drift. Being self-contained, the system can be used for long distance navigation over oceans and undeveloped areas of the globe.

Doppler navigation sensors are often integrated with other aircraft navigation systems. Alternatively, Doppler sensors are used in other specialised airborne applications, including weather radar and missile warning systems. Enhanced VOR ground installations also incorporate Doppler principles. In this chapter, we will review some basic scientific principles, look at Doppler navigation as a stand-alone system, and then review some of the other Doppler applications.

15.1 The Doppler effect

The 'Doppler effect' is named after Christian Doppler (1803–1853), an Austrian mathematician and physicist. His hypothesis was that the frequency of a wave apparently changes as its source moves closer to, or further away from, an observer. This principle was initially proven to occur with sound; it was subsequently found to occur with any wave type including electromagnetic energy. An excellent example of the Doppler effect can be observed when fast trains (or racing cars) pass by an observer.

To illustrate this principle, consider Figure 15.1, an observer located at a certain distance from a sound source that is emitting a fixed-frequency tone. As the train approaches the observer, the number of cycles 'received' by the observer is the fixed tone, plus the additional cycles received as a function of the train's speed. This will have the effect of increasing the tone (above the fixed frequency) as heard by the observer. At the instant when the train is adjacent to the observer, the true fixed-frequency will be heard. When the train travels away from the observer, fewer cycles per second will be received and the tone will be below the fixed-frequency as heard by the observer. The difference in tone is known as the **Doppler shift**; this principle is used in Doppler navigation systems. Doppler shift is, for practical purposes, directly proportional to the relative speed of movement between the source and observer. The relationship between the difference in frequencies and velocity can be expressed as:

$$F_D = \frac{vf}{c}$$

where F_D = frequency difference, v = aircraft velocity, f = frequency of transmission, and c = speed of electromagnetic propagation (3×10^8 metres/second).

15.2 Doppler navigation principles

Doppler navigation systems in aircraft have a focused beam of electromagnetic energy transmitted ahead of the aircraft at a fixed angle (theta, θ) as shown in Figure 15.2. This beam is scattered in all directions when it arrives at the surface of the earth. Some of the energy is received back at the aircraft. By measuring the difference in frequency between the transmitted and received signals, the aircraft's **ground speed** can be calculated. The signal-to-noise ratio of the received signal is a function of a number of factors including:

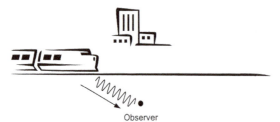

(a) Train moving towards the observer (more cycles in a given time therefore the observer perceives a higher pitch)

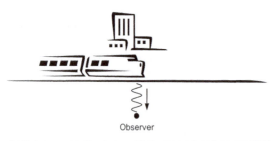

(b) Train nearest to the observer (observer perceives the exact pitch)

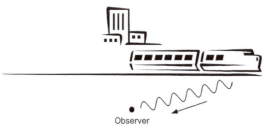

(c) Train moving away from the observer (less cycles in a given time therefore the observer perceives a lower pitch)

Figure 15.1   The Doppler effect

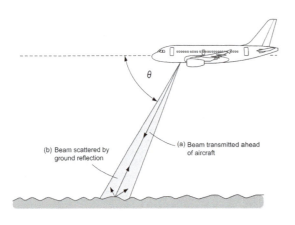

Figure 15.2   Doppler navigation principles

- Aircraft range to the terrain
- Backscattering features of the terrain
- Atmospheric conditions, i.e. attenuation and absorption of radar energy
- Radar equipment.

Note that the aircraft in Figure 15.2 is flying straight and level. If the aircraft were pitched up or down, this would change the angle of the beam with respect to the aircraft and the surface; this will change the value of Doppler shift for a given ground speed. This situation can be overcome in one of two ways; the transmitter and receiver can be mounted on a stabilised platform or (more usually) two beams can be transmitted from the aircraft (forward and aft) as shown in Figure 15.3. By comparing the Doppler shift of both beams, a true value of ground speed can be derived. The relationship between the difference in frequencies and velocity in an aircraft can be expressed as:

$$F_\mathrm{D} = \frac{2\cos\theta \times vf}{c}$$

where F_D = frequency difference, θ = the angle between the beam and aircraft, v = aircraft velocity, f = frequency of transmission, and c = speed of electromagnetic propagation (3×10^8 metres/second).

Note that a factor of two is needed in the expression since both the transmitter and receiver are moving with respect to the earth's surface. It can be seen from this expression that aircraft altitude is not a factor in the basic Doppler calculation. Modern Doppler systems (such as the CMC Electronics fifth generation system) operate up to 15,000 feet (rotary wing) and 50,000 feet (fixed wing).

Having measured velocity along the track of the aircraft, we now need to calculate **drift**. This can be achieved by directing a beam at right angles to the direction of travel, see Figure 15.4. Calculation of drift is achieved by utilising the same principles as described above. In practical installations, several directional beams are used, see Figure 15.5.

The calculation of ground speed and drift provides 'raw navigation' information. By combining these two values with directional information from a gyro-magnetic compass

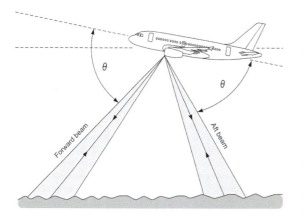

Figure 15.3 Compensation for aircraft pitch angle

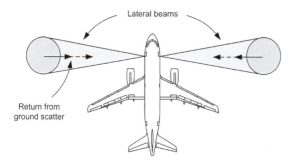

Figure 15.4 Measuring drift by Doppler shift

With no drift, the reflected return from ground scatter has no Doppler shift

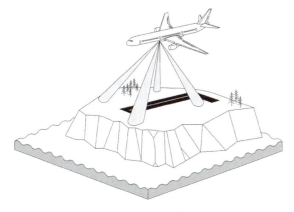

Figure 15.5 Measuring ground speed and drift using directional beams

system, we have the basis of a complete self-contained navigation system. By integrating the velocity calculations, the system can derive the distance travelled (along track) and cross track deviations. The Doppler system has a resolution of approximately 20–30 Hz (frequency difference) per knot of speed. Note that, in addition to ground speed and drift information, Doppler velocity sensors can also detect **vertical displacement** from a given point. Errors accumulate as a function of distance travelled; typical Doppler navigation system accuracy can be expressed in knots (V_t) as follows (data courtesy of CMC Electronics):

$$V_t = \sqrt{V_x^2 + V_y^2 + V_z^2}$$

The individual components of velocity along the x, y and z axes (ground speed, drift and vertical components) have accuracies given in Table 15.1 for both sea (Beaufort scale of 1) and land conditions.

When flying over oceans, the Doppler system will calculate velocities that include movement of the sea due to tidal effects, i.e. not a true calculation of speed over the earth's surface. These short-term errors will be averaged out over time. Doppler sensors are ideally suited for rotary wing aircraft that need to hover over an object in the sea, e.g. during search and rescue operations, see Figure 15.6. The surface features of water are critical to the received **backscatter**; this must be taken into account in the system specification. The 'worst case' conditions for signal to noise ratios are with smooth sea conditions; to illustrate this point consider the two reflecting surfaces illustrated in Figure 15.7. (Note that the reflecting surface of water would never actually be optically perfect, but smooth surface conditions do reduce the amount of scatter.) When hovering over water in search and rescue operations, Doppler systems

Table 15.1 Doppler navigation system component accuracy

Component	Land	Sea
V_x	0.3% V_t + 0.2 knots	0.25% V_t + 0.2 knots
V_y	0.3% V_t + 0.2 knots	0.25% V_t + 0.2 knots
V_z	0.2% V_t + 0.2 fpm	0.20% V_t + 0.2 fpm

have the distinct advantage of being able to track a vessel as it drifts with the tide. This reduces pilot workload, particularly if the Doppler system is coupled to an automatic control system.

Doppler system specifications for navigation accuracy are often expressed with reference to the Beaufort scale; this scale has a range of between 1 and 12. A sea state of 1 on the Beaufort scale is defined by a wind of between 1 and 3 knots with the surface of the water having ripples, but no foam crests.

Key point

Christian Doppler's hypothesis was that the frequency of a wave apparently changes as its source moves closer to, or farther away from, an observer.

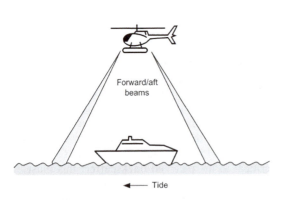

Forward/aft beams

◄──── Tide

(a) Aircraft tracks object using forward/aft beams

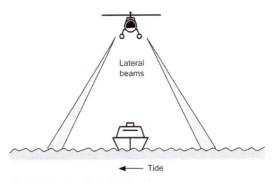

Lateral beams

◄──── Tide

(b) Aircraft tracks object using lateral beams

Figure 15.6 Using Doppler during hover

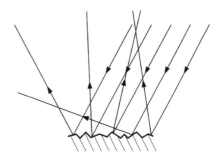

(a) Diffuse surface conditions

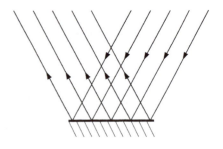

(b) Smooth surface conditions

Figure 15.7 Surface reflections

Key point

Doppler sensors are ideally suited for rotary wing aircraft that need to hover over an object in the sea, e.g. during search and rescue operations.

Test your understanding 15.1

What value of Doppler shift along the aircraft track would be measured if the radar beam were transmitted vertically down from the aircraft?

Test your understanding 15.2

What effect does increasing the frequency of a transmitted Doppler beam have on sensitivity of the frequency shift?

15.3 Airborne equipment overview

Doppler navigation systems use directional beams to derive ground speed and drift as previously described; these beams are arranged in a number of ways as illustrated in Figure 15.8. The fore and aft beams are referred to as a 'Janus' configuration (after the Roman god of openings and beginnings, Janus, who could face in two directions at the same time).

Three beams can be arranged in the form of the Greek letter lambda (λ). The four-beam arrangement is an X configuration; only three beams are actually required, the fourth provides a level of monitoring and redundancy. In the four-beam arrangement, the fore and aft signals are transmitted in alternative pairs.

Referring to the relationship:

$$F_D = \frac{2\cos\theta \times vf}{c}$$

it can be seen that the sensitivity of Doppler velocity calculations increases with the transmitted frequency; this means that a smaller antenna can be used.

The frequencies allocated to Doppler navigation systems are within the SHF range, specifically 13.25–13.4 GHz; some Doppler systems operate within the 8.75–8.85 GHz range.

(a) Four beam Janus X

(b) Three beam Janus λ

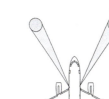

(c) Three beam Janus T

(d) Two beam non-Janus

Figure 15.8 Doppler beam arrangements

Test your understanding 15.3

How does a Doppler navigation system derive aircraft heading?

Test your understanding 15.4

What is the reason for having more than one Doppler beam transmission, e.g. the lambda configuration?

Test your understanding 15.5

What effect does the sea state have on the back scattering of a Doppler beam?

Key point

By measuring the difference in frequency between the transmitted and received signals, the aircraft's velocity in three axes can be calculated using the Doppler shift principle.

Key point

Doppler navigation systems are self-contained; they do not require any inputs from ground navigation aids. The system needs an accurate on-board directional input, e.g. from a gyrocompass.

The basic Doppler system comprises an antenna, transmitter and receiver. The antenna can be fixed to the airframe thereby needing corrections for pitch attitude (achieved via the Janus configuration). Alternatively the antenna could be slaved to the aircraft's attitude reference system. The antenna produces a very narrow conical- or pencil-shaped beam.

The Doppler navigation system has been superseded for commercial airline use by inertial and satellite navigation systems. Rotary wing aircraft, however, use Doppler sensors to provide automatic approach and stabilisation during hover manoeuvres; in this case the display would provide vertical displacement above/below the selected hover altitude and lateral/longitudinal deviation from the selected hover position.

15.4 Typical Doppler installations

Doppler principles can either be used in self-contained navigation systems, or as stand-alone velocity sensors. An early (1970s) version of a control display panel used on the MR1 Nimrod aircraft is illustrated in Figure 15.9. The stand-alone velocity sensor is in the form of a radar transmitter–receiver as illustrated in Figure 15.10, item 1. This sensor has a resolution of less than 0.1 knots and can be interfaced with other avionic systems and displays using data bus techniques. With increasing digital processing capability, the Doppler velocity sensor can be integrated with other navigation sensors to provide filtered navigation calculations. This subject is addressed in Chapter 16.

Typical self-contained Doppler navigation systems comprise the radar transmitter–receiver, signal processor, control display unit and steering indicator, as illustrated in Figure 15.10 (pictures and data courtesy of CMC Electronics). This navigation system transmits at 13.325 GHz using frequency modulation/continuous wave signals at a low radiated power of 20 mW.

Digital signal processing is used for continuous spectrum analysis of signal returns; this leads to enhanced tracking precision accuracy and optimises signal acquisition over marginal terrain conditions (sand, snow and calm sea conditions). Doppler systems compensate for attitude changes as described earlier; these manoeuvres can be

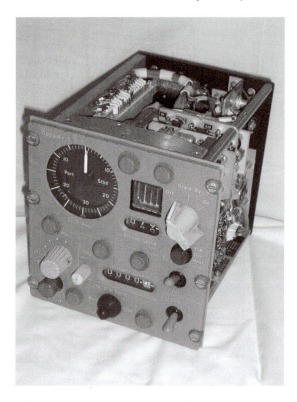

Figure 15.9 Doppler control display unit (1970's technology)

aggressive for certain helicopter operations. Changes of up to 60 degrees per second can be accommodated for pitch and roll excursions; the system can accommodate rate changes of up to 100 degrees per second in azimuth.

15.5 Doppler summary

In summary, Doppler navigation has a number of advantages:

- Velocity and position outputs from the system are provided on a continuous basis
- It requires no ground navigation aids, i.e. it is self-contained and autonomous
- Velocity outputs are very accurate
- Navigation is possible over any part of the globe, including oceans and polar regions
- The system is largely unaffected by weather (although certain rainfall conditions can affect the radar returns)

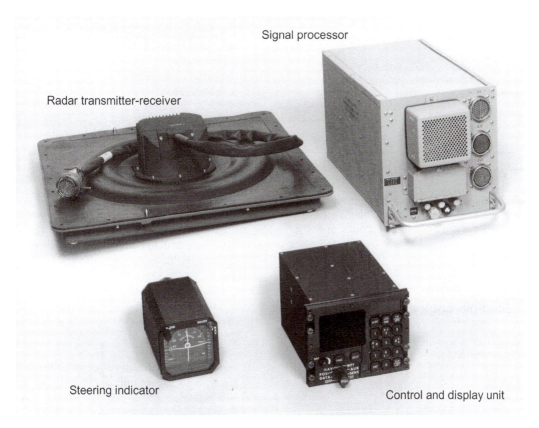

Signal processor

Radar transmitter-receiver

Steering indicator

Control and display unit

Figure 15.10 Doppler system (photo courtesy of CMC Electronics)

- The system does not require any preflight alignment.

The disadvantages of Doppler navigation are:

- It is dependent upon a directional reference, e.g. a gyro-magnetic compass
- It requires a vertical reference to compensate for aircraft attitude
- Position calculations degrade with distance travelled
- Short-term velocity calculations can be inaccurate, e.g. when flying over the tidal waters, the calculated aircraft velocity will be in error depending on the tide's direction and speed. (This effect will average out over longer distances, and can actually be used to an advantage for rotary wing aircraft)
- Military users have to be aware that the radar transmission is effectively giving away the location of the aircraft.

15.6 Other Doppler applications

In addition to self-contained navigation, the Doppler shift principle is also employed in several other aerospace systems. Missile warning systems and military radar applications include pulse-Doppler radar target acquisition and tracking; the Doppler principle is employed to reduce clutter from ground returns and atmospheric conditions.

In Chapter 10 reference was made to siting errors of conventional VOR ground stations. Many VOR stations now employ the Doppler principle to overcome these errors; these are referred to as Doppler VOR (DVOR) navigation aids; more details are provided in Chapter 10. Enhanced weather radar systems for commercial aircraft have the additional functionality of being able to detect turbulence and predict wind shear (see Chapter 20).

Key point

Doppler navigation system accuracy can be expressed in knots; errors accumulate as a function of distance travelled.

Test your understanding 15.5

When crossing over a coastal area, from land towards the sea, what effect will tidal flow have on the Doppler system's calculated ground speed and drift?

15.7 Multiple choice questions

1. Doppler navigation systems operate in which frequency range?
    (a) SHF
    (b) VHF
    (c) UHF.

2. When moving towards a sound source, what effect will Doppler shift have on the pitch of the sound as heard by an observer?
    (a) No effect
    (b) Increased pitch
    (c) Decreased pitch.

3. What effect does increasing the frequency of a transmitted Doppler beam have on sensitivity of the frequency shift?
    (a) Decreased
    (b) No effect
    (c) Increased.

4. Raw Doppler calculations include:
    (a) pitch and roll
    (b) directional information
    (c) ground speed and drift.

5. Velocity and position outputs from a Doppler navigation system are provided:
    (a) only when the aircraft is moving
    (b) on a continuous basis
    (c) only in straight and level flight.

6. The backscattering features of the terrain affect the Doppler navigation system's:
    (a) accuracy
    (b) signal to noise ratio
    (c) coverage.

7. Integrating Doppler ground speed calculations will provide:
    (a) distance travelled
    (b) drift angle
    (c) directional information.

8. Drift can be measured by directing a beam:
    (a) at right angles to the direction of travel
    (b) in line with the direction of travel
    (c) directly below the aircraft.

9. Doppler position calculations degrade with:
    (a) time
    (b) attitude changes
    (c) distance travelled.

10. When hovering directly over an object in the sea with a six-knot tide, the Doppler system will indicate:
    (a) six knots drift in the opposite direction of the tide
    (b) six knots drift in the direction of the tide
    (c) zero drift.

11. When hovering over water, the 'worst case' conditions for signal to noise ratios are with:
    (a) smooth sea conditions
    (b) rough sea conditions
    (c) tidal drift.

12. Doppler system beams in the lambda arrangement have beams directed in the following way:
    (a) forward and to each side of the aircraft
    (b) forward, aft and to one side of the aircraft
    (c) forward, aft and to each side of the aircraft.

13. Backscattering of a Doppler beam from the surface of water is:
    (a) low from a rough surface
    (b) low from a smooth surface
    (c) high from a smooth surface.

Chapter 16 Area navigation

Area navigation (RNAV) is means of combining, or filtering, inputs from one or more navigation sensors and defining positions that are not necessarily co-located with ground-based navigation aids. This facilitates aircraft navigation along any desired flight path within range of navigation aids; alternatively, a flight path can be planned with autonomous navigation equipment. Optimum area navigation is achieved using a combination of ground navigation aids and autonomous navigation equipment.

Typical navigation sensor inputs to an RNAV system can be from external ground-based navigation aids such as VHF omni-range (VOR) and distance measuring equipment (DME); autonomous systems include global satellite navigation or inertial reference system (IRS). Many RNAV systems use a combination of numerous ground-based navigation aids, satellite navigation systems and self-contained navigation systems. In this chapter, we will focus on area navigation systems that use VOR and DME navigation aids to establish the basic principles of RNAV. The chapter concludes with a review of Kalman filters and how RNAV systems are specified with a 'required navigation performance' (RNP).

16.1 RNAV overview

Two basic ground navigation aids that can be used for RNAV are VOR and DME; see Figures 16.1 and 16.2. RNAV is a guidance system that uses various inputs, e.g. VOR and/or DME to compute a position. The VOR system transmits specific bearing information, referred to as **radials**, see Figure 16.1. The pilot can select any radial from a given VOR navigation aid and fly to or from that aid.

Distance measuring equipment (DME) is a short-/medium-range navigation system based on secondary radar. Both VOR and DME are described in earlier chapters of this book. Conventional airways are defined by VOR and DME navigation aids, see Figure 16.3.

Since the VOR–DME systems are **line of sight**, the altitude of the aircraft will have a direct relationship with the range that the system can be used, see Figure 16.4. Using VOR–DME navigation aids imposes a limit on the working range that can be obtained. The maximum line-of-sight (LOS) distance between an aircraft and the ground station is given by the relationship:

$$d = 1.1\sqrt{h}$$

where d is the distance in nautical miles, and h is the altitude in feet above ground level (assumed to be flat terrain). The theoretical LOS range for altitudes up to 20,000 feet is given in Table 16.1. At higher altitudes, it is possible to receive VOR signals at greater distances but with reduced signal integrity. Although the actual range also depends on transmitter power and receiver sensitivity, the above relationship provides a good approximation.

The positions defined in an RNAV system are called **waypoints**; these are geographical positions that can be created in a number of ways. RNAV systems can store many waypoints in a sequence that comprises a complete route

Table 16.1 Theoretical LOS range

Altitude (feet)	Range (nm)
100	10
1,000	32
5,000	70
10,000	100
20,000	141

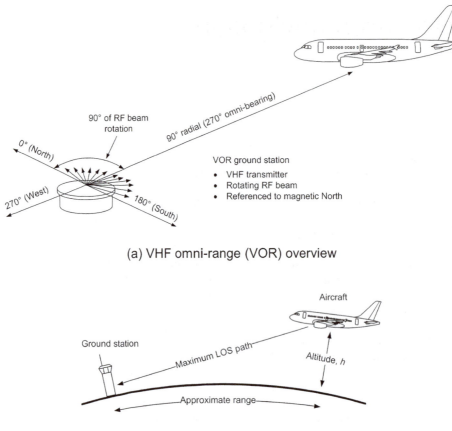

(a) VHF omni-range (VOR) overview

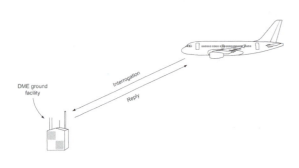

(b) VHF omni-range—line of sight

Figure 16.1 VOR principles

(a) Secondary radar used for DME   (b) DME transponder (right of photo)

Figure 16.2 DME principles

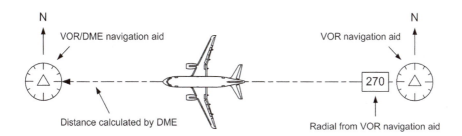

Figure 16.3(a) Aircraft flying along a conventional airway

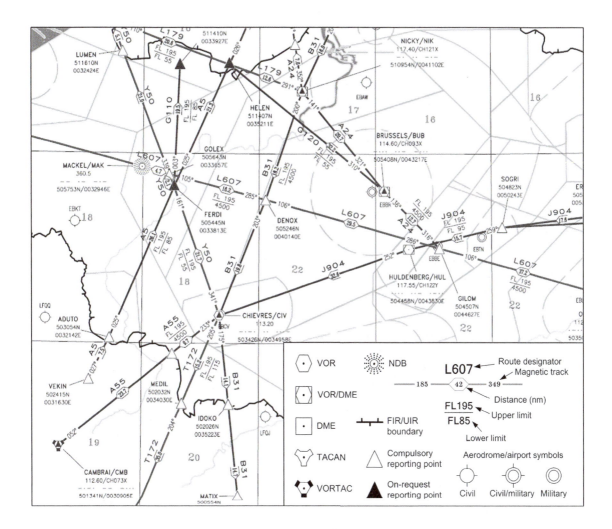

Figure 16.3(b) A typical airways chart

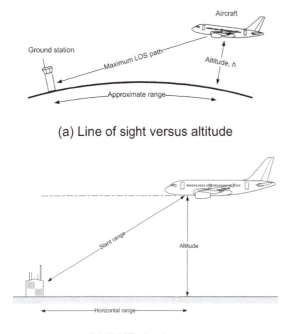

(a) Line of sight versus altitude

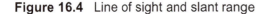

(b) DME slant range

Figure 16.4   Line of sight and slant range

from origin to destination. Creating waypoints that are not co-located with fixed ground aids provides a very flexible and efficient approach to flight planning. These waypoints are stored in a **navigation database** (NDB) as permanent records or entered by the pilot. Waypoints can be referenced to a fixed position derived from VOR and/or DME navigation aids, see Figure 16.5. The desired track between waypoints (Figure 16.6) is referred to as an RNAV **leg**. Each leg will have a defined direction and distance; a number of legs in sequence becomes the route. The advent of digital computers has facilitated comprehensive area navigation systems that use a combination of ground navigation aids and airborne equipment.

The features and benefits of RNAV are illustrated in Figure 16.7, these include:

- Customised and/or modified routes, e.g. to avoid congested airspace, or adverse weather conditions (Figure 16.7a)
- Optimising the route (Figure 16.7b) to bypass navigation aids ('cutting corners'), e.g. if VOR-C is out of range, the RNAV leg is created thereby shortening the distance flown

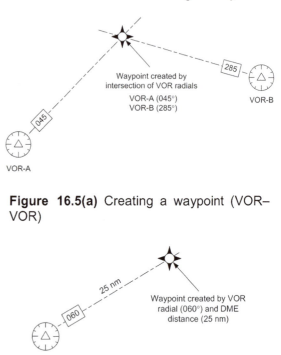

Figure 16.5(a) Creating a waypoint (VOR–VOR)

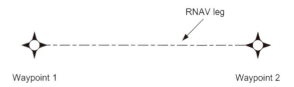

Figure 16.5(b) Creating a waypoint (VOR–DME)

Figure 16.6   Creating an RNAV leg

- Flying parallel tracks, i.e. with a specified cross track distance (Figure 16.7c). This provides greater utilisation of airspace, especially through congested areas
- Flying 'direct to' a VOR navigation aid (or waypoint) when cleared or directed by air traffic control (ATC), thereby shortening the distance flown (Figure 16.7d).

These features lead to a reduction of operating costs achieved by saving time and/or fuel. RNAV equipped aircraft are able to operate in flexible scenarios that are not possible with conventional airway routes, this leads to higher utilisation of

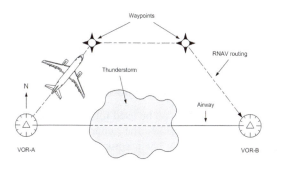

(a) Avoiding weather via RNAV routing

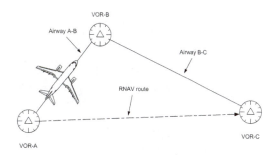

(b) Optimising a route to bypass a navigational aid

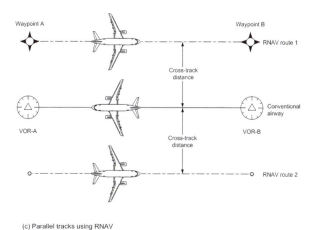

(c) Parallel tracks using RNAV

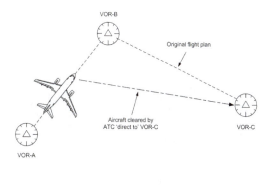

(d) RNAV 'direct to' clearance

Figure 16.7   Features and benefits of RNAV

the aircraft. VOR–DME defined airways were supplemented in the 1970s with RNAV routes, but this scheme has now been superseded (see 'Required navigation performance' at the end of this chapter).

16.2 RNAV equipment

In addition to the VOR and DME equipment described in previous chapters, an RNAV system also incorporates a control display unit, navigation instruments and a computer.

16.2.1 Control display unit (CDU)

Pilot inputs to the system are via a control display unit (CDU), see Figure 16.8(a). Typical CDU displays include:

- Present position in latitude and longitude
- Wind speed and wind direction
- Distance, bearing and time to the active waypoint.

The pilot can call up stored waypoints; alternatively, the pilot can create waypoints. The CDU can also be used for selecting the **direct to** feature. Navigation guidance information displayed on the CDU will also be displayed (via a 'Rad/Nav' switch) for the primary navigation instruments, e.g. the course deviation indicator (CDI). Guidance information on this instrument will include a continuous display of aircraft position relative to the desired track. Navigation sensor failure warnings will also be displayed on the primary navigation instruments. To achieve the maximum benefits of an RNAV system, outputs are coupled to the automatic flight

(a) RNAV control display unit (CDU)

(b) Course deviation indicator

Figure 16.8 RNAV control and display

Key point

RNAV systems use a combination of navigation system inputs.

Key point

RNAV equipped aircraft are able to operate in conditions and scenarios that would not have been previously possible, thereby obtaining higher utilisation of the aircraft.

---

control system (AFCS) by selecting 'NAV' on the AFCS mode control panel. Auto-leg sequencing with associated turn anticipation is possible within the control laws of the AFCS.

16.2.2 Navigation instruments

One instrument that can be used for the display of RNAV information is the course deviation indicator (CDI). This has a compass display and course selector as shown in Figure 16.8(b). The course selector (lower right-hand side of instrument) is set to the desired leg; a deviation pointer moves left or right of the aircraft symbol to indicate if the aircraft is to the right or left of the selected leg.

16.2.3 Computer

The RNAV computer is used to resolve a variety of navigation equations. In order to realise the benefits of RNAV, systems contain a **navigation database** (see below). Simple area navigation is achieved by solving geometric equations; the data required for these calculations is obtained from the relative bearings of VOR stations and/or distances from DME stations.

In navigation calculations, bearings are referred to as theta (θ) and distances as rho (ρ). The RNAV definition of a waypoint using a co-located VOR–DME navigation aid is illustrated in Figure 16.9. Accurate horizontal range can be calculated by the computer based on:

- DME slant range
- DME transponder elevation
- aircraft altitude.

This calculation provides the true range as illustrated in Figure 16.10. Transponder elevation is obtained from the computer's navigation database. Altitude information is provided by an encoding altimeter or air data computer (ADC). Cross track deviation, either intentional or otherwise, is calculated as shown in Figure 16.11.

Computers in more sophisticated systems are also able to **auto-tune** navigation aids to provide the optimum navigation solution. The system decides on whether to use combinations of VOR–VOR (theta–theta), VOR–DME (theta–rho) or DME–DME (rho–rho). Note that when two DME navigation aids are used, there is an ambiguous

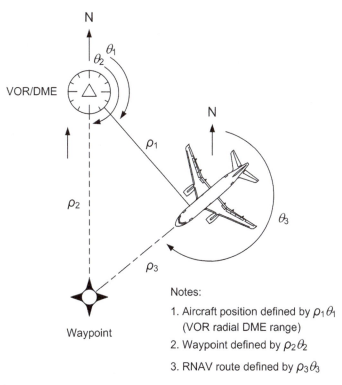

Notes:

1. Aircraft position defined by $\rho_1\theta_1$ (VOR radial DME range)
2. Waypoint defined by $\rho_2\theta_2$
3. RNAV route defined by $\rho_3\theta_3$

(a) RNAV triangulation

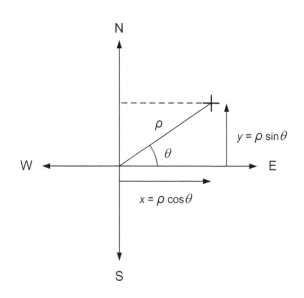

(b) RNAV calculation

Figure 16.9   RNAV geometry

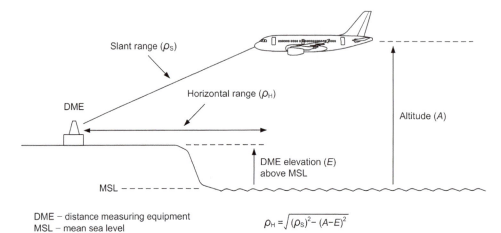

DME – distance measuring equipment
MSL – mean sea level

$$\rho_H = \sqrt{(\rho_S)^2 - (A-E)^2}$$

Figure 16.10 RNAV geometry—vertical profile

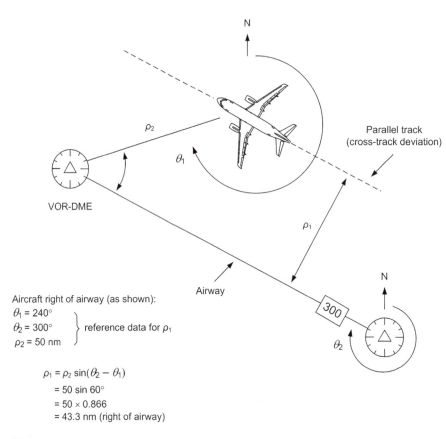

Aircraft right of airway (as shown):

$\theta_1 = 240°$
$\theta_2 = 300°$ $\Bigg\}$ reference data for ρ_1
$\rho_2 = 50$ nm

$\rho_1 = \rho_2 \sin(\theta_2 - \theta_1)$
$\quad = 50 \sin 60°$
$\quad = 50 \times 0.866$
$\quad = 43.3$ nm (right of airway)

Note:
If the aircraft was to the left of the airway, a negative value of ρ would be calculated – this is interpreted as being left of the airway

Figure 16.11 RNAV geometry—lateral profile

position fix, see Figure 16.12; this can be resolved in a number of ways, e.g. by tuning into a third DME navigation aid or tuning into a VOR station. Systems use algorithms to determine which combination of navigation aids to use; this will depend on signal strength and geometry. Four-dimensional waypoints can also be defined by specifying the required time of arrival over a three-dimensional waypoint. This is discussed further in the flight management system chapter. Other aircraft sensor inputs such as initial fuel quantity, fuel flow, airspeed and time provide the means of calculating range, estimated time of arrival (ETA), endurance etc. This data can be provided for specific waypoints or the final destination.

16.2.4 Navigation database (NDB)

The navigation database (stored within the RNAV computer's memory) contains permanent records for VOR, DME and VORTAC navigation aids. Table 16.2 illustrates the locations, identification codes, and navigation aid type for a typical European country. Details that are stored in the database include specific information for each navigation aid such as:

- Name
- Identification code
- Navigation aid type
- Latitude and longitude
- Elevation
- Transmission frequency.

The navigation database is updated every 28 days to take into account anything that has changed with a navigation aid, e.g. frequency changes, temporary unavailability etc. The pilot can enter new or modified details for navigation aids that might not be contained in the navigation database. Waypoints can either be entered as they appear on navigation charts, or the pilot can create them. The navigation database in more sophisticated RNAV systems will also include standard instrument departures (SIDs), standard terminal arrival routes (STARs), runway data and three-dimensional (latitude, longitude and altitude) waypoints to facilitate air traffic control requirements. Figures 16.13 and 16.14 give examples of SIDS and STARS. Note that since VOR and DME navigation aids have to be

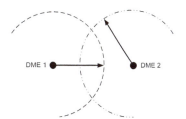

Figure 16.12 Ambiguous DME position fix

located on land, RNAV based on these navigation aids alone does not extend far beyond coastal regions. Referring to Figure 16.15, a combination of radio navigation stations located in a number of European countries provides a certain amount of navigation guidance in the North Atlantic, Norwegian Sea and North Sea. This diagram assumes a line-of-sight range of approximately 200 nm. The gaps in this radio navigation network can be overcome by the use of alternative navigation systems including: inertial navigation, Doppler, global satellite navigation systems and Loran-C; these are all described elsewhere in this book.

Table 16.2 Navigation aids in Belgium

Name	Identification	Type
Affligem	AFI	VOR–DME
Antwerpen	ANT	VOR–DME
Beauvechain	BBE	TACAN
Bruno	BUN	VOR–DME
Brussels	BUB	VOR–DME
Chievres	CIV	VOR
Chievres	CIV	TACAN
Costa	COA	VOR–DME
Flora	FLO	VOR–DME
Florennes	BFS	TACAN
Gosly	GSY	VOR–DME
Huldenberg	HUL	VOR–DME
Kleine Brogel	BBL	TACAN
Koksy	KOK	VORTAC
Liege	LGE	VOR–DME

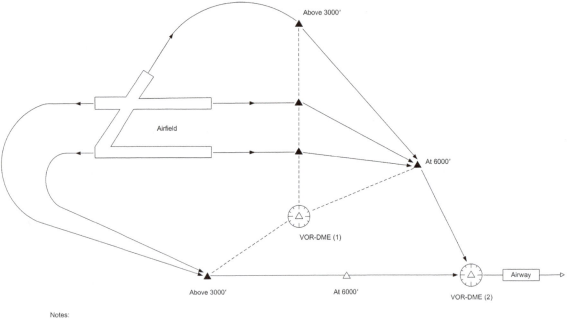

Notes:

1. In this illustration, each of the three runways has a specific departure route to the VOR-DME (2) navigation aid; the aircraft then joins the airways network
2. The SIDs are typically referenced to navigation aids, e.g. VOR-DME or marker beacons
3. There would also be published departure routes for aircraft joining airways to the south, east and north
4. Reporting points (triangles) are often specified with altitude constraints, e.g. at, below or above 3000'

Figure 16.13   Illustration of standard instrument departures (SID)

Test your understanding 16.1

Give (a) three features and (b) three benefits of RNAV.

Test your understanding 16.2

The navigation database contains permanent records for radio navigation aids. List the typical information that is stored for each one.

Test your understanding 16.3

What feature is used to select the best navigation aids for optimised area navigation?

16.3 Kalman filters

One essential feature of advanced RNAV systems is the use of **Kalman filters,** named after Dr Richard Kalman who introduced this concept in the 1960s. Kalman filters are optimal **recursive** data processing algorithms that filter navigation sensor measurements. The mathematical model is based on equations solved by the navigation processor. To illustrate the principles of Kalman filters, consider an RNAV system based on inertial navigation sensors with periodic updates from radio navigation aids. (Inertial navigation is described in Chapter 17.) One key operational aspect of inertial navigation is that system errors accumulate with time. When the system receives a position fix from navigation aids, the inertial navigation system's errors can be corrected.

The key feature of the Kalman filter is that it can analyse these errors and determine how they might have occurred; the filters are recursive, i.e. they repeat the correction process on a succession of navigation calculations and can 'learn' about

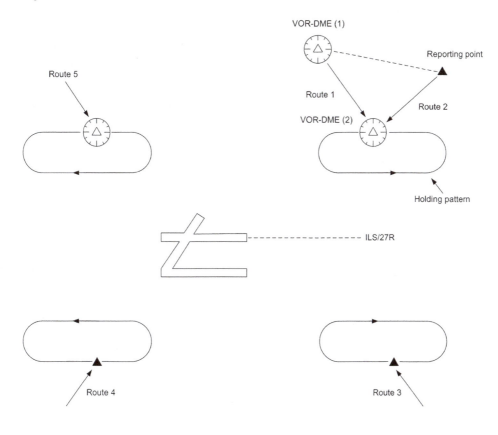

Notes:

1. In this illustration, each of the three arrival routes is associated with a navigation aid (VOR-DME) and reporting point (solid triangles)
2. Each arrival route is normally allocated a holding pattern
3. Minimum sector altitudes are published for each route
4. When cleared by ATC, the aircraft would leave the holding pattern and be given a heading to join the ILS for the active runway, e.g. 27R

Figure 16.14   Illustration of standard terminal arrival routes (STAR)

Test your understanding 16.4

What is the difference between a SID and STAR?

Key point

The RNAV navigation database is updated every 28 days to take into account anything that has changed with a navigation aid, e.g. frequency changes, temporary unavailability etc.

the specific error characteristics of the sensors used. The numerous types of navigation sensors employed in RNAV systems vary in their principle of operation as described in the specific chapters of this book. Kalman filters take advantage of the dissimilar nature of each sensor type; with repeated processing of errors, complementary filtering of sensors can be achieved.

Test your understanding 16.5

Explain the purpose of a Kalman filter.

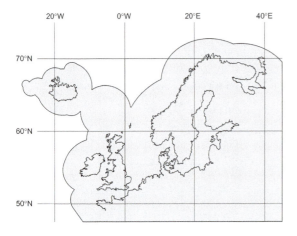

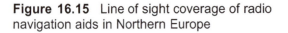

Figure 16.15 Line of sight coverage of radio navigation aids in Northern Europe

16.4 Required navigation performance (RNP)

Simple area navigation systems can use radio navigation aid inputs such as VOR and DME to provide definitions of waypoints as described in this chapter. Comprehensive area navigation systems use a variety of sensors such as satellite and inertial reference systems; these specific systems are addressed in more detail in subsequent chapters. The accuracy and reliability of area navigation systems has led to a number of navigation performance standards and procedures for the aircraft industry; these are known as **required navigation performance** (RNP). Various RNAV systems together with their associated RNP are evolving via individual aviation authorities. This is embraced by the generic term of **performance-based navigation (PBN)**. Factors that contribute to overall area navigation accuracy include:

- External navigation aids
- The aircraft's navigation equipment (including displays)
- Automatic flight control system (AFCS).

The International Civil Aviation Organisation (ICAO) has defined RNAV accuracy levels covering terminal, en route, oceanic and approach flight phases with specific navigation performance values between 1 and 10 nm. These values are expressed by a number, e.g. RNP-5. This indicates that (on a statistical basis) the aircraft's area navigation system must maintain the aircraft for 95% of the flight time within 5 nm of the intended flight envelope, i.e. either side of, and along the track. RNP-5 is used for basic RNAV (**BRNAV**) in Europe. It is not specified how this navigation performance should be achieved, or what navigation equipment is to be used. RNP for terminal operations is less than 1 nm; these systems require performance monitoring and alert messages to the crew in the event of system degradation.

Typical functions required of a BRNAV system include:

- Aircraft position relative to the desired track
- Distance and bearing to the next waypoint
- Ground speed, or time to the next waypoint
- Waypoint storage (four minimum)
- Equipment failure warnings to the crew.

Recommended BRNAV functions that maximise the capabilities of the system include:

- Roll commands to an automatic flight control system (AFCS)
- Aircraft position expressed as latitude and longitude
- A 'direct to' capability
- Navigation accuracy indication
- Automatic tuning of navigation aids
- Navigation database
- Automatic leg sequencing and/or turn anticipation.

Note that if an inertial reference system (IRS) is used as a sensor, the BRNAV system must have the capability of automatically tuning into radio navigation aids after a maximum period of two hours; this is because an IRS derived position will drift (see Chapter 17). If a global navigation satellite system (GNSS) is used as a sensor into the RNAV system, the GNSS must have fault detection software known as receiver autonomous integrity monitoring (RAIM), see Chapter 18. Single RNAV systems are permissible, however, the aircraft must be able to revert to conventional navigation using VOR, DME and ADF in the event of RNAV equipment failure.

In more remote areas, e.g. isolated oceanic regions where it is impossible to locate ground navigation aids, RNP-10 applies. This allows spacing of 50 nm between aircraft in place of 100 nm. The RNAV system now needs two independent long-range systems, e.g. IRS and/or GNSS. If using IRS as a sensor, the system has to receive a position fix with a specified period, typically 6.2 hours. The GNSS has to have fault detection and exclusion (FDE) capability, a technique used to exclude erroneous or failed satellites from the navigation calculations by comparing the data from six satellites.

Key point

Waypoints can be based on existing navigation aids and defined mathematically as:

- rho–theta (using one DME and one VOR navigation aid)

- rheta–theta (using two VOR navigation aids)

- rho–rho (using two DME navigation aids).

Key point

Auto-tuning of navigation aids is used by RNAV systems to select the best navigation aids for optimised area navigation.

Key point

Required navigation performance (RNP) is the performance-based successor to area navigation (RNAV).

Test your understanding 16.6

Explain why RNAV systems using VOR–DME are generally unavailable beyond land and its immediate coastal regions.

Test your understanding 16.7

Explain what is meant by RNP and why it is needed.

Test your understanding 16.8

Explain why an RNAV database needs to be updated every 28 days.

16.5 Multiple choice questions

1. Waypoints are defined geographically by:
   (a) latitude and longitude
   (b) VOR frequency
   (c) DME range.

2. SIDs are used during the following flight phase:
   (a) arrival
   (b) cruise
   (c) departure.

3. Accurate area navigation using DME–DME requires:
   (a) slant range
   (b) horizontal range
   (c) VOR radials.

4. Rho–theta is an expression for which area navigation solution?
   (a) DME–DME
   (b) VOR–DME
   (c) VOR–VOR.

5. Navigation legs are defined by:
   (a) speed and distance
   (b) bearing and distance
   (c) bearing and speed.

6. Specific information for each navigation aid is contained in the:
   (a) navigation database
   (b) control display unit
   (c) course deviation indicator.

7. Flying a parallel track requires a specified:
   (a) cross track deviation
   (b) bearing
   (c) distance to go.

8. A three-dimensional waypoint is defined by:
   (a) VOR–DME
   (b) latitude, longitude, altitude
   (c) rho–theta–rho.

9. Autotuning of navigation aids is used by
   RNAV systems to:
   (a) update the navigation database
   (b) create waypoints in the CDU
   (c) select the best navigation aids for
       optimised area navigation.

10. Cross track deviation is displayed on the CDU
    and:
    (a) RMI
    (b) DME
    (c) HSI.

11. The navigation database is normally updated:
    (a) at the beginning of each flight
    (b) every 28 days
    (c) when selected by the pilot.

12. A four-dimensional waypoint is defined by:
    (a) lateral position, altitude and time
    (b) latitude, longitude, altitude and speed
    (c) altitude, direction, speed and time.

13. VORTAC navigation aids comprise which
    two facilities:
    (a) VOR and DME
    (b) VOR and TACAN
    (c) TACAN and NDB.

14. RNP-2 requires that the aircraft:
    (a) uses a minimum of two different
        navigation sensor inputs
    (b) is maintained within two nautical miles of
        the specified flight path
    (c) is maintained within two degrees of the
        specified flight path.

15. An area navigation position calculated from
    two DME stations is referred to
    mathematically as:
    (a) theta–theta
    (b) rho–theta
    (c) rho–rho.

16. The feature marked X in Figure 16.16 is a:
    (a) VOR–DME
    (b) STAR
    (c) waypoint.

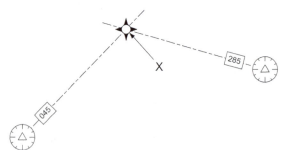

Figure 16.16   See Question 16

Chapter 17 Inertial navigation system

Inertial navigation is an autonomous dead reckoning method of navigation, i.e. it requires no external inputs or references from ground stations. The system was developed in the 1950s for use by the US military and subsequently the space programmes. Inertial navigation systems (INS) were introduced into commercial aircraft service during the early 1970s. The system is able to compute navigation data such as present position, distance to waypoint, heading, ground speed, wind speed and wind direction. It does not need radio navigation inputs and it does not transmit radio frequencies. Being self-contained, the system is ideally suited for long distance navigation over oceans and undeveloped areas of the globe. The reader should be aware that, as with many avionic systems, significant developments have occurred with inertial navigation systems in recent decades; the inertial system is often integrated with other avionic units and there are a variety of system configurations being operated. This chapter seeks to provide an introduction to the principles of inertial navigation together with some examples of typical hardware.

17.1 Inertial navigation principles

The primary sensors used in the system are **accelerometers** and **gyroscopes** (hereinafter gyro) to determine the motion of the aircraft. These sensors provide reference outputs that are processed to develop navigation data.

To illustrate the principle of inertial navigation, consider the accelerometer device illustrated in Figure 17.1; this is formed with a mass and two springs within a housing. Newton's second law of motion states that a body at rest (or in motion) tends to stay at rest (or in motion) unless acted upon by an outside force. Moving the accelerometer to the right causes a relative movement of mass to the left. If the applied force

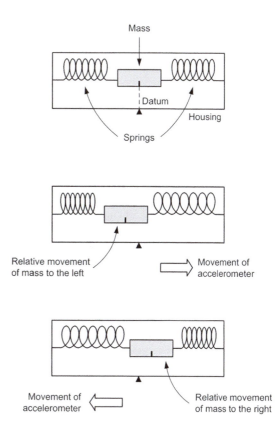

Figure 17.1 Accelerometer

is maintained, the mass returns to the neutral position. When the accelerometer is moved to the left, or brought to rest, the relative movement of the mass is to the right. The mass continues in its existing state of rest or movement unless the applied force changes; this is the property of **inertia**. Attaching an electrical pick-up to the accelerometer creates a transducer that can measure the amount of relative movement of the mass. This relative movement is in direct proportion to the acceleration being applied to the device, expressed in m/s^2. If we take this electrical output and mathematically **integrate** the value, we are effectively multiplying the acceleration output by time; this can be expressed as:

Time × acceleration = s × m/s^2 = m/s = *velocity*

If we now take this velocity output and mathematically integrate the value, we are once again multiplying the output by time; this can be expressed as:

Time × velocity = s × m/s = m = *distance*

In summary, we started by measuring acceleration, and were able to derive velocity and distance information by applying the mathematical process of integration. To illustrate this principle, consider a body accelerating at 5 m/s^2, after ten seconds the velocity of the body will be 50 m/s. If this body now travels at a constant velocity of 50 m/s for ten seconds, it will have changed position by 500 m.

Referring to the profile illustrated in Figure 17.2, at t_0, the accelerometer is at rest and its output is zero. When the accelerometer is moved during the time period $t_0 - t_1$, there is a positive output from the accelerometer; this is integrated to provide velocity and distance as illustrated in Figure 17.2. As the accelerometer reaches a steady velocity during the time period $t_1 - t_2$, the distance travelled increases. Acceleration increases and decreases during the journey (shown as positive and negative), until the destination is reached.

This accelerometer is providing useful velocity and distance information, but only measured in one direction. If we take two accelerometers and mount them on a **platform** at right angles to each other, we can measure acceleration (and subsequently velocity and distance information) in any **lateral direction**. Thinking of an aircraft application, if we can align the platform with a known reference, e.g. **true north,** the two accelerometers are then directed N–S and W–E respectively, see Figure 17.3. We now have the means of calculating our velocity and distance travelled in any lateral direction.

An illustration of how the basic navigation calculations are performed is given in Figure 17.4. In a practical inertial navigation system, there is very little actual movement of the mass. The relative displacement between the mass and

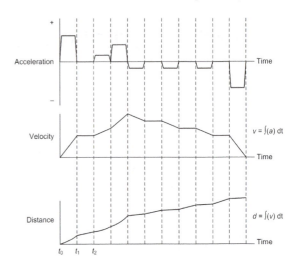

Figure 17.2   Integration profile

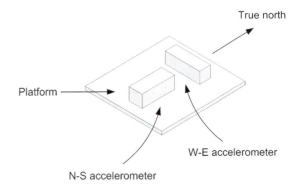

Figure 17.3   Platform NS-EW

housing is sensed by an electrical pick-off signal. A closed loop servomechanism feedback signal (proportional to acceleration) is then amplified and used to restrain the mass in the **null position**. The amount of feedback required to maintain the null position is proportional to the sensed acceleration; this becomes the accelerometer's output signal. Calculation of basic navigation data is illustrated in Figure 17.5. By combining two accelerometer outputs in the directions N–S and W–E, we can sum the vector outputs and calculate distance and velocity in the horizontal plane. By comparing the distance travelled with the starting position (see 'alignment process' in Section 17.4) we can calculate our present position.

Since the aircraft will be operating through a range of pitch and roll manoeuvres, it is vital that

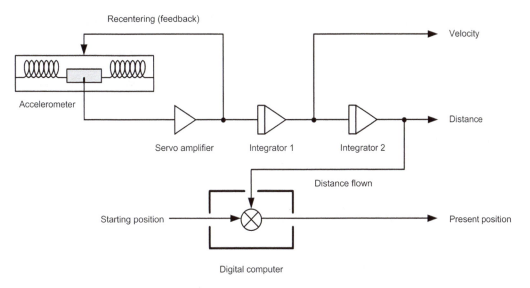

Figure 17.4 Navigation calculations (1)

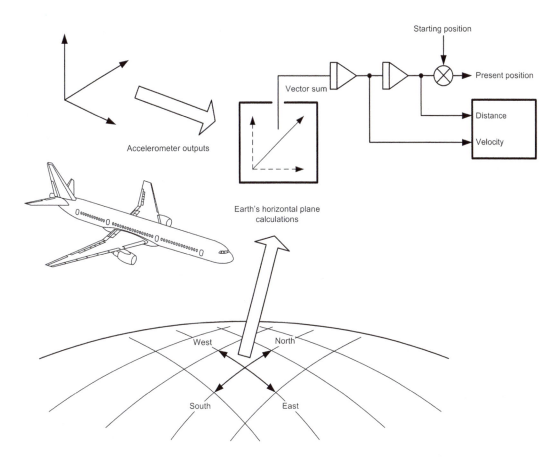

Figure 17.5 Navigation calculations (2)

when measuring acceleration in the N–S and W–E directions, we do not measure the effects of gravity. The original inertial navigation systems maintained a physical platform such that it was always aligned with true north, and always level with respect to the earth's surface. Electromechanical gyros and torque motors mounted within gimbals in each of the three axes achieved these requirements. The platform is aligned with true north and levelled at the beginning of the flight; this condition is maintained throughout the flight.

A by-product of aligning and levelling the platform is that attitude information is available for use by flight instruments and other systems. Modern day commercial aircraft inertial navigation systems are equipped with **strap-down** devices including solid-state gyros and accelerometers. The alignment process and attitude compensation is now achieved in the computer's software, i.e. there is no physical platform.

17.2 System overview

The key principles of inertial navigation are based on accelerometer and gyro references together with a navigation processing function. These can be combined within a single **inertial navigation system** (INS) and dedicated crew interface as illustrated in Figure 17.6. Alternatively, the accelerometers and gyros are contained within an **inertial reference unit** (IRU), the processing function and crew interface is then integrated within the flight management computer system (FMCS). Within the FMCS, the flight management computer (FMC) combines area navigation and performance management into a single system (described in Chapter 19).

Key point

The primary sensors used in the inertial navigation system are accelerometers and gyros to determine the aircraft's movement. These sensors provide outputs that are processed to provide basic navigation data.

17.3 System description

The components described in this section are for devices used in typical commercial transport aircraft; note that system architecture varies considerably with different aircraft types. A long-range aircraft will have three independent inertial reference systems, each providing its own navigation information.

The inertial navigation system can be considered to have three functions:

- References
- Processing
- Crew interface.

17.3.1 References

Accelerometers

These can be single or three axis devices; a typical single axis device is packaged in a 25×25 mm casing weighing 45 grams (see Figure 17.7). This contains a pendulum (**proof-mass**) that senses acceleration as previously described over the range ± 40 g; relative displacement between the pendulum and casing is sensed by a high gain capacitance pick-off and a pair of coils.

A closed loop servomechanism feedback signal (proportional to acceleration) is then amplified and demodulated. This feedback signal (analogue current or digital pulses) is applied to the coils to restrain the pendulum at the **null position**.

The feedback required to maintain the null position is proportional to the sensed acceleration; this becomes the accelerometer's output signal. Because of the high gain of the servomechanism electronics used, pendulum displacements are limited to microradians. An integral temperature sensor provides thermal compensation. The IRU contains three devices, measuring acceleration in the longitudinal, lateral and normal axes of the aircraft.

Key point

The inertial navigation system needs to establish a local attitude reference and direction of true north for navigation purposes. During this process, the aircraft should not be moved.

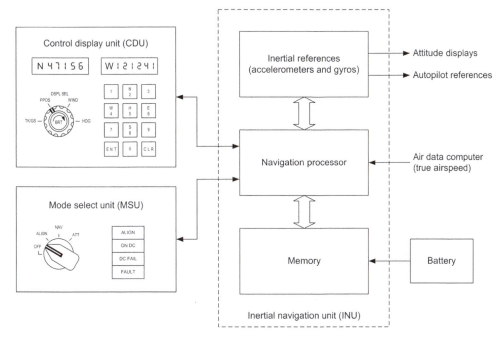

Figure 17.6(a) Inertial navigation system (general arrangement)

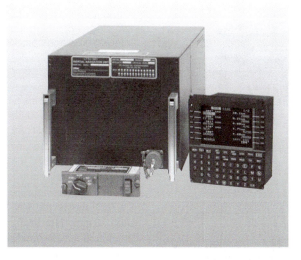

Figure 17.6(b) Inertial navigation system (courtesy of Northrop Grumman)

Key point

Synthesised magnetic variation can be obtained from inertial navigation systems meaning that remote sensing compass systems are not required.

Key point

By comparing the position outputs of three on-board inertial navigation systems, this also provides a means of error checking between systems.

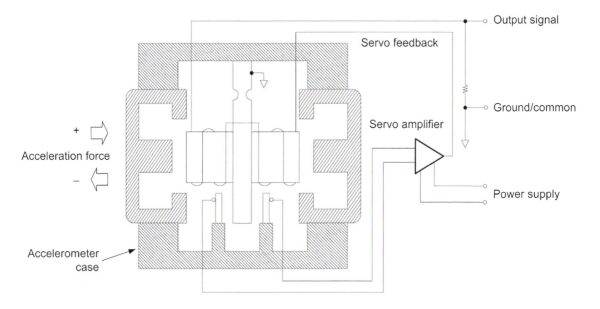

Figure 17.7(a)   Accelerometer arrangement

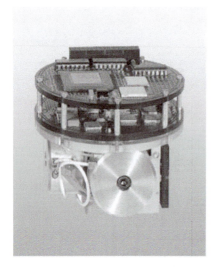

Figure   17.7(b)   MEMS   accelerometers mounted   above   the   fibre   optic   gyros (courtesy of Northrop Grumman)

Key point

Errors in the inertial navigation system are random and build up as a function of time; this applies even if the aircraft is stationary.

Test your understanding 17.1

The output from an accelerometer goes through two stages of integration; what does each of these integration stages produce?

Developments   in   micro-electromechanical systems (MEMS) technology has led to silicon accelerometers that are more reliable and can be manufactured onto an integrated circuit. MEMS is the integration of mechanical elements, sensors and electronics on a common silicon substrate through   micro-fabrication   technology.   Figure 17.7(b) shows the LN-200 inertial measurement unit   from   Northrop   Grumman;   the   micro-machined accelerometers are in the upper section of the unit. The lower section of the unit contains fibre optic gyros (see below). The entire unit weighs   less   than   750   grams   and   is   packaged within a 9 cm diameter housing.

Gyros

The original inertial navigation systems used electromechanical gyros; these were subsequently replaced   by   a   more   reliable   and   accurate technology: the **ring laser gyro** (RLG). Ring laser gyros use interference of a laser beam within an optic path, or ring, to detect rotational displacement. An IRU contains three such devices (see Figure 17.8) for measuring changes in pitch, roll and azimuth. (Note that laser gyros are not actually gyroscopes in the strict sense of the word—they are in fact sensors of angular rate of rotation about an axis.) Two laser beams are transmitted in opposite directions (contra-rotating) around a cavity within a triangular block of **cervit glass**; mirrors are located in two of the corners. The cervit glass (ceramic) material is very hard and has an ultra-low thermal expansion coefficient. The two laser beams travel the same distance, but in opposite directions; with a stationary RLG, they arrive at the detector at the same time.

The principles of the laser gyro are based on the **Sagnac effect**, named after the French physicist Georges Sagnac (1869–1926). This phenomenon results from interference caused by rotation. Interferometry is the science and technique of superposing (interfering) two or more waves, which creates a resultant wave different from the two input waves; this technique is used to detect the differences between input waves. In the aircraft RLG application, when the aircraft attitude changes, the RLG rotates; the laser beam in one path now travels a greater distance than the beam in the other path; this changes its phase at the detector with respect to the other beam. The angular position, i.e. direction and rate of the RLG, is measured by the **phase difference** of the two beams. This phase difference appears as a fringe pattern caused by the interference of the two wave patterns. The fringe pattern is in the form of light pulses that can be directly translated into a digital signal. Operating ranges of typical RLGs are $1000°$ per second in pitch, roll and azimuth. In theory, the RLG has no moving parts; in practice there is a device required to overcome a phenomena called **lock-in**. This occurs when the frequency difference between the two beams is low (typically 1000 Hz) and the two beams merge their frequencies. The solution is to mechanically oscillate the RLG to minimise the amount of time in this lock-in region.

Ring laser gyros are very expensive to manufacture; they require very high quality glass,

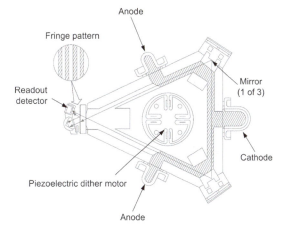

Figure 17.8(a)   Ring laser gyro arrangement

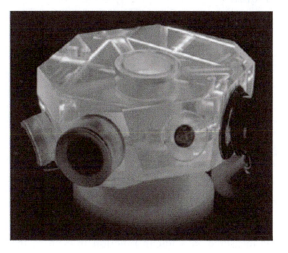

Figure 17.8(b)   Ring laser gyro (photo courtesy of Northrop Grumman)

cavities machined to close tolerances and precision mirrors. There are also life issues associated with the technology. A variation of this laser gyro technology is the fibre optic gyro (FOG), where the transmission paths are through coiled fibre optic cables packaged into a canister arrangement to sense pitch, roll and yaw, see Figure 17.9. The fibre optic gyroscope also uses the interference of light through several kilometres of coiled fibre optic cable to detect angular rotation. Two light beams travel along the fibre in opposite directions and produce a phase

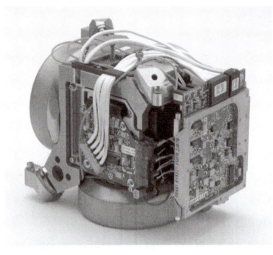

Figure 17.9   Fibre optic gyro assembly

shift due to the Sagnac effect. Fibre optic gyros
have a life expectancy in excess of 3.5 million
hours.

17.3.2 Inertial signal processing

The acceleration and angular rate outputs from
the IRU are transmitted via a data bus to the
navigation processor, in the flight management
computer (FMC). Aside from navigation
purposes, the IRU outputs are also supplied to
other systems, e.g. the primary flying display and
weather radar for attitude reference. Acceleration
is measured as a **linear function** in each of the
three aircraft axes; normal, lateral and
longitudinal. Attitude is measured as an angular
rate in pitch, roll and yaw.

These outputs are resolved and combined with
air data inputs to provide navigation data, e.g.
latitude, longitude, true heading, distance to the
next waypoint, ground speed, wind speed and
wind direction. The processor simultaneously
performs these navigation calculations using
outputs from all three accelerometers and angular
rate sensors; in addition to these calculations, the
processor has to compensate for three physical
effects of the earth.

- Gravity
- Rotation
- Geometry.

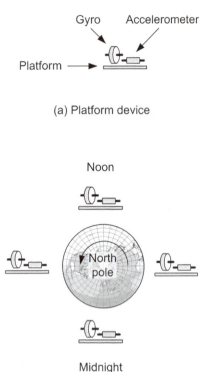

(a) Platform device

(b) Platform aligned with inertial space

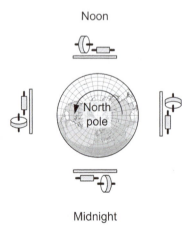

(c) Platform aligned with the earth's surface

Figure   17.10   Effect   of   gyro   and
accelerometer   alignment   relative   to   the
earth's rotation

Effects of gravity

The navigation processor needs to determine the relationship between the aircraft attitude and surface of the earth such that the accelerometers only measure aircraft motion, not **gravity**. Outputs from each of the laser gyros are angular rates of rotation about an axis. These outputs are integrated, i.e. multiplied by time, to provide measurements of pitch, roll and heading. To illustrate this principle, a yaw rate of 4.5 degrees/second over a ten second period equates to a heading change of 45°. If the aircraft's heading is now a constant 090°, and pitch/roll rates are zero, the only acceleration measured is along the longitudinal axis of the aircraft. With a constant velocity of 500 knots, after one hour the navigation processor calculates that the aircraft has changed position by 500 nm in an easterly direction.

Now consider the same scenario, but with the aircraft climbing with a 10 degrees nose up attitude. The processor needs to separate out vertical and lateral accelerations caused by gravity and motion of the aircraft respectively. The accelerometer is needed to measure motion parallel to the earth's surface but if the aircraft is pitching or rolling, it will not be able to distinguish between gravity and aircraft acceleration. The component of gravity has to be separated out of the measured acceleration.

Effects of the earth's rotation

The processor now has to take into account the effect of the **earth's rotation**. To illustrate this effect, consider a platform device with an accelerometer and gyro as shown in Figure 17.10 (a). In this schematic illustration, the gyro is used to maintain the platform in a stable position. As the earth rotates, the platform maintains its position with respect to inertial space (Figure 17.10(b)), however, it is moving relative to the earth's surface. The platform has to be aligned with the earth's surface (Figure 17.10(c)) so that we can use it practically for navigation purposes.

With a strap-down system, each laser gyro will measure the angular rate of rotation of the aircraft about an axis. Since each laser gyro is fixed in position within the IRU, with the aircraft on the ground, it will also measure the rotation of the earth in an easterly direction. This motion includes (a) the rotation of the earth about its own axis and (b) the orbit around the sun:

- 360° over 24 hours = 15°/hour
- 360° over 365 days = 0.04°/hour.

This **earth rate** of up to 15.04 degrees per hour depends on latitude. Earth rate is a component that is subtracted from any measurement of aircraft angular rate sensed in an easterly direction.

Effects of the earth's geometry

The final consideration for the processor to address is the **spherical geometry** of the earth. As the aircraft travels around the earth in straight and level flight (parallel to the surface) it actually describes an arc. The pitch laser gyro senses this as an angular rate with respect to **inertial space**. When integrated, this rate output is converted into a change of pitch attitude. Clearly no pitch change has actually occurred due to this **transport rate**; the processor needs to subtract this component from the pitch laser gyro measurement. To calculate transport rate, the distance travelled (described by an arc) and angle subtended from the earth's centre are divided by time. This relationship can be developed to relate lateral (tangential) velocity and angular rate. The navigation processor calculates transport rate from lateral velocity divided by an estimate of the earth's radius plus the aircraft altitude. Transport rate is then subtracted from any gyro output using a process known as **Schuler tuning** (after the Austrian physicist Max Schuler who solved the problem of accelerations due to the effect of ship manoeuvres on pendulum-based gyro-magnetic compass systems). Schuler tuning is achieved by feeding back aircraft rate terms such that the system is always aligned to the local vertical as the aircraft travels over the spherical earth.

17.3.3 Crew interface

A complete inertial navigation system (INS) is illustrated in Figure 17.11. The LTN-92 system (from Northrop Grumman) contains the inertial navigation unit (IRU), control display unit (CDU) and mode selector unit (MSU). The CDU is the crew's interface with the system; it used to enter data into the IRU, e.g. present position during the **alignment process**. It also provides warnings and

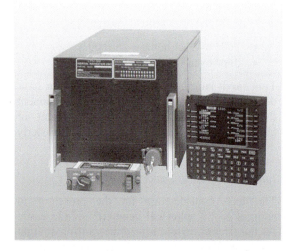

Figure 17.11 LTN92 system (photo courtesy of Northrop Grumman)

alerts back to the crew, e.g. if incorrect data has been inserted or if the system develops navigation errors. The MSU is used to turn the system on and to initiate the alignment process before selection of the navigation mode.

If the IRS is integrated with the flight management system (FMS) (as described in Chapter 19), crew interface with the IRU is via the FMS control display unit (Figure 17.12) and inertial reference mode panel (IRMP) on the overhead panel as shown in Figure 17.13.

The IRMP is used to initiate the alignment process before selection of the navigation mode. (NB Alignment must be achieved before moving the aircraft.) Four operating modes can be selected via the IRMP for each of the systems:

- Off
- Align
- Navigate
- Attitude.

When the system is selected from off to align, the initialisation process is started. Present position is entered via the CDU, this is checked for accuracy within the IRU. When the system is aligned (see Section 17.4) the navigation mode can be selected. In the event of navigation computer failure, the IRU can be selected to provide attitude references only for the flight instruments. Four annunciators are provided for each of the

systems. These provide the inertial reference system (IRS) status and fault indications as illustrated in Table 17.1.

The IRMP also has two alphanumeric displays; the data being displayed depends on what has been selected by a rotary switch. This displayed information is illustrated in Table 17.2. (A second rotary switch selects which system information is being displayed, e.g. left, centre or right.)

The IRMP normally displays data that has been entered via the control display unit (Figure 17.14); however, the IRMP's alphanumeric keyboard can also be used to enter data including latitude, longitude and magnetic heading.

Table 17.1 IRS status and fault indications

Caption	Colour	Purpose
Align	White	IRU is in the align mode, initial attitude mode, or powering down
On DC	Amber	IRU has switched to backup battery power
DC fail	Amber	DC power failure to the IRU
Fault	Amber	Built-in test has detected a failure, or certain alignment problems have occurred

Table 17.2 Inertial reference mode panel (IRMP) displays

Switch position	Left display	Right display
TK/GS	Track angle	Ground speed
PPOS	Latitude	Longitude
Wind	Wind angle	Wind speed
HDG	True heading	Blank

Test your understanding 17.2

What is the difference between an RLG and an FOG?

Figure 17.12 FMS control display unit (CDU)

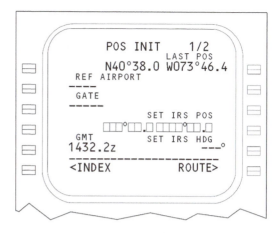

Figure 17.14 CDU position information

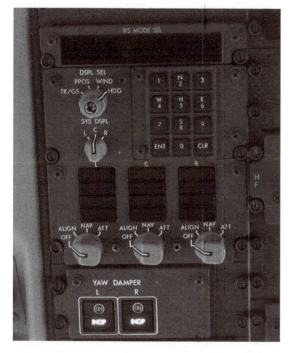

Figure 17.13 IRS panel in the overhead display of a Boeing 757

17.4 Alignment process

A fundamental requirement of inertial navigation is the initial alignment process; this is required to determine a **local vertical** and direction of **true north**. Alignment must be carried out with the aircraft on the ground and stationary. (Note that in certain cases, e.g. on the flight deck of an aircraft carrier, alignment has to be accomplished whilst the aircraft is moving. In this case, an external reference is required, e.g. the carrier's own inertial navigation system.) Systems using an electromechanical gimballed platform need this time for the gyros to level the platform with respect to the local vertical and align the platform in the direction of true north. For strap-down systems, there is no platform as such; however, the system still needs to establish a local attitude reference and direction of true north for navigation purposes.

For illustration purposes, a strap-down IRS using ring laser gyros (RLG) is described. With the aircraft's longitudinal axis lined up exactly with true north (Figure 17.15(a)), the roll RLG will sense an angular rate corresponding to the earth's rotation; the pitch RLG output (as a function of the earth's rotation) will be zero. If the aircraft's longitudinal axis were lined up exactly to the east (Figure 17.15(b)), the pitch RLG would correspond to the earth's rotation and the roll RLG (as a function of the earth's rotation) would be zero. Any aircraft position other than these two examples will provide **N–S and W–E components** of the earth's rotation enabling the system to determine the direction of true north. Furthermore, the reference system can estimate latitude and true heading by sensing these rotational vectors.

Referring to Figure 17.16, local vertical is computed by sensing gravity via the system's

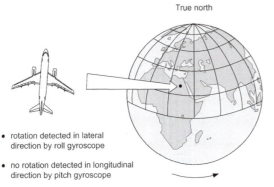

- rotation detected in lateral direction by roll gyroscope
- no rotation detected in longitudinal direction by pitch gyroscope
- IRS determines that the aircraft is aligned with true north

(a) Aircraft's longitudinal axis aligned with true north

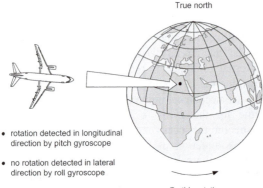

- rotation detected in longitudinal direction by pitch gyroscope
- no rotation detected in lateral direction by roll gyroscope
- IRS determines that the aircraft is aligned in easterly direction

(b) Aircraft's longitudinal axis aligned due east

Figure 17.15   Inertial system

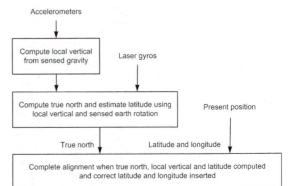

Figure 17.16   Computation of local vertical, true north and aircraft position by the navigation processor

three accelerometers. Utilising the local vertical and sensing the earth's rotation by the gyros allows the IRU to estimate latitude and compute the direction of true north. Once true north is established, the aircraft's present position can be entered; the system is now ready to navigate. Alignment is always initiated before departure and it is essential that the aircraft is not moved until alignment is completed. If the aircraft were moved, e.g. by a towing tug, the accelerometers would measure this, thereby corrupting the sensing of a local vertical. A warning (flashing 'align' light) is provided during the alignment

mode to indicate if:

- present position has not been entered
- there is a significant difference between the position entered and the last known position
- the aircraft has been moved.

If any of these events occur, the entire alignment process would have to be started again thereby causing a delay. The mode takes between 5 to 10 minutes to complete, depending on the system and latitude.

At the equator, earth rate is a maximum and the direction of true north can be determined relatively quickly. This process takes longer up to latitudes of 70 degrees, above which system accuracy and performance is degraded.

Once aligned, the inertial navigation computer is always referenced to true north. It is therefore possible to establish the variation of the earth's magnetic field by reference to a look-up table in the computer's memory.

The reader will be aware from Chapter 8 that **magnetic variation** is the difference between true north and magnetic north; this variation depends on where the observer is on the earth's surface. Magnetic variation also changes over the passage of time and so the computer's memory must be updated on a periodic basis.

The **synthesised** magnetic variation that can be obtained from inertial navigation systems means that remote reading compass systems are not required, thereby saving weight and system installation costs.

By entering the origin airport's position prior to departure, and then calculating distance travelled as shown in Figure 17.5, the navigation processor calculates the aircraft's present position and desired track to the destination at any given time. Waypoints can be entered into the memory for a given route, and the direction to that waypoint will be calculated and displayed. Additional information that can be supplied by the inertial navigation system is provided in Table 17.3, and illustrated in Figure 17.17.

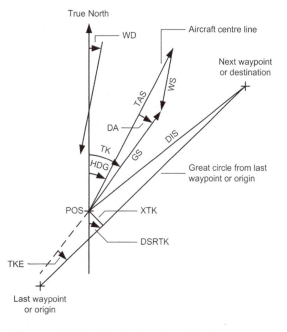

Figure 17.17 Navigation terminology

Test your understanding 17.3

How does an inertial navigation system derive the magnetic variation?

Table 17.3 Navigation terminology

Term	Abbreviation	Description
Cross track distance	XTK	Shortest distance between the present position and desired track
Desired track angle	DSRTK	Angle between north and the intended flight path of the aircraft
Distance	DIS	Great circle distance to the next waypoint or destination
Drift angle	DA	Angle between the aircraft's heading and ground track
Ground track angle	TK	Angle between north and the flight path of the aircraft
Heading	HDG	Horizontal angle measured clockwise between the aircraft's centreline (longitudinal axis) and a specified reference
Present position	POS	Latitude and longitude of the aircraft's position
Track angle error	TKE	Angle between the actual track and desired track (equates to the desired track angle minus the ground track angle)
Wind direction	WD	Angle between north and the wind vector
True airspeed	TAS	Measured in knots
Wind speed	WS	Measured in knots
Ground speed	GS	Measured in knots

17.5 Inertial navigation accuracy

The accuracy of an inertial navigation system depends on a number of factors including: the precision of the accelerometers and gyros; and the accuracy of alignment with respect to true north and local vertical. Errors in the system will build up as a function of time; typical errors of between one and two nautical miles per hour should be allowed for; however, smaller errors can be achieved.

These errors are random and start to accumulate from the moment that the navigation mode starts. When three systems (A, B and C) are aligned at the origin, their present positions are identical. As the flight progresses, see Figure 17.18, the three positions digress. By combining the present positions of three on-board systems it is possible to derive an optimised position (D) within the triangle of positions.

By incorporating other navigation sensor outputs into the navigation computer, e.g. two DME navigation aids, or global navigation sensors (see Section 17.7) it is possible to develop an updated and accurate position calculation. Furthermore, by comparing the position outputs of three on-board systems, we are also providing a means of error checking between systems. For example, if one system's position differs from the other two by a predetermined amount the crew can be alerted to this and they might decide to deselect the system.

17.6 Inertial navigation summary

Inertial navigation has a number of advantages and disadvantages compared with other systems. The disadvantages include:

- The position calculation degrades with time (even if the aircraft is not moving)
- The equipment is expensive
- Initial alignment is essential (this process is degraded at high latitudes, above 70 degrees)
- If the alignment process is interrupted, it has to be repeated leading to potential delays.

The advantages of inertial navigation include:

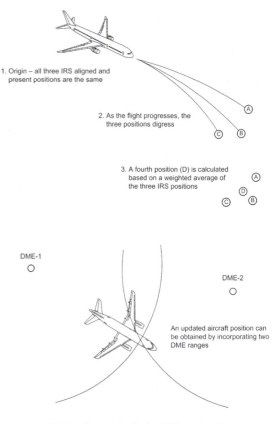

Figure 17.18   Position drift

- Instantaneous velocity and position outputs
- Autonomous operation, i.e. it does not rely on ground-based navigation aids
- Passive operation, i.e. it does not radiate signals and cannot be jammed
- The system can be used on a global basis and is unaffected by the weather.

17.7 System integration

There have been several references in this chapter to stand-alone inertial navigation systems, and those integrated with the flight management system. Many inertial systems are also integrated with global positioning systems and air data computers. An example of such a system is the Northrop Grumman global navigation air data inertial reference unit (GNADIRU) illustrated in Figure 17.19. This provides a powerful and accurate (RNP 0.1) navigation system and overcomes the problem of accumulated position errors. The system integrates inertial and global navigation satellite system (GNSS) measurements to provide highly accurate aircraft position with the high navigation integrity. Inertial sensing is based on state-of-the-art fibre optic gyros and micro-electromechanical systems silicon accelerometers. The system also provides air data information such as altitude, airspeed, angle of attack and other air data parameters. (GNSS principles are described in Chapter 18.)

Figure 17.19 Global navigation air data inertial reference unit (photo courtesy of Northrop Grumman)

Test your understanding 17.4

What are the sources of error in an inertial system?

Test your understanding 17.5

List (a) three advantages, and (b) three disadvantages of inertial navigation systems compared with other systems used for aircraft navigation.

17.8 Multiple choice questions

1. What output is produced from an accelerometer after the first integration process?
   (a) Acceleration
   (b) Velocity
   (c) Distance.

2. If the applied force on an accelerometer is maintained, the mass:
   (a) stays in the same position
   (b) moves in the direction of the force
   (c) returns to the neutral position.

3. During the 'align' mode, local vertical is sensed by:
   (a) accelerometers
   (b) gyros
   (c) the earth's rotation.

4. Establishing the orientation of true north is achieved by sensing:
   (a) local vertical
   (b) the earth's rotation
   (c) the earth's magnetic field.

5. Inertial navigation system errors are a factor of:
   (a) the aircraft's velocity
   (b) how long the system has been in the 'align' mode
   (c) how long the system has been in the 'navigation' mode.

6. Align mode is selected by the crew on the:
   (a) mode select unit
   (b) control display unit
   (c) inertial navigation unit.

7. During flight, with zero output from the accelerometers, the aircraft's ground speed and distance travelled are:
   (a) constant ground speed, increasing distance travelled
   (b) increasing ground speed, increasing distance travelled
   (c) decreasing ground speed, increasing distance travelled.

8. Magnetic north can be derived by an inertial
   reference system through:
   (a) knowledge of the present position and
       local magnetic variation
   (b) remote sensing of the earth's magnetic
       field
   (c) the earth's rotation.

9. Once aligned, the inertial navigation computer
   is always referenced to:
   (a) magnetic north
   (b) true north
   (c) latitude and longitude.

10. Errors in an inertial navigation system are:
    (a) random and build up as a function of time
    (b) fixed and irrespective of time
    (c) random and irrespective of time.

11. Magnetic variation depends on:
    (a) the location of magnetic north
    (b) the navigation computer's memory
    (c) where the observer is on the earth's
        surface.

12. Alignment of the inertial navigation system is
    possible:
    (a) at any time in flight
    (b) at any time on the ground
    (c) only when the aircraft is on the ground and
        stationary.

13. In 'attitude' mode, the inertial navigation
    system provides:
    (a) pitch and roll information
    (b) present position
    (c) drift angle.

14. The angle between the actual track and
    desired track is called:
    (a) track angle error
    (b) heading
    (c) drift angle.

15. During the alignment mode, a flashing align
    light indicates:
    (a) the system is ready to navigate
    (b) the aircraft was moved during align mode
    (c) the present position entered agrees with
        the last known position.

16. Referring to Figure 17.20, drift angle is
    defined as the:
    (a) difference between heading and ground
        track
    (b) angle between north and intended flight
        path
    (c) angle between north and the aircraft's
        flight path.

17. Referring to Figure 17.20, the shortest
    distance between the present position and
    desired track is:
    (a) desired track angle
    (b) ground track angle
    (c) cross track distance.

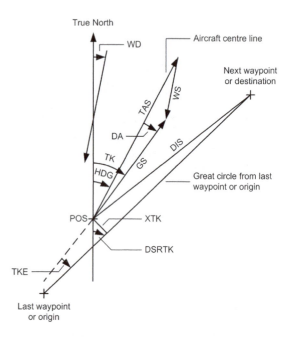

Figure 17.20   See Questions 16 and 17

Chapter 18 Global navigation satellite system

This chapter covers the subject of navigation using an artificial constellation of satellites. Global navigation satellite system (GNSS) is a generic reference for any navigation system based on satellites; the system in widespread use today is the United States' global positioning system (GPS). Other systems in operation include the Russian global navigation satellite system GLONASS that was established soon after GPS and the new European system Galileo. Several nations are developing new global satellite navigation systems; at the time of writing, GPS is the only fully operational system in widespread use throughout the world. For the purposes of explaining the principles and operation of GNSS, in this chapter we will refer to GPS. The chapter concludes with a review of augmentation systems used to increase GPS accuracy, availability and integrity for aircraft navigation, together with a brief insight into emerging technologies.

18.1 GPS overview

The US global positioning system (GPS) was initiated in 1973 and referred to as Navstar (**nav**igation **s**atellite with **t**iming **a**nd **r**anging). The system was developed for use by the US military; the first satellite was launched in 1978 and the full constellation was in place and operating by 1994.

GPS is now widely available for use by many applications including aircraft navigation. The system comprises a space segment, user segment and control segment. Twenty-four satellites (the **space segment**) in orbit around the earth send data via radio links that allows aircraft receivers (the **user segment**) to calculate precise position, altitude, time and speed on a 24-hour, worldwide, all weather basis. The principles of satellite navigation are based on radio wave propagation, precision timing and knowledge of each satellite's position above the earth; this is all monitored and controlled by a network of stations (the **control segment**).

18.2 Principles of wave propagation

The reader will have witnessed the effect of sound wave propagation by observing lightning and thunder during an electrical storm. If the storm is some distance away, there is a time delay between seeing the lightning flash and then hearing the thunder, see Figure 18.1. This delay is caused by the difference in time taken for the light and sound to travel from the lightning to the observer. The same principle applies when an electromagnetic wave is transmitted; except that the wave is propagated at the speed of light, 3×10^8 m/s (in a vacuum).

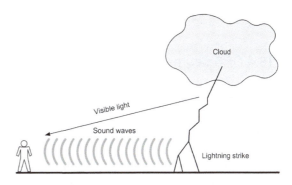

Figure 18.1 Delay in sound waves versus visible light

18.3 Satellite navigation principles

This property of wave propagation can be exploited for satellite navigation purposes. In the first instance, we need to know the exact position of a satellite in orbit above the earth. When this satellite transmits a radio wave to an observer on the earth's surface, the **time delay** between when the radio signal was transmitted and received provides the means of calculating the spherical **range** between the satellite and observer. (Note that the term range is used here when defining the distance from a target object.)

Consider an observer located at a point somewhere on the earth's surface receiving radio waves from a satellite (Figure 18.2). The range between the satellite and observer can be determined by the principle described above; however, this same range can occur at any position described by a circle around the globe. We can reduce this ambiguity through basic geometry by taking range measurements from a second satellite; this will now identify one of two positions on the earth's surface. By using a third satellite, we can remove all ambiguity and define our unique **two-dimensional position** on the earth's surface. Furthermore, a fourth satellite can be used to determine a three-dimensional position, i.e. latitude, longitude and **altitude**. Accuracy of the system depends on having good visibility of these satellites to provide angular measurements. Once the user's position has been calculated, the GPS receiver can derive other useful navigation information, e.g. track, ground speed and drift angle.

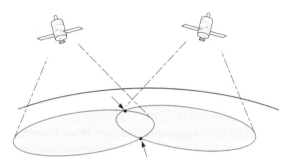

(a) Single satellite describes a circle on the earth's surface

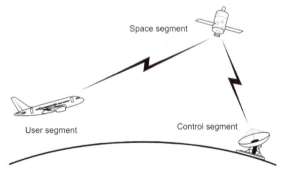

(b) Two satellites define two unique positions

Figure 18.2   Satellite ranging to determine position

18.4 GPS segments

Referring to Figure 18.3, the global positioning system (GPS) comprises three segments: space, ground and user. (Note that the GPS is being updated and modernized with next-generation satellites, alternative radio frequencies and higher specifications. The reader is encouraged to refer to recognised websites and relevant aircraft documentation for new developments.)

18.4.1 Space segment

There are a minimum of 24 (and up to 29) satellites in use, some are operational and others are used as backups. Each satellite is approximately 17 feet across (see Figure 18.4) and weighs approximately 2000 lb. The satellites are in orbit 10,900 nm (approximately 20,200 km) above the earth; this orbit provides optimum ground coverage with the least number of

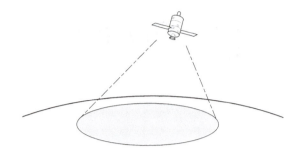

Figure 18.3   Global positioning system segments

satellites. Each satellite is installed with four **atomic clocks** that are extremely accurate, typically maintaining accuracy within three nanoseconds (3×10^{-9} seconds) per day. (Four clocks are installed for backup purposes in the event of failure.) The satellites are powered by the sun's energy via solar panels; nickel cadmium

Figure 18.4 Typical navigation satellite

batteries provide electrical power backup. Each satellite orbits the earth twice per day at an inclination angle of 55° with respect to the equatorial plane; there are six defined orbits each containing four satellites. Figure 18.5 provides an illustration of these orbital patterns. The net result of this orbital pattern is that a minimum of five satellites should be in view to a receiver located almost anywhere on the earth's surface. Satellites have a finite operational life, typically five to ten years.

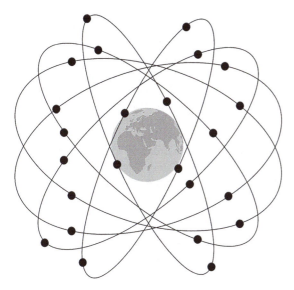

Figure 18.5 GPS space segment—six orbits, each with four satellites

Satellites also download **almanac** data; this is a set of orbital parameters status for all satellites in the constellation. The receiver uses almanac data during initial acquisition of satellite signals. **Ephemeris** data is also downlinked by each satellite; this data contains current satellite position and timing information.

18.4.2 Control segment

The control segment comprises one master control station (MCS) located at Schriever (formerly Falcon) Air Force Base in Colorado Springs, USA; five monitoring stations (located in Colorado Springs, Hawaii, Kwajalein, Diego Garcia and Ascension Island); and three ground antennas (located on Ascension Island, Diego Garcia and Kwajalein). The locations of the monitoring stations provide ground visibility for each satellite.

Although each satellite's clock is very accurate, the relative timing between satellites gradually drifts over time. The individual clocks are monitored and synchronised mathematically relative to Coordinated Universal Time (UTC) by

---

Key point

Three satellites are required to define a unique two-dimensional position on the earth's surface. A fourth satellite can be used to determine an aircraft's altitude.

Key point

The principles of satellite navigation are based on radio wave propagation, precision timing and knowledge of each satellite's position above the earth.

Test your understanding 18.1

How many satellites need to be in view to be able to calculate a two-dimensional position on the earth's surface?

---

the master station. (UTC is the basis for the worldwide system of time.)

Each of the monitoring stations tracks all satellites in view; ranging data and satellite health information is collected on a continuous basis. This data is processed at the MCS to establish precise satellite orbits and to update each satellite with its ephemeris (orbital) data. Updated data is transmitted to each of the satellites via one of the ground antennas.

18.4.3 User segment

GPS installed on an aircraft comprises two receivers and two antennas located in a forward position on the top of the fuselage, see Figure 18.6. Antennas are typically flat devices, $7 \times 5 \times 0.75$" with a single coaxial connector. Satellites that are less than $5°$ from the horizon are rejected as an inherent feature of the antenna's design. Other design features include the ability to reject signals that are reflected, e.g. from the sea by rejecting incorrectly polarised signals. The antennas receive signals directly from whichever GPS satellites are visible, i.e. within line of sight.

GPS receivers are often incorporated into multimode receivers (MMR) along with other radio navigation systems. In this chapter we shall refer to this item simply as the 'receiver', remembering that different aircraft types will have different configurations of equipment. The receiver contains RF filters, a quartz clock (to reduce equipment costs versus atomic clocks) and a processor.

The receiver and satellite generate identical pulse coded signals at precisely the same time (Figure 18.7); these signals are compared in the receiver to provide the basis of time delay (Δt) measurements. When the time delay from the satellite has been measured, it is compared with the known position and orbit of the satellite. This calculation provides a first **line of position** (LOP). Acquiring second and third satellites provides a unique position as previously described; however, the receiver needs to take into account its clock error (bias). Since the receiver's quartz clock is not as accurate as the each satellite's atomic clock, the clock error (bias) can be anticipated in the range calculations from four satellites. The **time bias error** means that the first LOP is not the true range; the

Figure 18.6   Location of GPS antennas

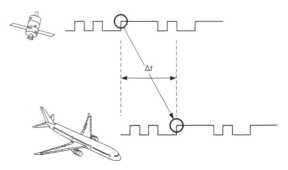

Figure 18.7   Pulse coded signals

calculated range is therefore a **pseudorange** (Figure 18.8), defined as true range ± the range associated with clock error. (Every microsecond of clock error represents a range of 300 metres.) Since the individual receiver's clock error is the same with respect to any satellite, using four satellites defines a precise and unique position as illustrated in Figure 18.9. Note that, since the satellites are in the order of 11,000 nm from the receiver, and are all in different orbits, we need to know the exact position of each satellite via its ephemeris data (transmitted as part of the message code).

Test your understanding 18.2

How many GPS satellites are there and how are they arranged into orbits?

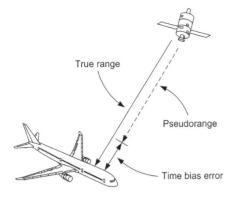

True range

Pseudorange

Time bias error

Figure 18.8   Illustration of pseudorange

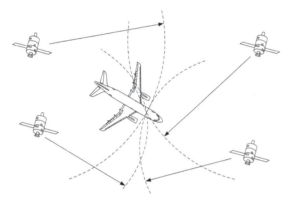

Figure 18.9   Pseudorange and position fixing with four satellites

---

Test your understanding 18.3

What is the purpose of the control segment?

Test your understanding 18.4

What is the difference between ephemeris and almanac data?

18.5 GPS signals

Each satellite transmits low power (20–50 watt) signals on two carrier frequencies: L1 (1575.42 MHz) and L2 (1227.60 MHz). Two carrier waves

are transmitted so that the effects of refraction through the ionosphere can be compared between the two signals, and corrections applied. These carrier frequencies are modulated with complex digital codes that appear like random electrical noise; these are called **pseudorandom** codes and they are a fundamental part of GPS. There are three sets of data to be modulated on the L1 and L2 carrier waves:

- Course acquisition (C/A) code
- Precise (or protected) P-code
- Navigation/system data.

The **coarse acquisition** (C/A) code is a pseudorandom string of digital data used primarily by commercial GPS receivers to determine the range of the transmitting satellite. The C/A code modulates the carrier wave at 1.023 MHz and repeats every 1 ms. The P-code (not available to civilian users) is modulated on both the L1 and L2 carriers at a frequency of 10.23 MHz. The P-code can be further encrypted as a Y-code to provide a high level of security for military users.

Data is exchanged between each satellite and the monitoring stations via uplink and downlink frequencies in the S-band (2227.5 and 1783.74 MHz respectively).

18.6 GPS operation

GPS has various levels of operation depending on how many satellites are in view. Three satellites provide a two-dimensional position fix; four satellites or more is desirable for optimum navigation performance. The receiver seeks out at least four satellites by monitoring their signal transmissions; this acquisition process takes about 15–45 seconds. To speed up the navigation process, the receiver can obtain an initial position

---

Key point

Transmission of GPS position and timing signals are sent to users in the UHF (L-band) of radar. These frequencies (1575.42 MHz and 1227.6 MHz) are designated L1 and L2.

fix from the inertial reference system; this allows the receiver to search for satellites that should be in view. In the event of poor satellite coverage for defined periods (typically less than 30 seconds) the system uses other navigation sensor inputs to enter into a dead reckoning mode. For prolonged periods of poor satellite reception, the system re-enters the acquisition mode.

18.6.1 Selective availability

Selective availability (SA) is a feature of GPS that intentionally introduces errors (typically 10 meters horizontally, and 30 meters vertically) into the publicly available L1 signals. This is a political strategy that denies any advantage for hostile forces acting against the USA. The highest GPS accuracy was available (in an encrypted form) for the US military, its allies and US government users. During the 1990s, a number of political factors were mounting in the USA:

- The shortage of military standard GPS units during the 1990s Gulf War
- The widespread availability of civilian products
- The FAA's long-term desire to replace ground navigation aids with GPS.

This led to the decision by US President Bill Clinton in 2000 allowing all users access to the L1 signal without the intentional errors.

18.6.2 GPS accuracy, errors and augmentation systems

Navigation errors can arise from poor satellite visibility or less than optimum geometry from the satellites that are visible. Accuracy of ephemeris data (i.e. each satellite's positional information) is fundamental to the accuracy of the system.

There are external effects that will affect the GPS signal, introduce errors and subsequently affect accuracy. Multipath ranging errors can be caused by reflections of the GPS signals from mountains and tall buildings. Atmospheric conditions in the ionosphere and troposphere will affect GPS signals, these errors can be predicted to a certain extent and therefore correction factors can be built in. The ionosphere will refract the satellites' signals; however, since two frequencies are transmitted (L1 and L2), the time difference between when these signals are transmitted and received can be compared, and correction factors applied.

Calculating ranges from the intersection of two range measurements (whether satellite or ground navigation aids) requires optimum geometry. If the angle between the two satellites viewed by the receiver is acute, this does not provide an accurate position fix. In satellite navigation, this is referred to as **geometric dilution of precision** (GDOP). The closer two satellites are, when viewed from the aircraft, the greater is the GDOP. This **dilution of precision** (DOP) can be broken down into specific components:

- PDOP (position DOP based on geometry only)
- HDOP (horizontal contribution to PDOP)
- VDOP (vertical contribution to DOP)
- TDOP (range equivalent of clock bias).

Almanac data within the receiver, together with ephemeris data from the satellite, is used to assist the receiver in acquiring specific satellites for optimum geometry.

The aforementioned errors are all unintentional. There is, however, the ongoing concern of intentional interference known as **spoofing**, i.e. the deliberate attempt to disrupt GPS signals. The Federal Aviation Administration (FAA) and other authorities are constantly testing the quality of GPS signals and working on ways to mitigate such threats. There are several schemes in place or proposed to improve system accuracy, integrity, and availability including:

- **Differential GPS** (DGPS) for marine users of GPS, this is maintained by the US Coast Guard
- **Wide area augmentation system** (WAAS) for aviation users, this is maintained by the FAA
- **Local area augmentation system** (LAAS) for aviation users, this is maintained by the FAA
- **European geostationary navigation overlay service** (EGNOS): this is a joint project of the European Space Agency (ESA), the European Commission (EC) and Eurocontrol.

All these augmentation systems operate on the principle of numerous ground stations in known geographical positions receiving GPS signals. Correction signals are then sent to users in a variety of ways. The **wide area augmentation system** (WAAS) was developed specifically for aviation users and is intended to enable GPS to be used in airspace that requires high integrity, availability and accuracy. WAAS improves a GPS signal accuracy of 20 metres to approximately 1.5 metres (typical) in both the horizontal and vertical dimensions. WAAS is based on a network of reference stations around the world that monitors GPS signals and compares them against the known position of the reference stations. These reference stations collect, process and transmit this data to a master station. Updated data is then sent from the master station via an uplink transmitter to one of two geostationary satellites; the aircraft receiver compares this with GPS data and messages are sent to the crew if the GPS signal is unreliable.

A further development of GPS augmentation for aircraft is the **local area augmentation system** (LAAS). This facility is located at specific airports and is intended to provide accuracy of less than one meter. Receiver stations are located in the local airport vicinity and these transmit integrity messages to the aircraft via VHF data links (VDL). The intention is for augmented GPS to gradually replace ground-based navigation aids, ultimately leading to global navigation satellite landing system (GLS) to replace the instrument landing system (ILS) for precision approaches and landings.

The GPS navigation receiver can also be installed with error detection software known as **receiver autonomous integrity monitoring** (RAIM). Monitoring is achieved by comparing the range estimates made from five satellites. In addition to this, failed satellite(s) can be excluded from the range estimates by comparing the data from six satellites. This technique is called **fault detection and exclusion** (FDE).

18.6.3 GPS airborne equipment

GPS can be used in isolation, or with other airborne systems to provide differing levels of operation. Referring to Table 18.1, the level of integration determines if the GPS can be used for oceanic, en route, terminal area or non-precision approach. In addition to position calculations, GPS can provide **derived navigation data**:

- Track (from taking several position fixes)
- Ground speed (from calculating the distance between fixes over a period of time)
- Drift angle (from the difference between heading and track).

Global navigation systems for general aviation are often integrated with ILS–VOR and VHF communication systems, see Figure 18.10. This is a self-contained panel mounted device. Text is displayed on the screen for selected frequencies, distances, bearings etc. Graphics are used to provide a multi-function type display, e.g. for navigation references, weather and traffic warnings (see Chapters 20 and 22).

Table 18.1 Classification of GPS integration

Class	Integration capability
A	GPS sensor and navigation capability (including RAIM)
B	Data sent to an integrated area navigation system, e.g. flight management system
C	Output guidance sent to an autopilot or flight director

18.7 Other GNSS

The Russian global navigation satellite system (**GLONASS**) features 24 satellites orbiting at a lower altitude of 19,100 km in three orbital planes; three satellites are in orbit as spares. The Russian defence organization owns the system and civilian usage is managed by the Russian Space Agency. At the time of writing there is limited take-up of GLONASS outside of Russia for civilian applications compared with the worldwide acceptance and usage of GPS. Several satellites have exceeded their design life thereby reducing system capability; these are being replaced on a progressive basis.

Galileo is a European system that is intended to be compatible with, but more advanced than,

GPS or GLONASS. The system is based on 30 satellites in a higher (23,000 km) orbit; the satellites form three orbital planes each comprising 10 satellites. With this higher orbit is an increased time to circle the earth; 14 hours. The ground system comprises two control centres, five monitoring and control stations and five uplink stations. The system is planned to be in operation during 2008. The European geostationary navigation overlay service (EGNOS) is the first phase of Galileo. EGNOS utilises a network of ground stations and three geostationary satellites to provide increased accuracy, integrity and reliability of any global satellite navigation system.

In addition to these systems, Japan is planning its own satellite navigation system; other nations are either joining or forming partnerships. Some novel ideas include optimising orbits such that the satellite(s) remain visible over certain areas of the globe for longer periods to obtain maximum usage. Estimates vary, but it is conceivable that over 100 navigation satellites could be in orbit over the next 20 years.

Since the two original global navigation systems were established, there remains a political debate about the deployment of additional systems. This debate is fuelled by a number of factors, e.g. national security aspects

Figure 18.10 Integrated GPS control panels

(remember that GPS was originally established as a military asset).

Another aspect to consider is the industrial advantage that comes with satellite technology; the mass market is hungry for hand-held GPS receivers and in-car satellite navigation systems. Finally, from the commercial aircraft viewpoint, if there are numerous systems in place, there is the added 'complication' of what equipment will be approved in each and every nation. The political and economic aspects of these factors are beyond the scope of this book; however, the reader is encouraged to monitor events through the press and other media.

Test your understanding 18.5

How does WAAS increase GPS integrity, availability and accuracy?

18.8 The future of GNSS

Given the above, the long-term intention of the aviation community is to rationalise the air traffic management through increased use of GNSS; this will be realised with the various augmentation systems discussed in this chapter and no doubt the additional satellite constellations.

There are programmes in place in the USA to eventually replace ground-based radio navigation aids including non-directional beacons (ADF) and en route navigation aids (VOR). DME navigation aids will be retained for a longer period, with the possibility of relocating some of these facilities. Automatic approach and landing trials are under way using satellite derived navigation references.

It is clear that any GNSS is vulnerable to disruption; this can be either a deliberate attempt to interfere with the transmissions, or as a result of atmospheric conditions. With increased dependence on GNSS, the impact of any disruption is significant. The solution to this is to have an alternative navigation system working alongside GPS as a back-up, e.g. DME–DME, inertial navigation systems, or eLoran. All of these systems are described elsewhere in this book. Operators also have the flexibility offered by area navigation, i.e. using a combination of satellite navigation and other navigation sensors (see Chapter 16). The reader is encouraged to read the industry press and monitor developments of this subject.

Key point

In the event of poor satellite coverage, the aircraft's navigation system automatically selects other navigation sensors and enters into a dead reckoning mode.

18.9 Multiple choice questions

1. Ephemeris data refers to the satellite's:
   (a) orbital position
   (b) current status
   (c) frequency of radio transmission.

2. GPS accuracy and integrity for en route operation can be increased by:
   (a) local area augmentation system (LAAS)
   (b) wide area augmentation system (WAAS)
   (c) differential GPS (DGPS).

3. The GPS orbital pattern is such that a minimum of how many satellites should be in view to a receiver?
   (a) Five
   (b) Four
   (c) Three.

4. Selective availability is a feature of GPS that:
   (a) applies correction factors to known causes of error
   (b) intentionally introduces errors
   (c) determines which users can receive signals.

5. In the event of poor satellite coverage, the system:
   (a) automatically selects other navigation sensors and enters into a dead reckoning mode
   (b) continues using the same satellites
   (c) automatically selects other satellites.

6. To speed up the satellite acquisition process, the aircraft receiver can obtain an initial position fix from the:
   (a) flight management system
   (b) internal clock
   (c) inertial reference system.

7. The deliberate attempt to disrupt GPS signals is known as:
   (a) spoofing
   (b) selective availability
   (c) satellite acquisition.

8. Fault detection is achieved by comparing the position calculations made from how many satellites?
   (a) Five
   (b) Four
   (c) Six.

9. The GPS navigation concept is based upon calculating satellite:
   (a) speed
   (b) altitude
   (c) range.

10. Multi-path reflections of GPS signals are caused by:
    (a) mountains and tall buildings
    (b) atmospheric conditions
    (c) poor satellite visibility.

11. The local area augmentation system (LAAS) provides integrity messages to the aircraft via:
    (a) geostationary satellites
    (b) VHF data links
    (c) the GPS satellites.

12. How many GPS satellites need to be in view to be able to define a unique two-dimensional position on the earth's surface?
    (a) Two
    (b) Three
    (c) One.

13. Failed satellite(s) can be excluded from the navigation calculations by comparing the data from how many satellites?
    (a) Four
    (b) Five
    (c) Six.

14. During prolonged periods of poor satellite
    reception, the aircraft receiver:
    (a) enters into a dead reckoning mode
    (b) re-enters the acquisition mode
    (c) rejects all satellite signals.

15. In the diagram shown in Figure 18.11, which
    feature represents the control segment?
    (a) A
    (b) B
    (c) C.

16. In the diagram shown in Figure 18.12, X
    represents:
    (a) the actual range
    (b) the pseudorange
    (c) the distance error.

17. GPS satellites occupy orbits at a typical
    altitude of:
    (a) 20,000 km
    (b) 120,000 km
    (c) 200,000 km.

18. GPS transmissions are in the:
    (a) C-band
    (b) L-band
    (c) X-band.

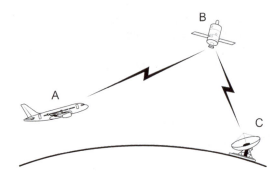

Figure 18.11   See Question 15

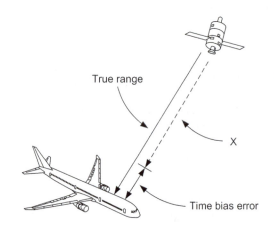

Figure 18.12   See Question 16

Flight management systems

The term 'navigation' can be applied in both the lateral and vertical senses for aircraft applications. Lateral navigation (LNAV) is effectively the area navigation function described in Chapter 16. Vertical navigation (VNAV) is concerned with optimising the performance of the aircraft to reduce operating costs. This has been traditionally achieved by the flight crew (particularly the flight engineer) making reference to data contained within charts, tables and performance manuals.

Aircraft performance data is based on a number of factors including aircraft weight, altitude and outside air temperature. Since these factors are constantly changing, the task of calculating optimum engine thrust limits, aircraft speed and altitude has gradually been automated with the advent of performance management systems. During the 1980s, lateral navigation and performance management functions were combined into a single system known as the flight management system (FMS). Various tasks previously performed by the crew can now be automated with the intention of reducing crew workload. In this chapter we will review the principles of flight management systems and explore some the key features and benefits.

19.1 FMS overview

The flight management system (FMS) combines area navigation and performance management into a single system. The two primary components of the system are the flight management computer (FMC) and control display unit (CDU). Primary aircraft interfaces with the FMC are the inertial reference system and automatic flight control system, including the autothrottle. Flight management systems were introduced at a time of rising operating costs; the contributing factors to these costs include fuel

and time. The cost of fuel is self-evident; the cost of time includes aircraft utilisation, e.g. if the aircraft is being leased on a cost per flying hour basis. Reducing aircraft speed will decrease fuel burn, but this leads to a longer flight time and increased 'cost of time'. Flying faster will reduce the cost of time but increase fuel burn.

Four-dimensional navigation is possible with flight management systems. The aircraft's latitude, longitude, altitude and arrival time requirements can be planned, calculated and subsequently predicted on an ongoing basis. Each airline will have its own financial model in terms of fuel and time costs; the FMS can be customised accordingly and expressed as a **cost-index**; this is entered into system within the range 0–100 to represent the extremes of minimum fuel through to minimum time. In order to perform the key functions of area navigation and performance management, the system interfaces with many other systems on the aircraft.

Flight management systems were the first examples of integrated multi-mode avionics. On transport category aircraft, the FMS integrates many systems including radio navigation systems, inertial navigation systems, global positioning systems, and centralised maintenance monitoring.

19.2 Flight management computer system (FMCS)

The two primary components of the system are the FMC and CDU; these are a subset of the FMS referred to as the flight management computer system (FMCS).

19.2.1 Flight management computer

The FMC contains an operational program, navigation database and performance database.

We have already come across the navigation database (NDB) in Chapter 16. The FMC's **navigation database** (see Table 19.1) is a comprehensive version of what has already been discussed in area navigation systems. The **performance database** (PDB) contains a detailed model of the aircraft's aerodynamic characteristics. This includes the aircraft's speed and altitude capabilities together with operating limits for both normal operation and abnormal conditions, e.g. engine failure. Engine parameters are also stored in the PDB, these include fuel flow and thrust models for the type of engine installed on the aircraft. Note that aircraft can be certified to fly with more than one engine type; these are all stored in the PDB.

An important feature of the FMC are the **program pins**. Rather than producing many different FMC software configurations for each aircraft type and each engine combination, one

FMC part number can be installed with software covering a number of aircraft and engine types in the PDB. The FMC (like most avionic computers) is installed in the equipment rack and connects to the wiring looms via pins/sockets at the rear of the computer.

Program pins are used to select various software options within the computer; these are connections that are made to the connector either to ground, 28 V DC power supply or not connected. Logic circuits inside the computer are thereby set into predetermined configurations depending on how the program pins are configured. For example, a program pin could be connected to ground for one engine type, and set to 28 V DC for another engine type. When the FMC is installed, it effectively recognises which engine type is installed and the relevant engine software is used. The same FMC installed on another aircraft with different engine type will recognise this via the program pin(s) and utilise the relevant engine software.

Certain functions are fixed and cannot be changed, e.g. the aircraft type/model. Other program pins are airline options; examples of these options are the use of metric or imperial units, e.g. Centigrade or Fahrenheit, pounds (×1000) or kilograms (×1000).

19.2.2 Control display unit

The CDU is the primary interface between the crew and FMC. It is designed such that data entry and displays are in the language used by ATC. The location of a CDU on a typical transport

Table 19.1 Navigation database

Content	Details
Radio navigation aids	VOR, DME, VORTAC, ADF identification codes, frequencies, locations, elevations
Waypoints	Names and locations, pre-planned within company routes
Airports and runways	Locations, ILS frequencies, runway identifiers, lengths
Standard instrument departures (SIDs)	Published departure procedures including altitude restrictions
Standard terminal arrival routes (STARs)	Published arrival procedures including altitude restrictions
En route airways	Navigation aid references, bearings, distance between navigation aids
Holding patterns	Fix point, inbound course, turn direction
Company routes	A combination of all the above, as specified by the airline

Figure 19.1 Location of FMCS control and display unit

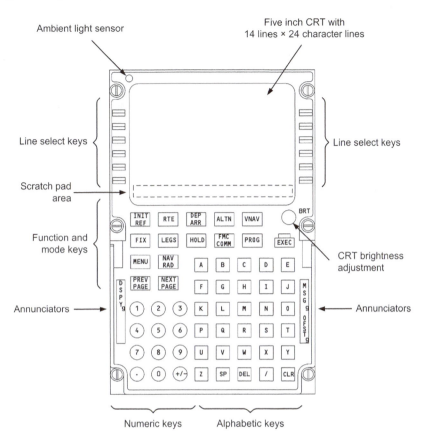

Figure 19.2   Location of FMCS control and display unit

aircraft is shown in Figure 19.1. The CDU comprises a variety of features, referring to Figure 19.2 these include the:

- data display area (typically a cathode ray tube—CRT)
- line-select keys (LSK)
- function and mode keys
- alpha-numeric key pad
- warning annunciators.

The display area is arranged in the form of chapters and pages of a book. When the system is first powered up, the CDU displays the IDENT page, see Figure 19.3.

The 'IDENT' page contains basic information as stored in the FMC including aircraft model, engine types etc. Other pages are accessed from this page on a menu basis using the line-select keys, or directly from one of the function or mode select keys.

Figure 19.3 'IDENT' page displayed on system power-up

19.3 System initialisation

Before the system can be used for lateral and vertical navigation, the FMC needs some basic initialisation data. Certain information required by the system has to be entered by the crew; other information is stored as a default and can be overwritten by the crew. To simplify the process, information to be entered by the crew is displayed in **box prompts**. Information displayed as default information is displayed as **dash prompts**. There are a number of ways that individual pages can be accessed, and there is a variety of information on each page. The description below illustrates an initialisation procedure, starting with position initialisation through to performance initialisation, following a logical process. During this initialisation description, we will be making reference to **fields**; these are specific areas on the CDU screen where data is either displayed and/or entered. In the following text, each **line-select key** (LSK) will be referred to by its location: left/right of the display and 1–6 from top to bottom.

19.3.1 Position initialisation

When the system is powered up, the IDENT page is displayed, see Figure 19.4 (more details about this page are provided after this section). Pressing LSK-6R, identified POS INIT for **position**

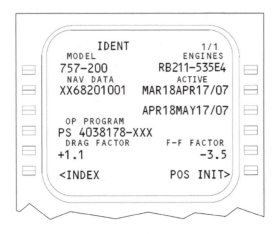

Figure 19.4 'IDENT' page—the system is prompting the selection of 'position initialisation' (POS INIT on LSK-6R)

initialisation, displays the POS INIT page, see Figure 19.4. The information needed at this point (indicated by box prompts) is present position for inertial reference system (IRS) alignment. Position can be entered in a number of ways, but let's assume at this point that we want to load present position by manually keying in latitude and longitude. Using the alphanumeric keys, latitude and longitude are entered via the key pad, entries appear in the bottom of the display (referred to as the **scratch pad**). When this data is confirmed in the scratch pad, LSK-4R (adjacent to the position boxes) is pressed and the scratch pad data replaces the 'Set IRS position' boxes. Present position is automatically transferred to the IRS and the next stage of initialisation is prompted by LSK-6R; this leads to the next section of initialisation for the desired ROUTE.

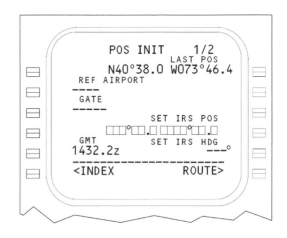

Figure 19.5 Position initialisation ('POS INIT') page

19.3.2 Route selection

The ROUTE page requires that an origin and destination be entered; these are entered (via the scratch pad) to replace the box prompts adjacent to LSK-1L (origin) and LSK-1R (destination), see figure 19.6(a). Origins and destinations are identified using the International Civil Aviation Organisation (ICAO) four-letter codes, e.g. London Heathrow is EGLL, New York Kennedy international airport is coded KJFK. This system is used in preference to the International Air Transport Association (IATA) three-letter codes,

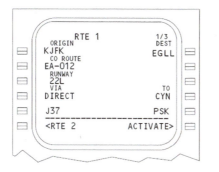

(a) Route page – origin and destination entered

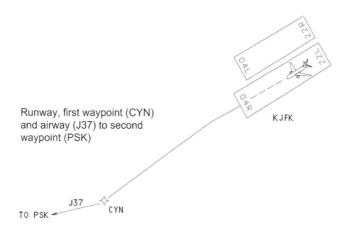

Runway, first waypoint (CYN)
and airway (J37) to second
waypoint (PSK)

(b) Route page – departure details

Figure 19.6 Route page details

e.g. LHR and JFK, since some of these three-letter codes are duplicated for some airfields. Note that most airlines have predetermined company routes, these are stored in the navigation database and can be entered (as a code) via LSK-2L. There may be more than one route between the origin and destination; when the company route code is entered into an appropriate field, this will automatically enter the origin and destination together with all en route waypoints. Specific departure details, e.g. runway and initial departure fix, can also be contained within the company route as illustrated in Figure 19.6(b). Once the route is activated (LSK-6R), the bottom

right field changes to PERF INT for **performance initialisation**.

19.3.3 Performance initialisation

The system requires gross weight (GW) or zero fuel weight (ZFW), reserve fuel, cost index and cruise altitude. Required entries are indicated as before with box prompts, see Figure 19.7. Note that since the total fuel onboard (52.3 tonnes in this example) is known by the FMC (via an input from the fuel quantity system) entering ZFW will automatically calculate GW and vice versa. Cost index can be entered manually, or it may be contained within the company route. All other entries on the page are optional; entry of data in these fields will enhance system performance. Once the performance initialisation details are confirmed, the system is ready for operation. Further refinement of the flight profile can be made by entering other details, e.g. take-off settings, standard instrument departures, wind forecasts etc.

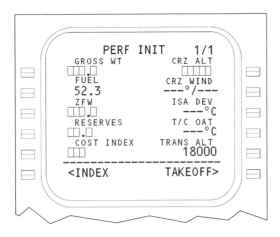

Figure 19.7 Performance initialisation ('PERF INIT') page

19.4 FMCS operation

The flight management computer system (FMCS) calculates key performance data and makes predictions for optimum operation of the aircraft based on the cost-index. We have already reviewed the system initialisation process, and this will have given the reader an appreciation of how data is entered and displayed. The detailed operation of a flight management system is beyond the scope of this book; however, the key features and benefits of the system will be reviewed via some typical CDU pages. Note that these are described in general terms; aircraft types vary and updated systems are introduced on a periodic basis. CDU pages can be accessed at any time as required by the crew; some pages can be accessed via the line-select keys as described in section 19.3; some pages are accessed via function/mode keys. The observant reader may

> **Key point**
>
> The FMS comprises the following subsystems: FMCS, AFCS and IRS.

> **Key point**
>
> The page automatically displayed upon FMC power-up is the identification page; this confirms that the FMC has passed a sequence of self-tests.

> **Key point**
>
> Required entries into the CDU are indicated by box prompts; optional entries are indicated by dashed line prompts.

> **Key point**
>
> To define the destination airport on the FMC route page requires entry of the airfield's four-character identifier.

have already noticed that in the top right of each CDU page is an indication of how many sub-pages are available per selected function.

19.4.1 Identification page ('IDENT')

This page is automatically displayed upon power-up; aside from displaying a familiar page each time the system is used, this also serves as confirmation that the FMC has passed a sequence of built-in test equipment (BITE) self-tests including: memory device checks, interface checks, program pin configuration, power supplies, software configurations and microprocessor operation. Information displayed on this page includes aircraft and engine types, navigation database references and the operational program number. By reference to the relevant aircraft documentation, one FMC part number could be fitted to a number of different aircraft types. Each aircraft type will have different aerodynamic characteristics and these differences will be stored in the FMC's memory. The FMC recognises specific aircraft types by program pins contained within the aircraft connector, see Figure 19.8. Given aircraft types can be operated with different engine models; these are recognised by using specific program pins. Furthermore, airlines have the option on the units used within the system, e.g. temperature in Centigrade or Fahrenheit, weights in kilograms or pounds etc. These are also determined by program pins.

The navigation database (NDB) is identified by when it becomes effective, and when it expires. Referring to Figure 19.4, the active (current) database is adjacent to LSK-2R. The updated database is adjacent to LSK-3R; this is selected on the changeover date (April 18 in this illustration). A comprehensive range of information is contained in the NDB as detailed in Table 19.1; note that this is an indicative list since databases are usually customised for individual airlines. The synergy of integrated avionic systems can be demonstrated by FMC database information also being displayed on the EHSI (Figure 19.9 is displaying a number of airports contained in the database).

19.4.2 Progress page

There are many pages available to the crew for managing and modifying data required by the system depending on circumstances. One of the pages used to monitor key flight information is the progress page, see Figure 19.10. By describing the information on the progress pages, the reader will gain an appreciation of the features and benefits of the flight management computer system.

The progress page can be accessed via the PROG key on the CDU. There are no entries required on this page; it is used for information only. The top line of the page displays details for the previous waypoint (CYN) in the active flight plan; name, altitude, actual time of arrival and fuel. The next three lines display details for the active waypoint (ENO), next waypoint (GVE) and final destination (KATL). Details include:

Figure 19.8 Program pins located in the computer's connector

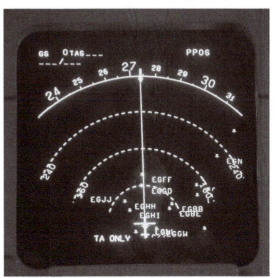

Figure 19.9 Airports within the navigation database displayed on the EHSI

Table 19.2 Typical CDU pages

Page title	Full title	Purpose of page
IDENT	Identification	Verifies aircraft model, active database, operational program number, engine type(s)
POS INIT	Position initialisation	Present position required by entering data using one of three methods: latitude/longitude coordinates via the keypad, line selection of last position, line selection of departure gate coordinates
RTE	Route	Entry of route details, either by company route code, or manual construction
CLB	Climb	Selection of desired climb mode, e.g. economy, maximum rate, maximum angle, selected speed, engine out
CRZ	Cruise	Selection of desired cruise mode, e.g. economy, long-range cruise, engine out, selected speed
DES	Descent	Selection of desired descent mode, e.g. economy, selected speed
DIR INTC	Direct intercept	Used to select a waypoint that will be flown directly towards from the present position
RTE LEGS	Route legs	Used for confirming and modifying en route details, e.g. waypoint identification, course and distance to waypoints, speed and altitude constraints (see Figure 19.11)
DEP ARR	Departure/Arrival	Provides access to the navigation database for origin or destination SIDS, STARS and specific runways
RTE HOLD	Holding pattern	Review or revision of holding pattern details, e.g. fix point, turn direction, inbound course, leg time and target speed
PROG	Progress	In-flight status of progress along route (see separate notes provided)
N1 LIMIT	N1 limit	The N1 limit is automatically selected and controlled by the FMC. This page provides a range of manually selected N1 limit options including go-around, maximum continuous, climb and cruise
FIX INFO	Fix	Used to create fix points on the current flight leg from known waypoints using radials and distances from the waypoint

distance to go (DTG), estimated time of arrival (ETA) and predicted fuel. The fifth line gives selected speed, predicted time and distance to an altitude change point, e.g. top of climb (T/C) as illustrated in Figure 19.10. The last line of the page is providing navigation source information. In this case, the FMC selected inertial reference system (number 3) is being updated by two DME navigation aids ENO and MLC; these are being tuned manually and automatically as indicted by the letters M and A next to the navigation aid identifier.

The second progress page contains a variety of useful information, e.g. wind speed and direction (displayed with associated head, tail and cross wind components), cross track (XTK) error,

vertical track (VTK) error, true airspeed (TAS), static air temperature (SAT) and various fuel quantity indications.

19.4.3 Legs page

Figure 19.11 provides an illustration of how en route lateral and vertical profiles are integrated within the FMC. In this example, the aircraft is flying towards waypoint CYN on a track of 312°. There is an altitude constraint of 6000 feet over CYN, climb speed is 250 knots. This combined lateral and vertical profile is depicted by the tracks, distances, speeds and altitudes for each waypoint. This level of detail also applies for **standard instrument departures** (SIDs) and

(a) Progress page (page 1/2)

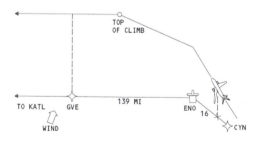

(b) Progress page (page 2/2)

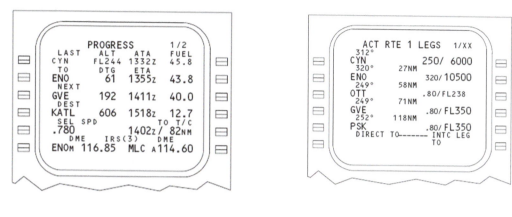

(a) Legs page

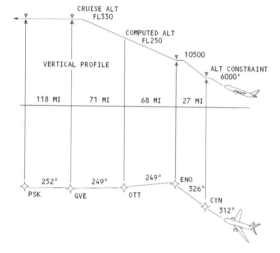

(b) Lateral and vertical profiles

Figure 19.11 Legs page and associated flight profiles

(c) Flight profile

Figure 19.10 FMCS progress pages and flight profile

standard terminal arrival procedures (STARs), see Figures 19.12 and 19.13.

19.4.4 Other CDU pages

A detailed review of every page available on the CDU is beyond the scope of this book; however, a summary of typical pages is provided in Table 19.2. Note that this table is provided for illustration purposes. Aircraft types vary together with the type and model of FMC installed.

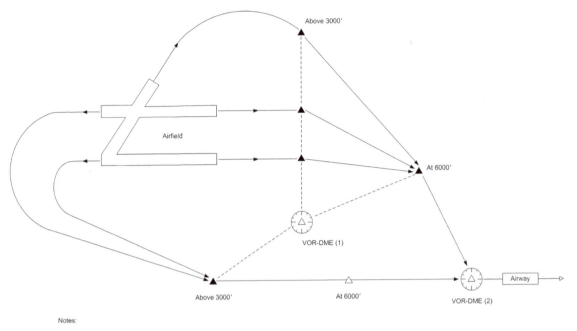

Figure 19.12   Standard instrument departure (SID)

Key point

The highlighted waypoint on the progress page is the active waypoint.

Key point

Alerting messages require attention from the crew before guided flight can be continued.

19.5 FMS summary

As we have seen, the FMCS performs all the calculations and predictions required to determine the most economical flight profile, either for minimum fuel, or minimum time (or indeed some point in between depending on the operator's financial and commercial models).

When coupled to the automatic flight control system, with vertical and lateral navigation modes engaged, the flight crew act as managers monitoring and entering data as required. Much of the data presented on the CDU is also displayed on the primary flying displays; aircraft with electronic flight instruments have the advantage in that information is displayed with coloured symbols to identify key features of the flight plan, e.g. navigation aids, airfields and descent points.

Test your understanding 19.1

(a) What is the meaning of four-dimensional navigation?

(b) How can you confirm that the FMC has passed its power-up test?

(c) What is the significance of box and dash prompts on the CDU?

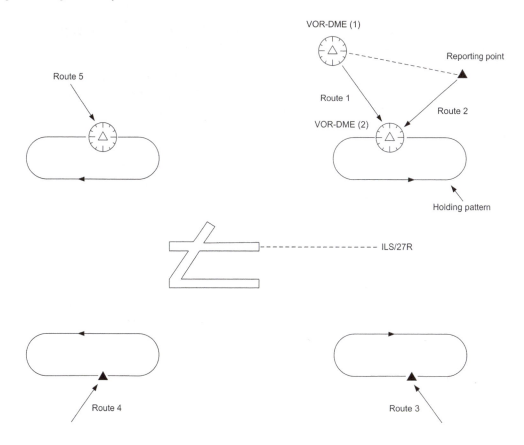

Notes:

1. In this illustration, each of the three arrival routes is associated with a navigation aid (VOR-DME) and reporting point (solid triangles)
2. Each arrival route is normally allocated a holding pattern
3. Minimum sector altitudes are published for each route
4. When cleared by ATC, the aircraft would leave the holding pattern and be given a heading to join the ILS for the active runway, e.g. 27R

Figure 19.13 Standard terminal arrival routing (STAR)

Test your understanding 19.2

What is the purpose of program pins?

Test your understanding 19.3

Where would you confirm details of each of the following: navigation database, operational program, aircraft and engine type?

19.6 Multiple choice questions

1. To define the destination airport on the FMC route page requires entry of the airfield's:
   (a) three-character identifier
   (b) four-character identifier
   (c) latitude and longitude.

2. The page automatically displayed upon FMC power-up is the:
   (a) identification page
   (b) navigation database
   (c) position initialisation page.

3. Program pins are defined by the:
   (a) FMC operational program
   (b) navigation database
   (c) aircraft wiring at the FMC connector.

4. Information entered into the CDU scratch pad
   is displayed in the:
   (a) lowest section of the display
   (b) box prompts
   (c) dash prompts.

5. Minimum flight time would be achieved with
   a cost-index of:
   (a) zero
   (b) 100
   (c) 50.

6. Aircraft and engine type can be confirmed on
   the:
   (a) progress page
   (b) identification page
   (c) position initialisation page.

7. The use of metric/imperial units is determined
   by:
   (a) the part number of the FMC
   (b) program pins
   (c) dashed line entries.

8. Required entries into the CDU are indicated
   by:
   (a) box prompts
   (b) dashed lines
   (c) highlighted text.

9. 'Not in database' is an example of an:
   (a) alert message
   (b) advisory message
   (c) active waypoint.

10. Display of the identification page after power-
    up confirms the:
    (a) IRS is aligned
    (b) navigation sources in use
    (c) FMC has passed its BITE check.

11. The FMC recognises specific aircraft types
    by:
    (a) CDU entry
    (b) program pins
    (c) the navigation database.

12. SIDs in the navigation database refer to:
    (a) arrivals
    (b) en route navigation
    (c) departures.

13. The EXEC key lights up when:
    (a) data is entered for initialisation/changes
    (b) advisory messages are displayed
    (c) incorrect data has been entered.

14. Alerting messages require attention from the
    crew:
    (a) before guided flight can be continued
    (b) when time is available
    (c) at the completion of the flight.

15. The highlighted waypoint on the progress
    page is the:
    (a) previous waypoint
    (b) active waypoint
    (c) destination.

Chapter 20   Weather radar

Weather radar (WXR) was introduced onto passenger aircraft during the 1950s for pilots to identify weather conditions and subsequently reroute around these conditions for the safety and comfort of passengers. Extreme weather conditions are a major threat to the safe operation of an aircraft. Approximately 33% of accidents are weather related; flight crews need to be aware of these conditions and understand the consequences. In the age of digital data communications, aircraft systems, e.g. aircraft communication addressing and reporting system (ACARS), can receive and transmit information about prevailing weather conditions. The onboard weather radar, however, provides the crew with their main source of identifying extreme weather conditions. A secondary use of weather radar is a terrain-mapping mode that allows the pilot to identify features of the ground, e.g. rivers, coastlines and mountains. Various features are being added to weather radar systems to provide many benefits including enhanced displays and improved turbulence detection. In this chapter we will review some basic radar principles, and examine the principles of weather radar including the detection of severe turbulence and lightning.

20.1 System overview

The word radar is derived from **ra**dio **d**etection **a**nd **r**anging; the initial use of radar was to locate aircraft and display their range and bearing on a monitor (either ground based or in another aircraft). This type of radar is termed **primary radar**; energy is directed via an antenna to a 'target'; this target could be an aircraft, the ground or specific weather conditions. In the case of weather radar, we want to detect the energy reflected back from the contents of a cloud, or from **precipitation**, see Figure 20.1. The latter may defined as the result of water vapour condensing in the atmosphere that subsequently

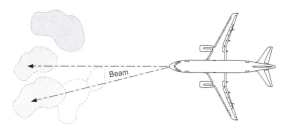

(a) Microwave energy directed via radar beam

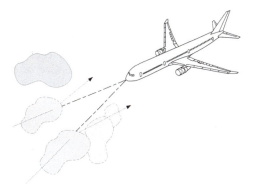

(b) Reflected energy from the contents of a cloud

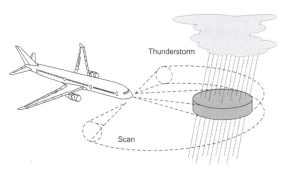

(c) Reflected energy from precipitation

Figure 20.1   Weather radar principles

falls to the earth's surface. Precipitation can occur in many different forms including: rain, freezing rain, snow, sleet, and hail. Weather radar operates either in the C-band (4–8GHz) or X-band (8–12.5 GHz); these two bands have their advantages and disadvantages for use in weather radar applications. C-band **microwave** energy pulses can penetrate through heavy precipitation, thereby providing **weather detection**, enabling the pilot to determine more details of the weather pattern. X-band microwave energy pulses can provide good resolution of images; however, this means that they can only be used for **weather avoidance**. Higher frequencies require a smaller antenna; for this reason, larger passenger aircraft use X-band radar.

The range of a weather radar system is typically 320 miles. Microwave energy pulses are reflected from the moisture droplets and returned to the radar antenna. The system calculates the time taken for the energy pulses to be returned; this is displayed as an image on a dedicated weather radar screen, or the image can be integrated with the electronic flight display system. The strength of the returned energy is measured and used to determine the size of the target. Higher moisture content in a cloud provides higher returned energy. The antenna is **scanned** in the lateral plane to provide directional information about the target.

20.2 Airborne equipment

The typical weather radar system comprises one antenna in the nose cone, two transceivers in the equipment bay, two control panels and two displays in the flight deck, see Figure 20.2.

20.2.1 Antenna

Microwave signals are transmitted and received via the antenna. Early versions of the antenna were in the form of a parabolic dish; however, the current versions are the planar array flat-plate type, see Figure 20.3(a). The flat-plate antenna projects a more focused beam than the parabolic type; this is due to the reduction in side-lobes as illustrated in Figure 20.4. The antenna comprises a flat steerable plate with a large number of radiating slots, each equivalent to a half-wave dipole fed in phase. The antenna is mounted on the forward pressure bulkhead behind the radome;

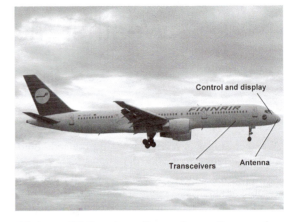

Figure 20.2   Location of weather radar equipment

Figure 20.3(a)   Weather radar antenna

Figure 20.3(b)   Antenna waveguide

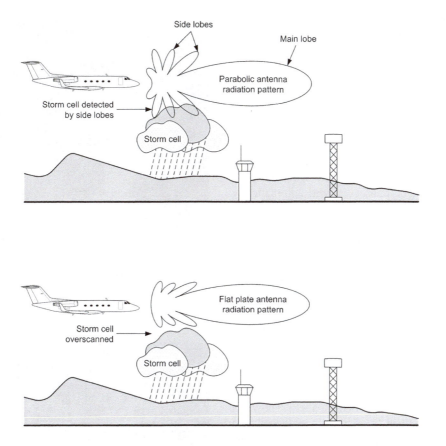

Figure 20.4 Parabolic and flat-plate antenna radiation patterns

this is a streamlined piece of structure constructed of materials that have **low attenuation** of the radar signals. The mechanical condition of the radome is very important to the effectiveness of the weather radar system, e.g. de-lamination will affect signal attenuation.

The antenna automatically traverses from left to right on a repetitive basis to be able to **scan** the weather patterns ahead of the aircraft. To investigate cloud formations, the pilot can also tilt the antenna up or down to provided different viewing perspectives. The reference position is to scan the antenna so as to provide images across the horizon; inputs from the aircraft's attitude reference system are used to provide the stabilisation. Motors are used as part of a drive mechanism to traverse the antenna in azimuth and to tilt the antenna in pitch. Synchro transmitters are used to relay the various positions of the antenna back to the transceiver.

Energy pulses are carried between the antenna and transceiver via a **waveguide**, see Figure 20.3 (b). This is because losses in a coaxial cable would be high at frequencies above 3 GHz, and prohibitive at frequencies above 10 GHz. Coaxial cables are also limited in terms of the peak power handling capability. Waveguides have their disadvantages; they are bulky, expensive and require more maintenance. Manufactured from aluminium alloy, in a hollow rectangular form, they have dimensions closely matched to the wavelength of the system. Chapter 2, Section 2.12 provides more details on waveguides.

Key point

The energy radiated from a weather radar system is hazardous and could cause injury.

20.2.2 Transceiver

The transceiver is a combined transmitter and receiver, Figure 20.5; antenna power output is in the order of 5–10 kW. Modern transceivers are solid-state devices, incorporating video processing for the display and stabilisation signals for the antenna. Since the energy received from a given size of water droplet varies with range, the energy returns from closer ranges will be higher than those received from droplets further away. The transceiver will automatically compensate for returns from targets that are near or far from the aircraft. This is achieved by altering the gain as a function of time from when the energy pulse is transmitted. Pulses of radar energy are transmitted on a repetitive basis; the interval between pulses depends on the range selected by the crew. Time has to be allowed for the energy pulse to be reflected from water droplets at the limit of the selected range before the next pulse is transmitted.

20.2.3 Control panel

A typical weather radar control panel is shown in Figures 20.6(a) and (b). This allows the pilot to select the left or right transceiver, select the weather radar mode, manually tilt the antenna and select the gain of the system.

20.2.4 Display

The basic display used for primary radar systems is the **plan-position-indicator** (PPI). As the beam sweeps from side to side, a radial image on the display (synchronised with each sweep) moves across the display. The image on the display depends on the amount of energy returned from the target. Original weather radar systems had dedicated monochrome displays based on a cathode ray tube (CRT); these have evolved over the years into full colour displays, often integrated with other electronic flight instruments. The full benefits of a weather radar system can be appreciated when the system is used on an aircraft with an electronic flight instrument system (EFIS) display, Figure 20.7. A symbol generator is used to provide specific weather radar images as determined by the transceiver. An electronic display control panel

Figure 20.5 Weather radar transceiver

Figure 20.6(a) Location of weather radar control panel

Figure 20.6(b) Typical weather radar control panel

allows each pilot to select the range of weather radar in increments of 10, 20, 40, 80, 160 and 320 miles.

The electronic display is overlaid onto the map mode allowing the pilot to relate the aircraft's heading with the weather images. These images are colour coded to allow the pilot to assess the severity of weather conditions. Colours (ranging from black, green, yellow, red and magenta) are used to indicate rainfall rates that can be interpreted as a level of turbulence.

20.3 Precipitation and turbulence

For a more detailed understanding of weather radars, factors that affect precipitation, turbulence and the formation of clouds are now considered.

20.3.1 Cloud formation

For the comfort and safety of the passengers, we want the weather radar system to detect the turbulence resulting from precipitation that leads to severe weather conditions, i.e. thunderstorms, such that these can be avoided if possible. Precipitation may defined as the result of water vapour **condensing** in the atmosphere that subsequently falls to the earth's surface. This can occur in many different forms including: rain, freezing rain, snow, sleet, and hail. Clouds are the visible accumulation of particles of water and/or ice in the atmosphere; their formation changes on a continuous basis, often resulting in no more than a 'light shower' of rain. Under certain atmospheric conditions, clouds become **large** and **unstable** leading to hazardous flying conditions. The flight crew needs to have accurate and up to date forecasts of en route weather conditions; this includes details of cloud classifications as detailed in Table 20.1.

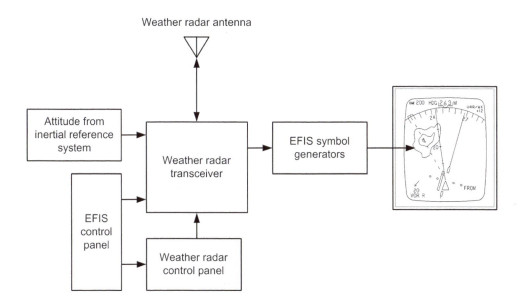

Figure 20.7 Weather radar EFIS display

Table 20.1 Classification of clouds

Name	Base (AMSL*)	Appearance
Cirrus (Ci)	> 20,000 feet	Fibrous and detached, mainly ice crystals
Cirrocumulus (Cc)	> 20,000 feet	Thin layers or patches without shading
Cirrostratus (Cs)	> 20,000 feet	Transparent, whitish veil that can cover the sky
Altocumulus (Ac)	> 6,500 feet	Patchy groups of white/grey layers
Altostratus (As)	> 6,500 feet	Greyish/blue fibrous sheets
Nimbostratus (Ns)	< 6,500 feet	Dark grey layers, covering the sky, hiding the sun
Stratocumulus (Sc)	< 6,500 feet	Grey/white patches with dark rounded features
Stratus (St)	< 6,500 feet	Grey layers, uniform base
Cumulus (Cu)	< 6,500 feet	Dense, vertical shapes developing in mounds
Cumulonimbus (Cb)	< 6,500 feet	Heavy, dense and towering. Upper portion fibrous

* Above mean sea level

Within the above classification, precipitation varies with each cloud type (see Figure 20.8):

- **Altocumulus**: precipitation does not actually reach the ground
- **Altostratus**: precipitation is in the form of rain or snow
- **Nimbostratus**: the cloud base is diffuse with continuous rain or snow
- **Stratocumulus**: light rain, drizzle or snow
- **Stratus**: drizzle
- **Cumulus**: rain or snow showers
- **Cumulonimbus**: lightning, thunder, hail. Associated heavy showers of rain/snow.

It can be seen from the above that **cumulonimbus** formations present the greatest hazard to aircraft, and maximum discomfort for passengers. We need to understand the nature of **thunderstorms** that contain heavy rainfall and turbulence.

20.3.2 Thunderstorms

Three conditions are needed to create thunderstorms: instability within the atmosphere, high moisture content and a catalyst to start the air rising. Air can be forced to rise in the atmosphere from a number of causes as illustrated in Figure 20.9:

- **Frontal**: when opposing warm and cold air masses combine
- **Convective**: the ground being heated by the sun
- **Orographic**: movement of air over the terrain.

Referring to Figure 20.10, the life cycle of a thunderstorm develops in three stages. During the first towering **cumulus stage**, warm, moist air containing water vapour rises up to higher altitudes. When the dew point is reached it cools down and the moisture content condenses into water droplets thereby creating clouds. As a result of the condensation process, latent heat is released causing the air to become warmer and drier and thereby less dense. This air rises as an **updraught** over one or two miles diameter due to convection. At this stage, air is drawn into the cell horizontally at all levels causing the updraught to become stronger with altitude. Water vapour is carried up to higher altitudes where it combines to form larger water droplets. This first stage of a thunderstorm develops over approximately 20 minutes.

When the water droplets are sufficiently large enough, they are too heavy to be supported by the updraught, and are released as rainfall. As the water droplets fall, they draw in the

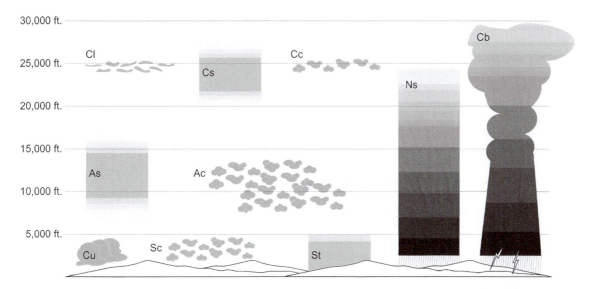

Figure 20.8   Classification of clouds

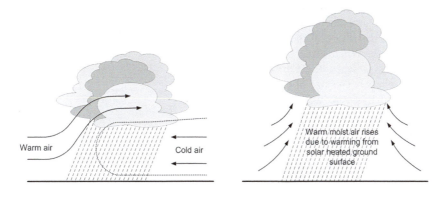

(a) Frontal   (b) Convective

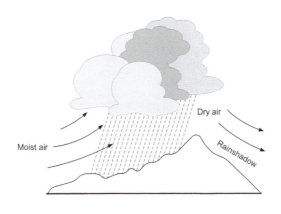

(c) Orographic

Figure 20.9   Causes of rising air in the atmosphere

surrounding air causing a **downdraught**. The air temperature of the downdraught is cold compared with the updraught; heavy rain falls from the base of the cloud. The updraught continues to carry the remaining water droplets up to very high altitude; this can be up to 50,000 feet in the tropics (well into the troposphere). At these altitudes, strong winds in the upper atmosphere (jet streams) carry the top of the cloud away to form the characteristic 'anvil' shape. This is the **mature stage** of the thunderstorm, and can last up to 40 minutes.

These two air masses are now moving in opposite directions forming localised cells of air movement; the relative movement is known as **wind shear**, and is the basis of **turbulence**. This wind shear is characterised by the relative speed and direction of the two air movements. The amount of increase in temperature and size of air mass being heated determines the rate at which wind shear, and hence turbulence, occurs. These factors will determine the nature of the boundary between the air masses. When these boundaries are very distinct, this is when the rainfall and turbulence is most severe. The size of droplets in thunderstorms and other severe weather is large compared with general rainfall; this produces much larger radar returns as described in Section 20.3.3.

In addition to the upward and downward movement of air within the cell of a thunderstorm, droplets also travel laterally depending on prevailing conditions. New cells are formed at the edges of the cloud; they may not contain rain but they often create turbulence.

The **final stage** of the thunderstorm is when the updraughts weaken thereby reducing the supply of warm, moist air containing water vapour. Downdraughts continue over a broader base, with less intensity and the cycle begins to dissipate. Temperatures within the cloud balance with the surrounding air, and the once towering cumulonimbus formation collapses.

20.3.3 Detection of water droplets

The antenna scans forward and to each side of the aircraft with a conical, or **pencil** beam of microwave energy, see Figure 20.11. When the weather radar energy pulses reach a water droplet,

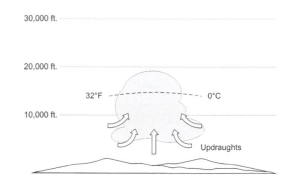

(a) Towering cumulus stage

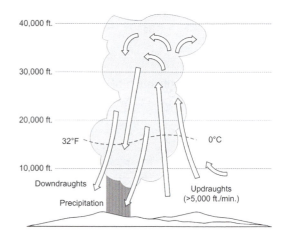

(b) Mature stage

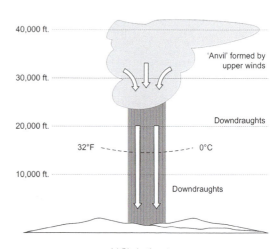

(c) Dissipating stage

Figure 20.10   Stages in the life cycle of a thunderstorm

the energy is absorbed, refracted or reflected from the front convex portion back to the antenna, as illustrated in Figure 20.12. Note that the water droplet is assumed to be spherical; it is unlikely to be a perfect sphere within a thunderstorm; however, the principles illustrated in Figure 20.12 are good approximations. The water droplet diameter affects the amount of energy returned to the antenna. With larger droplets, i.e. heavier rainfall, more energy is reflected back to the antenna. Smaller droplets from cloud and fog return significantly lower reflections. This relationship between individual size, rainfall rate and reflected energy is the basis of detecting the severity of the storm. Weather radar wavelengths (25 mm at 12 GHz) are larger than any water droplet sizes (refer to Table 20.2). The scattering of electromagnetic radiation by particles such as raindrops, i.e. particles smaller than the wavelength of the radiated energy is characterised by a phenomenon called **Rayleigh scattering**. This is named after John William Strutt (Lord Rayleigh 1842–1919), an English physicist. The intensity of the returns is given by the Rayleigh scattering equation; in this equation, the intensity of the returns varies as the sixth power of the droplet diameter. This means that returns from water vapour within fog or clouds is small compared with raindrops.

20.3.4 Hailstones

Hailstones are formed in cumulonimbus clouds and, in addition to the turbulence, they will cause physical damage to the aircraft. They are formed in strong thunderstorms with significant updraughts as irregular lumps (or pellets) of ice; this occurs when **super-cooled water** makes contact with an object such as dust or ice particles. Super-cooled water exists at temperatures below freezing point; this is because water needs a nuclei to form ice crystals. The water freezes on the surface of the nuclei, and then grows in size by forming layers. When the hailstone is too heavy to be supported in the updraught of the storm it will fall out of the cloud. Referring to Table 20.2, the sizes of hailstones are starting to approach the wavelength of the radar energy (25 mm at 12 GHz); hailstones have been recorded up to 60 mm in

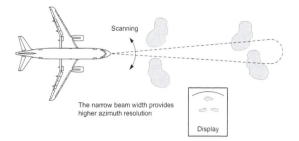

Figure 20.11 Pencil beam and azimuth resolution

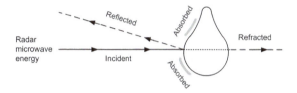

Figure 20.12 Effect of microwave energy on a water droplet

Key point

Microwave energy pulses from the weather radar are reflected from the moisture droplets and returned to the radar antenna.

Key point

X-band (8–12.5 GHz) energy pulses provide good resolution of images for the purposes of weather avoidance; they require a smaller antenna compared with C-band radar (4 to 8 GHz).

Key point

The weather radar antenna automatically traverses from left to right on a repetitive basis to be able to scan the weather patterns ahead of the aircraft.

---

diameter. When the wavelength of the radar energy is smaller than the target diameter, the hailstone acts as a lens, focussing energy onto the concave internal surface and reflecting higher levels (compared with water droplets) of energy back to the antenna.

20.3.5 Turbulence

Turbulence can be inferred from the measurement of precipitation, both in terms of the type of precipitation (rain or hail), droplet size and precipitation rate. Approximate dimensions (given as their diameter) of various weather targets are given in Table 20.2.

20.3.6 Predictive wind shear

Wind shear can occur in both the vertical and horizontal directions; this is particularly hazardous to aircraft during take-off and landing. Specific weather conditions known as **microbursts** cause short-lived, rapid air movements from clouds towards the ground. When the air from the microburst reaches the ground it spreads in all directions, this has an effect on the aircraft depending on its relative position to the microburst. Referring to figure 20.13, when approaching the microburst, it creates an **increase** in headwind causing a temporary increase of airspeed and lift for an aircraft approaching the cloud; if the pilot were unaware of the condition creating the increased airspeed, the normal reaction would be to reduce power. When flying through the microburst, the

Table 20.2 Weather radar targets and approximate dimensions

Target	Diameter
Hailstones (large)	>20 mm
Hailstones (medium)	>7.5 mm
Snowflakes (large)	>10 mm
Snowflakes (medium)	>5 mm
Rain (droplets)	<5 mm
Rain (drizzle)	0.5 mm

aircraft is subjected to a downdraught. As the aircraft exits the microburst, the downdraught now becomes a tail wind, thereby **reducing** airspeed and lift. This complete sequence of events happens very quickly, and could lead to a sudden loss of airspeed and altitude. In the take-off and climb-out phase of flight, an aircraft is flying just above stall speed; wind shear is a severe threat. During approach and landing, engine thrust will be low; if a microburst is encountered, the crew will have to react very quickly to recognise and compensate for these conditions.

Modern weather radar systems are able to detect the **horizontal** movement of droplets using Doppler shift techniques. Doppler is usually associated with self-contained navigation systems, and this subject is described in a separate chapter. The Doppler effect can be summarised here as: '...the frequency of a wave apparently changes as its source moves closer to, or farther away from an observer'. This feature allows wind shear created by microbursts to be detected.

Referring to Figure 20.14, the microwave energy pulses from the antenna are reflected by the water droplets as in the conventional weather radar system. Using the Doppler shift principle, the frequency of energy pulse returned by droplets (B) moving toward the aircraft will be at a higher frequency than the transmitted frequency. The frequency of energy pulse returned by droplets moving away from the aircraft (A) will be at a lower frequency than the transmitted frequency. These Doppler shifts in frequency are used to determine the direction and velocity of the air movement resulting from a microburst.

Visual and audible warnings of wind shear conditions are provided to the crew. The visual warnings are given on the weather radar and navigation displays using a wind shear icon and message, together with warning lights on the glare shield. Audible warnings are provided as computer generated voice alerts over the cockpit speakers, typically 'wind shear ahead'. The system automatically configures itself for the phase of flight; it is normally inhibited below 50 feet radio altitude during take-off and landing. During an approach, the system is activated below 2500 feet radio altitude.

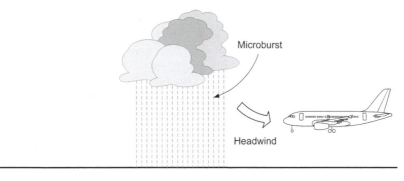

(a) An aircraft entering a microburst encounters headwinds that increase airspeed and lift

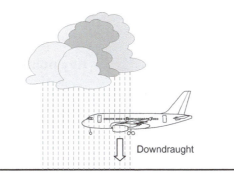

(b) The aircraft flies through the headwind and encounters a downdraught

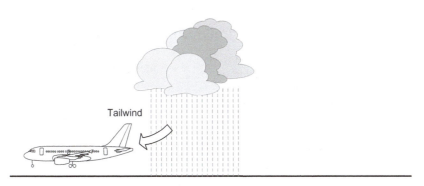

(c) The aircraft exits the microburst and encounters a tailwind, reducing airspeed and lift

Figure 20.13 Microburst and associated air movements

Test your understanding 20.1

In which bands of radar frequencies does weather radar operate?

Test your understanding 20.2

Explain what happens when radar energy reaches a water droplet.

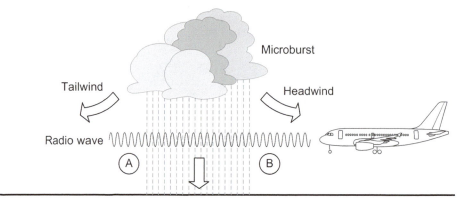

Figure 20.14   Doppler frequency shift between A and B indicates a wind shear condition

20.3.7 Terrain mapping

A secondary use of the weather radar system is for terrain mapping, e.g. identifying rivers, coastlines and mountains. This mode of operation is selected by the crew on the control panel. Returns from the various ground features are different, just as they are for precipitation. These variations are interpreted by the system and displayed using various colours. Since the energy of the return signal depends on the reflectivity of the terrain and angle at which the beam meets with the terrain, the gain control is used to provide the optimum display. If the gain is too low or too high, the images will be unclear.

The weather radar pencil beam is not suitable for terrain mapping since it does not cover a sufficiently large area. A fan-shaped beam (Figure 20.15) provides optimum coverage of the terrain along the intended track of the aircraft. The system can have the facility to reshape the beam depending on the selected mode. Modern systems achieve terrain mapping by sweeping the pencil beam at incremental vertical angles of the antenna to build up the overall display.

Key point

During a thunderstorm the effects of turbulence can be experienced up to 10 nm from the centre of the storm cell; in extreme cases this can be up to 20 nm.

Key point

Thunderstorms develop within cumulonimbus cloud formations and present severe hazards to aircraft:

- Reduced visibility from ground level up to 50,000 feet

- Turbulence and wind shear causing handling problems

- Hailstones up to 60 mm in diameter causing structural damage

- Lightning strikes causing, inter alia, structural damage

- Interference with navigation and communications equipment.

Key point

The temperature increase and size of air mass being heated determines the rate at which wind shear, and hence turbulence, occurs.

Test your understanding 20.3

What effect will radome delamination have on the weather radar signal?

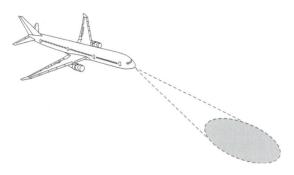

Figure 20.15 Terrain mapping

Key point

Predictive wind shear uses the principle of Doppler shift; '...the frequency of a wave apparently changes as its source moves closer to, or farther away from, an observer'.

Key point

In the take-off and climb-out phases of flight, an aircraft is flying just above stall speed; wind shear causes temporary changes in lift and is therefore a threat to aircraft safety.

20.4 System enhancements

Various features are being added to the basic weather radar systems to provide many benefits including enhanced displays, improved turbulence detection, and integration with other systems.

The weather radar system described so far is two dimensional, i.e. it provides range and directional information. **Three-dimensional** weather radar provides volumetric information relating to the weather pattern. The antenna scans 90 degrees either side of the aircraft centreline, from ground level up to 60,000 feet and up to 320 miles ahead of the aircraft. Energy returns are stored in a computer and used to build a **volumetric model** of the airspace ahead of the aircraft. Unwanted ground returns (**clutter**) are

filtered out in the computer using an on-board **terrain database**; this provides a clearer image of weather patterns. Some systems refine this de-cluttering by taking the earth's curvature into account. Weather radar software has been developed with the ability to predict storm cloud formations by analysing cloud growth characteristics, thereby providing an increased notification of turbulence. The three-dimensional image of the weather pattern is presented on the display both in terms of forward and side views. The rate of thunderstorm development can also be modelled within the weather radar system providing crews with increased notice of anticipated turbulence.

Knowledge of how thunderstorms develop over land versus oceans can be built into the system to modify the weather radar display. Different geographical areas of the world generate different types of storm with their associated turbulence; these characteristics can be built into the system's software model. These models are used to characterise the different type of storms that develop in the northern and southern hemispheres, taking into account the variations that occur with latitude. Turbulence warnings can also be customised for specific aircraft types and the various phases of flight.

20.5 Lightning detection

A complementary technology used for the detection of storms is lightning detectors. These are inexpensive and lightweight, making them very attractive for general aviation (especially for single-engine aircraft, where there is no space for a radome). Lightning detection system comprises an antenna, processor and display; the system weight is approximately 5 kg versus 15 kg for a weather radar system. The system monitors electrical activity within a storm (whereas weather radar detects precipitation as described in this chapter).

When clouds are developing as described earlier, lightning is not produced in the early stages; weather radar will indicate a developing storm before a lightning detector can. Weather radar can experience attenuation, i.e. where nearby precipitation can mask precipitation further away. Lightning detectors provide confirmation when a cloud has developed into a thunderstorm.

Lightning might occur in areas outside of the precipitation area detected by radar. Electrical activity can originate in the anvil of the thundercloud or on the outside edges of the precipitation area. Transport aircraft use weather radar in preference to lightning detectors because not all clouds develop into thunderstorms; weather radar detects smaller storms, i.e. without lightning, that also cause turbulence.

Test your understanding 20.4

What is the significance of the black, green, yellow, red and magenta images on a weather radar display?

Test your understanding 20.5

How is wind shear created and how are microbursts detected?

20.6 Multiple choice questions

1. Weather radar operates in which bands of radar frequencies?
   (a) C- and X-band
   (b) L-band
   (c) HF.

2. Larger passenger aircraft use X-band radar because it can:
   (a) determine more details of the weather pattern
   (b) penetrate through heavy precipitation
   (c) provide good resolution of images, and requires a smaller antenna.

3. What effect will radome delamination have on the weather radar signal?
   (a) none
   (b) decreased attenuation
   (c) increased attenuation.

4. Air can be forced to rise in the atmosphere from a number of causes including when opposing air masses combine; this is called:
   (a) convection
   (b) frontal
   (c) orographic.

5. What effect does increased water droplet diameter have on the amount of energy returned to the antenna?
   (a) increased
   (b) no effect
   (c) decreased.

6. The most severe weather radar images are colour coded:
   (a) black
   (b) magenta
   (c) green.

7. Predictive wind shear uses the principle of:
   (a) Doppler shift
   (b) detecting rain drop size
   (c) detecting rain drop shape.

8. Weather radar energy pulses are transmitted at rates that vary with selected:
   (a) range
   (b) mode
   (c) antenna tilt angle.

9. Compared with the transmitted frequency, energy pulses returned by droplets moving away from the aircraft will be at:
   (a) lower frequencies
   (b) higher frequencies
   (c) the same frequency.

10. As the aircraft travels away from a microburst, the downdraught affects airspeed and lift by:
    (a) reducing airspeed and reducing lift
    (b) increasing airspeed and reducing lift
    (c) reducing airspeed and increasing lift.

11. Weather radar energy pulses returned from a water droplet have been:
    (a) absorbed
    (b) refracted
    (c) reflected.

Chapter 21 Air traffic control system

The purpose of the air traffic control (ATC) system is to enable ground controllers to maintain safe separation of aircraft, both on the ground and in the air. In addition to this, the controllers are managing the flow of traffic in a given airspace. The system is based on secondary surveillance radar (SSR) facilities located at strategic sites, on or near airfields. Ground controllers use the SSR system to identify individual aircraft on their screens. The basic system is referred to as the ATC radar beacon system (ATCRBS); the updated version, Mode S, improves surveillance with a high integrity digital data link. This chapter describes the various modes of ATC operation and concludes with a review of automatic dependent surveillance-broadcast (ADS-B) and communications, navigation and surveillance/air traffic management (CNS/ATM).

21.1 ATC overview

We have seen examples of primary and secondary radar systems in previous chapters. To reiterate; with primary radar, high energy is directed via an antenna to **illuminate** a 'target'; this target could be an aircraft, the ground or water droplets in a cloud. In the case of ATC primary radar, the energy is reflected from the aircraft's body to provide range and azimuth measurements. ATC's primary radar system places the target(s) on a **plan position indicator** (PPI). Primary surveillance radar (PSR), see Figure 21.1, has its disadvantages; one of which is that the amount of energy being transmitted is very large compared with the amount of energy reflected from the target. Small targets, or those with poor relecting surfaces, could further reduce the reflected energy. Natural and man-made obstacles such as mountains and wind farms also shield the radar signals. **Secondary surveillance radar** (SSR) transmits a specific low energy signal (the interrogation) to a known target. This signal is

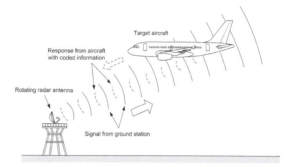

Figure 21.1 Primary surveillance radar (PSR)

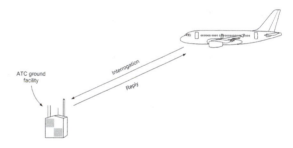

Figure 21.2 Secondary surveillance radar (SSR)

analysed by a **transponder** and a new (or secondary) signal, i.e. not a reflected signal, is sent back (the reply) to the origin, see Figure 21.2. Secondary radar was developed during the Second World War to differentiate between friendly aircraft and ships: this system was called Identification Friend or Foe (IFF). In the air traffic control system, the primary and secondary radar antennas are mounted on the same rotating assembly, thereby providing a coordinated system. The complete system is illustrated in Figure 21.3.

The ATC system operates on two frequencies within the L-band of radar:

- **Interrogation** codes on a 1030 MHz carrier wave
- **Reply** codes on a 1090 MHz carrier wave.

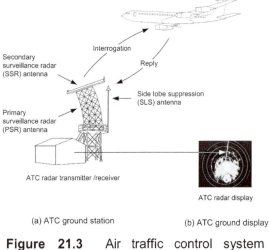

(a) ATC ground station   (b) ATC ground display

Figure 21.3   Air traffic control system overview

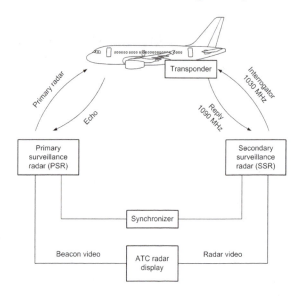

Figure 21.4   Combined PSR and SSR

The primary radar system provides a single icon per aircraft on the ATC controller's display; this means that each icon will look similar, depending upon the amount of reflected energy. As a consequence, an aircraft would have to change direction in order for it to be uniquely identified. By implementing the SSR transponder system, each icon can be identified via a unique **four-digit code** (allocated by ATC for each flight). Using SSR also means that the effects of clutter (from trees, buildings and hills etc.) are not displayed on the controller's screen. With an uncluttered screen, and each aircraft readily identified, more aircraft can be allowed into the controlled airspace. The combined PSR/SSR system is illustrated in Figure 21.4. Developments of the ATC transponder system have provided additional functionality allowing details such as flight number and altitude to be displayed on the controller's screen. Emergency codes can be sent in the event of radio failure or hijacking. The reader will appreciate that it is essential for an aircraft operating in controlled airspace to be equipped with an ATC transponder.

21.2 ATC transponder modes

SSR systems have been developed for both military and commercial aircraft applications; a summary of commercial aircraft modes is as follows:

Mode A: In this transponder system, the pilot selects the four-digit code on the ATC control panel prior to each flight. The SSR system confirms this aircraft's azimuth on the controller's screen with an icon confirming that the aircraft is equipped with a transponder. If the controller needs to distinguish between two aircraft in close proximity an identity code will be requested; the pilot pushes a switch on his ATC control panel, and this highlights the icon on the controller's screen. Since each aircraft is allocated with a unique code, only one icon per aircraft will be highlighted; this unique identification is referred to as a **squawk** code. Each of the four digits ranges from 0 to 7, these are then coded as octal numbers for use by the transponder. (This system is called Mode 3 for military users.)

Mode C: Azimuth is now augmented by pressure altitude; this is displayed on the controller's screen, adjacent to the aircraft icon thereby providing three-dimensional information. Altitude can be taken from the pilot's altimeter from an encoder that sends parallel data (in Gillham/Gray code) to the transponder. This coded data is in 100-foot increments. Aircraft with air data computers will send altitude to the transponder in serial data form, typically ARINC 429.

Mode S (select): In addition to the basic identification and altitude information, Mode S includes a data linking capability to provide a cooperative surveillance and communication system. Aircraft equipped with Mode S transponders allow specific aircraft to be interrogated; this increases the efficiency of the ATC resources. To illustrate this point, when aircraft equipped with Mode A or C transponders are interrogated, all aircraft with this type of transponder will send replies back to the ground station. This exchange occurs each time an interrogation signal is transmitted. Imagine a room full of people; the question is asked: 'please state your name and location in the room'. The person asking the question could become overwhelmed with the replies. If the question was posed in a different way, i.e. on a **selective** basis: 'Mike, where are you?' followed by: 'David, where are you?', the replies are only given by the person being addressed. The Mode S system has a number of advantages:

- Increased traffic densities
- Higher data integrity
- Efficient use of the RF spectrum
- Reduced RF congestion
- Alleviation of Mode A and C code shortages
- Reduced workload for ground controllers
- Additional aircraft parameters available to the ground controller.

Mode S transponders only send a reply to the first interrogation signal; the ground station logs this aircraft's address code for future reference. Mode S provides additional surveillance capability into controlled airspace; this is being introduced on a progressive basis. Aircraft equipped with Mode S transponders are also able to communicate directly with the Mode S transponders fitted to other aircraft; this is the basis of the traffic alert and collision avoidance system (TCAS) and will be described in the next chapter. (Note that Modes B and D are not used by commercial aircraft.)

21.3 Airborne equipment

Commercial transport aircraft are installed with two ATC antennas, a control panel and two transponders as illustrated in Figure 21.5. Since the ATC system and distance measuring equipment (DME) operates in the same frequency range, a mutual suppression circuit is utilised to prevent simultaneous transmissions.

21.3.1 Control panel

This is often a combined air traffic control and traffic alert and collision avoidance system control (TCAS) panel, see Figure 21.6 (refer to the next chapter for detailed operation of TCAS). The four-digit aircraft identification code is selected by either rotary switches or push buttons, and displayed in a window. Altitude reporting for Mode C transponders can be selected on or off. When requested by ATC, a momentary make switch is pressed; this transmits the selected code for a period of approximately 15 to 20 seconds. Table 21.1 illustrates the codes that are used in emergency situations.

Table 21.1 Emergency ATC transponder codes

Code	Meaning
7700	General air emergency
7600	Loss of radio
7500	Hijacking

Figure 21.5 Airborne equipment

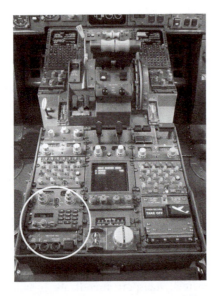

Figure 21.6(a) ATC control panel location

Figure 21.6(b) ATC control panel

21.3.2 Transponder

The aircraft transponder (Figure 21.7) provides the link between the aircraft and ground stations. General aviation products have a combined panel and transponder to save space and weight. These can be Mode S capable for IFR operations.

The ground station SSR antenna is mounted on the antenna of the primary radar surveillance system, thereby rotating synchronously with the primary returns. The airborne transponder receives interrogation codes on a 1030 MHz carrier wave from the ground station via one of two antennas located on the airframe. These signals are then amplified, demodulated and decoded in the transponder. The aircraft reply is coded, amplified and modulated as an RF transmission reply code on a 1090 MHz carrier wave. If the transponder is interrogated by a TCAS II equipped aircraft, it will select the

appropriate antenna to transmit the reply. This technique is called **antenna diversity**; this enhances visibility with TCAS-equipped aircraft flying above the host aircraft.

21.4 System operation

Although SSR has many advantages over primary radar, the smaller antenna's radiation pattern contains substantial **side-lobes**. These side-lobes can generate false returns (Figure 21.8), and so a method of coding the interrogation signals via pulse techniques is employed. The solution is to superimpose an omnidirectional pattern from a second antenna onto the directional beam. Suppressing these side-lobes is discussed in the following sections on interrogations and replies.

21.4.1 Mode A and C interrogation

Interrogation is based on a three-pulse format as illustrated in Figure 21.10; each pulse is 0.8 μs wide. Two pulses (P1 and P3) are transmitted on the rotating antenna thereby producing a **directional** signal. A third pulse (P2) is transmitted on the fixed antenna that radiates an **omnidirectional** signal. The purpose of the P2 pulse is described in the Mode A reply section. Referring to Figure 21.10, P1 and P2 have a 2 μs interval; P3 is transmitted at an interval of 8 μs for Mode A and 21 μs for Mode C interrogations (see Figure 21.11). This spacing between P1 and P3 therefore determines the type of interrogation signal (Mode A or C). The pulse repetition frequency (PRF) of interrogation signals is unique to each ground station; a typical PRF is 1200 interrogation signals per second. Replies are sent by the aircraft at the same PRF.

21.4.2 Mode A reply

A given aircraft's transponder will receive maximum signal strength each time the ground station's directional beam passes, i.e. once per revolution. Since P2 is transmitted from the fixed omnidirectional antenna, it is received with constant signal strength; but with lower amplitude than P1/P3. When the aircraft's transponder receives the maximum P1/P3 signal

Figure 21.7 ATC transponder

position X) at 1.45 μs intervals within F1 and F2. The twelve pulses are grouped into four groups of three; each group represents an octal code. Each of the four groups is labelled A, B, C and D; single pulses within the group carry a numerical weighting of 1, 2 and 4. For illustration purposes, Table 21.2 shows how group A pulses represent the octal values between 0 and 7.

When a pulse occurs in group A, this represents the value 1, 2 or 4 depending on the position of the pulse. When a pulse is not transmitted in the allocated time frame, this represents a value of zero. With four groups of data, the **octal numbers** between 0000 and 7777_8

strength, i.e. when the rotating antenna is directed at the aircraft, they are received at higher amplitude than P2. Referring to Figure 21.9, an aircraft not within the main-lobe of the directional beam would receive a P2 pulse from the omnidirectional antenna at higher amplitude than the P1/P2 pulses. The transponder recognises this as a side-lobe signal and suppresses any replies until 25 to 45 μs after P2 is received. This is called **side-lobe suppression** (SLS), a technique ensuring that only the main lobe of the rotating antenna is being replied to and not a side-lobe. The physical arrangement and antenna patterns are illustrated in Figures 21.3 and 21.9 respectively.

The Mode A reply is the ATC code allocated to that flight, formed into a series of pulses. This reply is framed between two pulses (F1 and F2) that have a time interval of 20.3 μs as illustrated in Figure 21.12. Data to be transmitted is coded by twelve pulses (plus an unused 'spare' pulse in

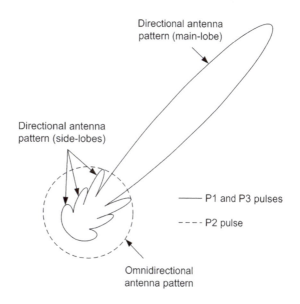

Figure 21.9 Side-lobe suppression (SLS)

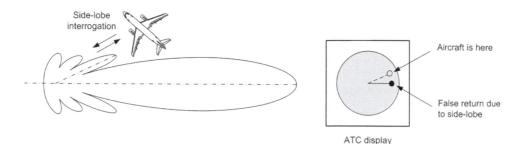

Figure 21.8 False returns from side-lobes

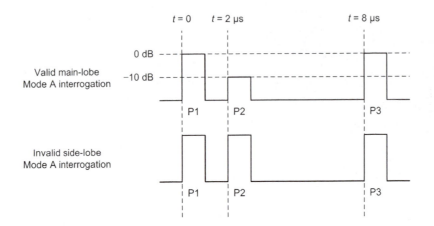

Figure 21.10 Interrogation signal validity

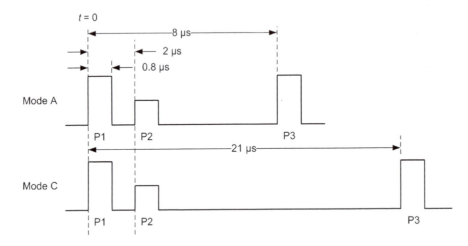

Figure 21.11 Pulse format for Mode A and Mode C interrogations

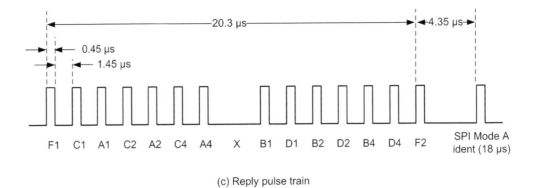

(c) Reply pulse train

Figure 21.12 Mode A/C reply pulse train (ATC code or altitude)

can be transmitted; this corresponds to the ATC code allocated to the flight and selected by the crew on their ATC control panel. (4096 codes are possible using these four octal digits.)

The final part of the aircraft reply is a pulse that is sent after F2; this pulse occurs 4.35 μs after F2, but it is only sent when an 'ident' is requested by ATC. The flight crew send this **special position identity** (SPI) pulse by pressing a momentary make switch on the control panel. The SPI pulse is sent for a period of 15 to 20 seconds after the switch has been pressed; this highlights the aircraft icon on the controller's screen.

21.4.3 Mode C reply

The aircraft's altitude is encoded by the transponder and transmitted as binary coded octal (BCO) (in 100 foot increments) as described for Mode A replies. The reply will also contain a code representing the aircraft's altitude; this is referenced to standard pressure, 1013.25 mB if the aircraft is above the transition altitude, see Figure 21.12. (The transponder sends Mode A and C replies on an alternating basis.) Figure 21.13 illustrates how an ATC identification code of 2703 is combined with an altitude code representing 1,900 feet.

Since all SSR transmissions are on the same frequencies (interrogation on 1030 MHz and replies on 1090 MHz), problems can occur when aircraft are within range of two or more ground stations. Several replies could be sent by an aircraft to each ground station that sends an interrogation signal; these undesired replies are known as non-synchronised garble, or **false replies from unsynchronised interrogator transmissions** (FRUIT). Note that **FRUIT** is sometimes written as false replies uncorrelated in time. When interrogations are received simultaneously, the transponder will reply to as many ground stations as possible.

If two or more aircraft are in close proximity, e.g. in a holding pattern, and within the ground station's directional antenna beam width, it is possible that their individual replies overlap at the ground station's computer, see Figure 21.14. The situation where replies are received from two or more interrogators answering the same interrogation is referred to as **synchronized**

Table 21.2 Illustration of Group A pulses

A1	A2	A3	Octal value
0	0	0	0
0	0	1	1
0	1	0	2
0	1	1	3
1	0	0	4
1	0	1	5
1	1	0	6
1	1	1	7

Note: 0 = no pulse transmitted
        1 = pulse transmitted

garbling. To resolve this, the controller can request the flight crew on a specific aircraft to provide an ident pulse. The problem is that ATCRBS for Modes A/C requires many interrogations to determine the position (**range and azimuth**) of an aircraft; this requires increased capacity of the ATCRBS infrastructure. The increasing density of aircraft within a given air space leads to false replies as the ground station saturates with garbling conditions. The solution to this is the Mode S (select) system.

21.4.4 Mode S operation

Although Mode S communication is very different to that of Modes A/C, both types of equipment operate on the same frequencies. The system is two-way compatible in that aircraft equipped with Mode A/C transponders will respond with ident and altitude data if interrogated by a Mode S ground station. The principles of Mode S are illustrated in Figure 21.15. Individual interrogations are sent to specific aircraft; only the transponder on this aircraft sends a reply. This reply contains additional information, e.g. selected altitude and flight number. Directional and omnidirectional beam patterns are transmitted as illustrated in Figure 21.16. Unlike ATCRBS, Mode S uses a **monopulse** SSR; this reduces the number of

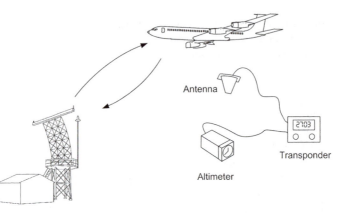

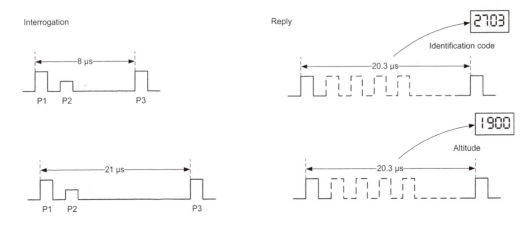

Figure 21.13(a)   Mode C principles

Figure 21.13(b)   Mode C replies

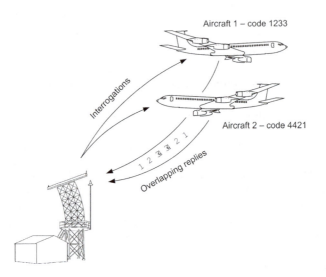

Figure 21.14   Synchronized garbling (from two aircraft within the antenna beamwidth)

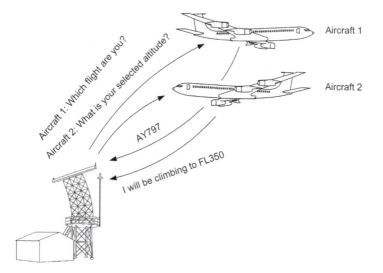

Figure 21.15 Principles of Mode S

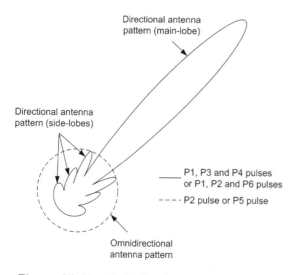

Figure 21.16 Mode S antenna pattern

interrogations required to track a target. In theory, monopulse radar only requires one reply to determine the target's azimuth (direction and range).

Two interrogation uplink formats (UF) are transmitted as shown in Figure 21.17; these are the **all-call** and **roll-call** interrogations. The two interrogations are transmitted on an alternating basis and differentiated by the width of a P4 pulse; this is either 0.8 µs or 1.6 µs. The shorter pulse is used to solicit replies from Mode A/C

transponders; they reply with their ATC code and altitude as before. Mode S transponders will not reply to this interrogation. When the P4 pulse is 1.6 µs, Mode S equipped aircraft will reply with their unique address. These replies are stored by the Mode S system as unique identifiers for each specific aircraft, see Figure 21.18(b).

The Mode S discrete addressed interrogation uplink format (UF) is shown in Figure 21.17(b). Pulse P1 and P2 both have the same amplitude and are part of the directional antenna's main-lobe. This pair appears as suppression pulses to Mode A/C transponders, so they do not reply. Mode S transponders then seek the start of the P6 data pulses; this is formed by a pattern of phase reversals that form a series of logic 1/0. **Phase-shift keying** (PSK) is a modulation technique that shifts the phase by minus 90 degrees for a logic one, and +90 degrees for a logic zero. Each data pulse's duration is 0.25 µs; the pulse's phase is sampled at these intervals. A reference pulse of 1.25 µs duration is used to indicate the start of the data word. The word length of P6 (56 or 112 bits) depends on the transponder type.

The Mode S reply is sent via the 1090 MHz carrier wave, as illustrated in Figure 21.18(a); this contains a four-pulse preamble, starting with two pairs of synchronising pulses followed by a block of data pulses (either 56 or 112 bit blocks). Using **pulse position modulation** (PPM), each data bit is allocated a 1 µs time interval, divided into two

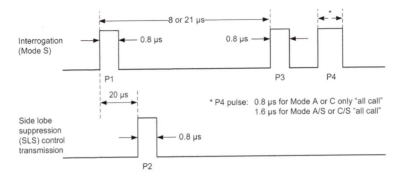

Figure 21.17(a)   Mode S 'all-call' interrogation uplink format

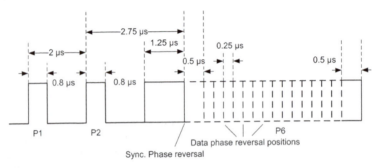

Figure 21.17(b)   Mode S discrete interrogation signal uplink format

halves. If the first half of this interval contains a pulse, this represents logic 1; if the second half of the interval contains a pulse this represents logic 0. Note that both states are indicated by the presence of a pulse.

Each Mode S-equipped aircraft has a unique address allocated to it by ICAO via the individual national registration authorities; the **aircraft address** (AA) is a 24-bit code that cannot be changed. Each national authority allocates a header code within the 24 bits, e.g. the UK code is 01000. A 24-bit code of all zeros is not valid; all ones are used for the all-call interrogation.

Individual interrogators are also coded, and this is a key feature of Mode S. **Interrogator codes** (IC) comprise 15 interrogator identifier (II) and 63 surveillance identifier (SI) codes. The use of these interrogator codes ensures unambiguous data exchange between interrogators and transponders. The Mode S all-call request acquires Mode S equipped aircraft entering a given airspace. The aircraft transponder replies with its unique aircraft address (AA). Lock-out protocols are used to suppress further replies by aircraft to subsequent requests from the

same interrogator. Transponders are thereby conditioned for replying to a specific interrogator identified by the interrogator codes (IC); the transponder will subsequently ignore requests from other interrogations. Following the all-call lock-out, the interrogator will address individual aircraft transponders on a **selective** basis. Only the selected aircraft transponder will reply; interrogators will ignore replies not intended for them.

Mode S is being introduced on a progressive basis via a transitional phase of equipment standards. The two standards are: **elementary surveillance** (ELS) and **enhanced surveillance** (EHS). Data sent by each of these two standards is shown in Tables 21.3 and 21.4 respectively.

21.4.5 Mode S transponders

Aircraft are being equipped on a progressive basis with transponders that meet the necessary functionality for the category of operations. There are four levels of transponder; these levels specify the datalink capability as detailed in Table 21.5.

Mode S transponders are installed with 255

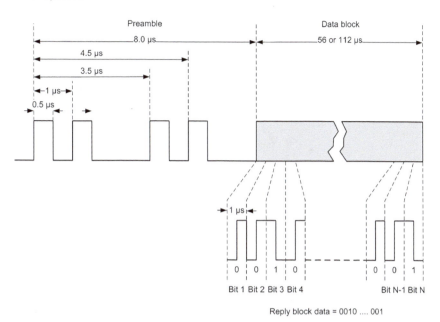

Figure 2.18(a) Mode S reply downlink format

Test your understanding 21.1

On which frequencies are ATC interrogations and transmitted ?

Test your understanding 21.2

Explain the term 'FRUIT' in the context of air traffic control.

Table 21.3 ELS transponder replies

Data	Notes
24-bit aircraft address	Allocated to individual aircraft by ICAO via the national registration authority
SSR Mode 3/A	Range and azimuth measurements; selective addressing
SSR Mode C	25 foot altitude resolution (reduced from 100 foot)
Flight status	Ground/airborne. This includes the squawk ident function in the form of downlink aircraft communications (DAP)
Data link capability report	This information is extracted when the transponder ID is first acquired by the interrogator, see Table 21.5
Common usage GICB report	Ground initiated Comm B (GICB), see Table 21.5
Aircraft identification	Call sign or registration number; selected by the flight crew (sometimes referred to as the flight ID). This will eventually replace the existing 4096 ATC codes
ACAS active resolution advisory report	Airborne collision avoidance system

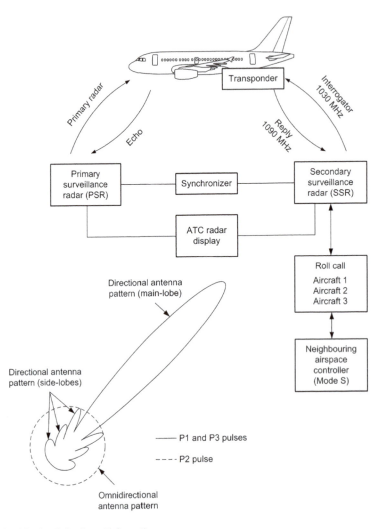

Figure 21.18(b)   Mode S 'roll-call' function

Table 21.4 EHS transponder replies (in addition to ELS)

Data	Notes
Selected altitude	Typically from the autopilot mode control panel
Roll angle	
Track angle rate	True airspeed as an alternative
True track angle	
Ground speed	
Magnetic heading	
Indicated airspeed	Mach number as an alternative
Rate of climb/descent	

Table 21.5 Mode S transponder datalink capability levels

Level	Type	Detail
1	Comm A	Mode A, C and S surveillance, without a datalink capability
2	Comm B	Level 1, plus standard length transmitting and receiving 112-bit messages
3	Comm C	Level 2, plus receiving of 16 linked 112 extended length messages (ELM)
4	Comm D	Level 3, plus transmitting and receiving 16 linked 112 ELM

data registers, each containing 56 bits. These registers are automatically loaded with specific aircraft derived data. When interrogated, the transponder registers are extracted and sent as the reply messages. Referring to Table 21.5, the minimum level of transponder capability for elementary surveillance is Comm B. The registers are referred to as Comm B data selectors (BDS); each register contains either specific or common information. Data can be transmitted or received by level 2 (Comm B) transponders; this is via a downlink format (DF) or uplink format (UF). Alphanumeric strings of data are sent as **downlink aircraft parameters** (DAP). Certain BDS registers in the transponder are used for specific parameters, other registers have common usage. These registers are checked for timely updates. A report on the status of these updates is provided when requested by the ground station: ground initiated Comm B (GICB).

Test your understanding 21.3

Explain the principles of pulse position modulation.

Test your understanding 21.4

What are the three emergency ATC codes?

Test your understanding 21.5

What are the differences between ATC transponder Modes A, C and S?

21.5 Automatic dependent surveillance-broadcast (ADS-B)

21.5.1 Introduction to ADS-B

The development of Mode S has led onto two related systems: traffic alert and collision avoidance systems (TCAS) and automatic dependent surveillance-broadcast (ADS-B). Both systems can exchange data directly between aircraft. TCAS is a **surveillance** system that provides warnings directly to the crew when other navigation systems (including ATC) have failed to maintain safe separation of aircraft. ADS-B is an emerging technology for **air traffic management** (ATM) that is intended to replace conventional secondary surveillance radar. TCAS is addressed in the Chapter 22; ADS-B is described below.

Automatic dependent surveillance-broadcast (ADS-B) is intended eventually to replace conventional ground-based ATC radar systems. The system also provides surveillance in remote areas where ground radar coverage is not possible, e.g. over oceans. ADS-B forms part of the FAA's next generation air transportation system (NGATS). It will revolutionise how pilots obtain traffic and weather information. The intention is to increase air navigation safety by providing crews with real-time information about other traffic; this makes it possible for the crew to be responsible for their own aircraft's separation and collision avoidance. The system is **automatic** in that no interrogation is required to initiate a transponder broadcast from the aircraft; this type of unsolicited transmission is known as a **squitter**.

ADS-B utilises conventional global navigation satellite system (GNSS) and onboard broadcast equipment for communication via satellites

Key point

SSR has the following advantages compared with primary radar:

- Low power transmitter

- Superior returns from the target in terms of signal strength and integrity

- Transmissions and returns can be coded to include data

- Smaller antennas.

Key point

Altimeters are used in some aircraft to provide an encoded digital output (barometric altitude to the transponder) in Gillham code. This is a modified form of Gray code, where two successive values differ in only one digit. This code prevents spurious outputs from the analogue encoder. To illustrate how this code is used, a four bit parallel output would count from zero to seven as follows: 0000, 0001, 0011, 0010, 0110, 0111, 0101, 0100.

Key point

The 100 foot resolution used in many Mode C transponders is being updated to 25 foot resolution for Mode S; serial data from the encoder will be required to achieve this.

making it **dependent**. Air traffic coordination is thereby provided though **surveillance** between aircraft; the system has a range of approximately 150 nm.

A significant benefit of ADS-B is the estimated 90% cost saving compared with replacing ageing SSR system infrastructures. Other benefits include greater access to optimum routes and altitudes; this leads to reduced fuel consumption and greater utilisation of aircraft. The system provides **real-time data** for both flight crews and air traffic controllers. Data is exchanged between aircraft and can be independent of ground

equipment. Since SSR is based on range and azimuth measurements, the accuracy of determining an aircraft's position reduces as a function of range from the antenna. Two aircraft in close proximity, but some range from an SSR ground station, can exchange data via ADS-B and calculate their relative positions more accurately. ADS-B is being proposed with three methods of exchanging data:

- Mode S transponder extended squitter (ES)
- VHF digital link (VDL)
- Universal access transceiver (UAT).

There are advantages and disadvantages for these methods; each technology is competing with the other, driven by technical, operational and political factors. A combination of the above is being introduced on a progressive basis to serve the needs of general aviation and commercial air transport. There are examples of where Mode S and VDL have been integrated into single ground stations. The reader is encouraged to monitor developments via the industry press.

21.5.2 Extended squitter (ES)

The Mode S method extends the information already described above for enhanced surveillance. Extended squitter (ES) messages include aircraft position and other status information. The advantage of using ES is that the infrastructure exists via Mode S ground stations and TCAS-equipped aircraft. Note that Mode S provides only unidirectional communications.

21.5.3 VHF digital link (VDL)

VDL utilises the existing aeronautical VHF frequencies to provide bi-directional communications; digital data is within a 25 kHz bandwidth. This protocol is based on a technique called 'self organising time division multiple access' (STDMA). VDL is suited for short message transmissions from a large number of users over longitude range. The system utilises conventional global navigation satellite system (GNSS) to send messages of up to 32 bytes at 9.6 kbps. The system can manage 9,000 32-byte messages per minute. The system is self-organising, therefore no master ground station is required.

21.5.4 Universal access transceivers (UAT)

In order to illustrate the principles of ADS-B, the universal access transceiver (UAT) is described in more detail. UAT uses conventional global navigation satellite system (GNSS) technology and a relatively simple broadcast communications link, see Figure 21.19 for typical ADS-B architecture. The 978 MHz universal access transceiver (UAT) receives inputs from a global navigation satellite system (GNSS), combines this data with other parameters, e.g. airspeed, heading, altitude and aircraft identity, to facilitate the **air traffic management**, see Figure 21.20. Flight information services-broadcast (FIS-B), such as weather and other non-ADS-B radar traffic information services-broadcast (TIS-B), can also be uplinked. This data is transmitted to aircraft in the surrounding area, and to ground receivers that distribute the data in real time via existing communication infrastructures. The system allows operations in remote and/or mountainous areas not covered by ground radar. Trials have been conducted in the Yukon-Kuskokwim delta (Alaska), to provide radar-like surveillance (this is an area where secondary radar cannot be deployed).

Referring to Figure 21.21, aircraft A is in a remote area, and has flown beyond the range of a ground-based transceiver (GBT). This aircraft continues to broadcast its ADS-B data; however, no other aircraft is within air-to-air range. Air traffic control and other ADS-B-equipped aircraft receive aircraft A's data via **satellite** link. Furthermore, aircraft A receives all other ADS-B aircraft positions (latitude and longitude) together with routine ATC data, e.g. weather updates via satellite link until it flies within range of a GBT. In Figure 21.21, aircraft B and C broadcast and receive data via the nearest GBT. (Data courtesy of ADS-B Technologies, LLC.)

21.6 Communications, navigation and surveillance/air traffic management (CNS/ATM)

This subject is derived from the numerous disciplines and technologies required to enable aircraft to navigate, and air traffic control to manage, airspace. In many parts of the world, including the USA and Europe, dense traffic flows are currently being managed; other parts of the world are seeing continuing increases in air traffic. Today's ATC infrastructure, including operating methods and equipment, cannot possibly manage the predicted demands of air traffic management. It is vital that **global standards** are developed and implemented for the delivery of a safe, efficient and economic air navigation service provision.

Navigation is not always about flying great circle routes for the shortest distance between two points, e.g. tailwinds should be exploited and headwinds avoided. This requires real-time weather information for pilots and controllers. Close cooperation is also required with airports to ensure efficient arrivals and departures thereby minimising delays. Air traffic management addresses traffic flow at the optimum speed, height and route to minimise fuel consumption.

Numerous enabling factors for CNS/ATM will lead to higher navigation accuracy at lower cost (not just the cost of fuel, but also the impact of air travel on the **environment**). Area navigation already provides a flexible and efficient means of en route and terminal area operations in place of airway routings.

The FAA is proposing a new generation of VHF datalink mode 2 (VDL-2) equipment enabling **controller-pilot datalink communications** (CPDLC). This will supplement company or engineering information currently handled by the lower capacity ACARS data link system. CPDLC features include four-dimensional navigation management, en route clearances and traffic flow rerouting. The phased introduction of CPDLC will see existing voice communications for air traffic control purposes only used as a backup.

Key point

ADS-B provides real-time data for both flight crews and air traffic controllers; data is exchanged between aircraft and is independent of ground equipment. The system allows operations in remote and/or mountainous areas not covered by ground radar.

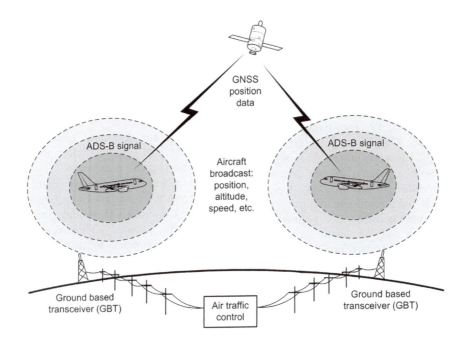

Figure 21.19   Typical ADS-B architecture

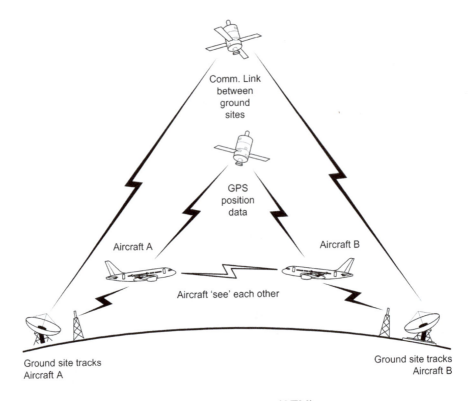

Figure 21.20   ADS-B used for air traffic management (ATM)

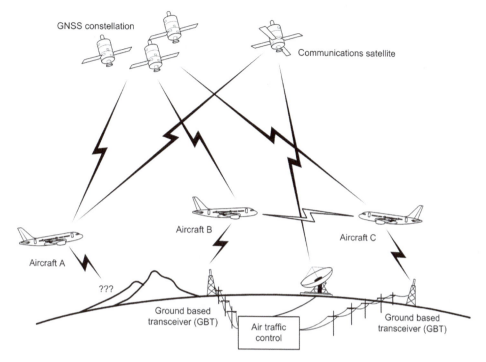

Figure 21.21   ADS-B operations in remote areas

Key point

Side-lobe suppression is achieved by transmitting directional and omnidirectional pulses.

Key point

ATC and DME transponders operate in the same frequency band; their transmissions have to be coordinated.

Key point

Mode S eliminates synchronous garbling, increases the capacity of a given air space and improves surveillance accuracy.

Key point

The phased introduction of CPDLC will see existing voice communications for air traffic control purposes only used as a backup.

Key point

Enhanced Mode S provides data for the current state of motion of the aircraft, together with the aircraft's vertical intention, i.e. selected altitude.

Key point

The demands on ATC resources are further reduced by Mode S since replies are only sent during the initial 'all-call' broadcast, and thereafter only when specifically requested.

Key point

The Mode S reply is in a different form to that of Modes A or C, and is only sent to a Mode S interrogation.

Test your understanding 21.6

Who allocates the following: ATC identity codes, 24-bit aircraft address codes?

Test your understanding 21.7

What is the difference between elementary and enhanced surveillance?

21.7 Multiple choice questions

1. Side-lobes are inherent in which part of the SSR interrogation?
   (a) Omnidirectional antenna
   (b) Rotating antenna
   (c) Transponder reply.

2. The transponder always suppresses any reply when:
   (a) P2 amplitude is =P1/P3
   (b) P2 amplitude is ≤P1/P3
   (c) P2 amplitude is ≥P1/P3

3. Compared with primary radar, the transmission power used by secondary surveillance radar is:
   (a) higher
   (b) the same
   (c) lower.

4. ATCRBS ident codes are formatted in which numbering system?
   (a) Binary
   (b) Octal
   (c) Decimal.

5. Directional and omnidirectional pulses are transmitted by the SSR for:
   (a) ident and altitude data
   (b) DME suppression
   (c) side-lobe suppression.

6. Mode C (altitude information) is derived from an:
   (a) altimeter or air data computer
   (b) ATC transponder
   (c) ATC control panel.

7. The special position ident (SPI) pulse is transmitted for a period of:
   (a) indefinite time
   (b) 15 to 20 seconds
   (c) 8 to 21 μs.

8. Side-lobe suppression is achieved by transmitting:
   (a) directional (P1/P3) and omnidirectional (P2) pulses
   (b) directional (P2) and omnidirectional (P1/3) pulses
   (c) directional (P1) and omnidirectional (P2/3) pulses.

9. The transponder code of 7700 is used for:
   (a) general air emergency
   (b) loss of radio
   (c) hijacking.

10. ATC interrogations and replies are transmitted on the following frequencies:
    (a) interrogation on 1030 MHz, replies on 1090 MHz
    (b) interrogation on 1090 MHz, replies on 1030 MHz
    (c) interrogation on 1030 MHz, replies on 1030 MHz.

11. The 'Mode S all-call' interrogation will cause Mode A and C transponders to reply with their:
    (a) ident and altitude data
    (b) aircraft address code
    (c) special position ident (SPI) pulse.

Chapter 22

Traffic alert and collision avoidance system

With ever increasing air traffic congestion, and the subsequent demands on air traffic control (ATC) resources, the risk of a mid-air collision increases. This risk led to the concept of airborne collision avoidance systems (ACAS) being considered as early as the 1950s. Several ACAS technologies have been developed; this chapter focuses on the traffic alert and collision avoidance system (TCAS). TCAS is an automatic surveillance system that helps aircrews and ATC maintain safe separation of aircraft. It is an airborne system based on secondary radar that interrogates and replies directly between aircraft via a high-integrity data link. The system is functionally independent of ground stations, and alerts the crew if another aircraft comes within a predetermined time to a potential collision. Airborne collision avoidance is a complex task; it has taken years to develop. It is important to note that TCAS is a backup system, i.e. it provides warnings and guidance when other navigation systems (including ATC) have failed to maintain safe separation of aircraft. This chapter provides an overview of TCAS, and describes how the system contributes to the safe operation of aircraft.

22.1 Airborne collision avoidance systems (ACAS)

There are five different types of ACAS technology in use, or being planned:

Passive receivers. These units are intended for general aviation use; they monitor ATC transponder signals in the immediate area, and provide visual or audible signals to warn of nearby traffic. They have a range of approximately six miles and monitor 2,500 feet above or below the host aircraft. The receiver is normally located on the aircraft's glareshield. It has an internal antenna, which can lead to intermittent coverage depending on how and where the unit is positioned. Passive systems provide approximations of where another aircraft is, and its position relative to the host aircraft. Some passive devices provide vertical trend information, e.g. indicating if the other aircraft is climbing or descending.

Traffic information system (TIS). This uses the host aircraft's Mode S transponder (refer to Chapter 21) to communicate with the ground-based secondary surveillance radar (SSR) network. Traffic information can be obtained within a five-mile radius, 1,200 feet above or below the host aircraft. This traffic information is provided on a (near) real-time basis. The attraction of TIS is that aircraft hardware and software are minimal since the system 'feeds' off ground station computations. TIS is unavailable outside of areas covered by SSR and will be superseded by a system called automatic dependent surveillance-broadcast (ADS-B).

Traffic advisory system (TAS). The host aircraft's TAS actively monitors the airspace seeking nearby transponder-equipped aircraft and provides relevant traffic information via a display and audio warning. TAS uses active interrogation of nearby transponders to determine another aircraft's position and movement. The system can track up to 30 aircraft with a range of up to 21 nm, 10,000 feet above or below the host aircraft.

Traffic alert and collision avoidance system (TCAS). This is the industry standard system mandated for use by commercial transport aircraft, and the main subject of this chapter. Two types of TCAS are in operation, TCAS I and II. Both systems provide warnings known as 'advisories' to alert the crew of a potential collision. **TCAS I** assists the crew in visually locating and identifying an **intruder** aircraft by issuing a **traffic advisory** (TA) warning. **TCAS II** is a collision avoidance system and, in addition to traffic advisories, provides vertical flight manoeuvre guidance to the crew. This is in the form of a **resolution advisory** (RA) for **threat traffic**. A resolution advisory will either increase or maintain the existing vertical separation from an intruder aircraft. If two aircraft in close proximity are equipped with TCAS II, the flight manoeuvre guidance is coordinated between both aircraft. A third type of system (TCAS III) was intended to provide lateral guidance to the crew; however, this has been superseded by a new concept: automatic dependence surveillance-broadcast (ADS-B).

Automatic dependent surveillance-broadcast (ADS-B). This system is intended eventually to replace conventional ground-based ATC radar systems. The system also provides surveillance in remote areas where ground radar coverage is not possible, e.g. over oceans. ADS-B forms part of the FAA's next generation air transportation system (NGATS) and it will revolutionise how pilots obtain traffic and weather information. The intention is to increase air navigation safety by providing crews with real-time information about other traffic; this makes it possible for the crew to be responsible for their own aircraft's separation and collision avoidance. The system is **automatic** in that no interrogation is required to initiate a transponder broadcast from the aircraft; this type of unsolicited transmission is known as a **squitter**. ADS-B utilises conventional global navigation satellite system (GNSS) and onboard broadcast equipment for communication via satellites making it **dependent**. Air traffic coordination is thereby provided though **surveillance** between aircraft. ADS-B is further described in Chapter 21 (ATC).

22.2 TCAS overview

Secondary surveillance radar (SSR) transmits a specific low energy signal (the interrogation) to a known target. This signal is analysed and a new (or secondary) signal, i.e. not a reflected signal, is sent back (the reply) to the origin, see Figure 22.1. In the TCAS application, interrogations and replies are sent directly between the on-board ATC transponders, see Figure 22.2. The TCAS computer interfaces with the ATC transponder and calculates the time to a potential collision known as the **closest point of approach** (CPA).

TCAS creates a **protected volume of airspace** around the host aircraft, see Figure 22.3; this is based on altitude separation and a calculated time to the CPA. The Greek letter **tau** (T) is the symbol used for the approximate time (in seconds) to the CPA, or for the other aircraft reaching the same altitude. This protected volume of airspace is determined as a function of time (tau) for both range and vertical separation:

$$\text{Range tau} = \frac{3,600 \times \text{slant range (nm)}}{\text{closing speed (knots)}}$$

$$\text{Vertical tau} = \frac{\text{altitude separation (feet)} \times 60}{\text{combined vertical speed (fpm)}}$$

TCAS interrogates other aircraft within this protected airspace and obtains their flight path details, i.e. range, altitude and bearing. This data is analysed along with the host aircraft's flight path. If there is a potential conflict between flight paths, a visual and audible warning is given to the crew. This warning depends on the type of equipment installed in the host and other traffic as shown in Figure 22.4.

Tau is programmed for varying sensitivity levels determined by altitude bands as illustrated in Table 22.1. For each altitude band, there is a different sensitivity level and corresponding value of tau for traffic and resolution advisories (TA and RA respectively). Higher sensitivity levels provide a larger protected volume of airspace. If closure rates are low, modifying the range

---

Key point

TCAS is an airborne system based on secondary radar that interrogates and replies to other aircraft; the system utilises the aircraft's Mode S transponders, and is functionally independent of the aircraft navigation systems and ground stations.

Key point

The closest point of approach is derived as a function of time, referred to as tau.

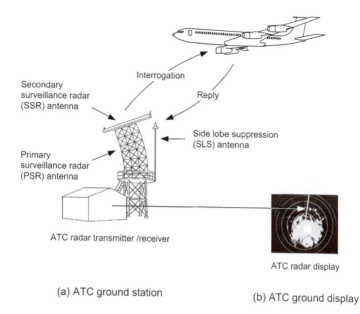

Figure 22.1 Secondary surveillance radar

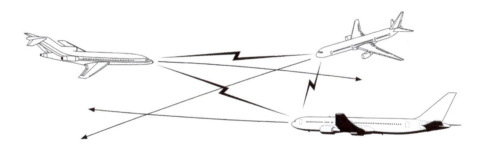

Figure 22.2 Airborne equipment—data link communication

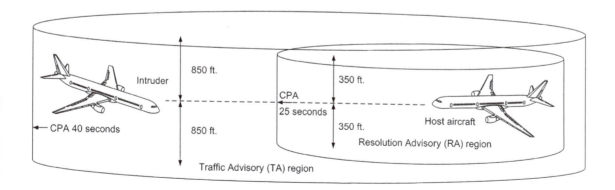

Figure 22.3 Protected airspace volume

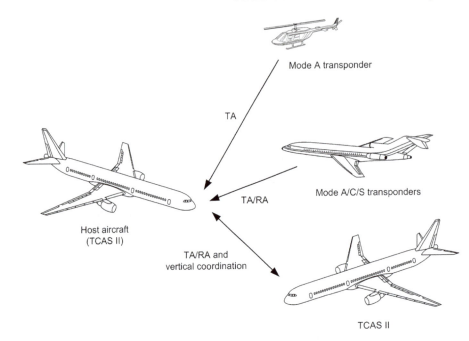

Figure 22.4 Variations of warning given to host aircraft depending on intruder's equipment

Table 22.1 TA/RA sensitivity levels

Altitude (feet) *	Sensitivity levels	Tau (seconds)		DMOD (nm)		Altitude Threshold (feet)	
		TA	RA	TA	RA	TA	RA
0 to 1,000	2	20	None	0.30	None	850	n/a
1,000 to 2,350	3	25	15	0.33	0.20	850	300
2,350 to 5,000	4	30	20	0.48	0.35	850	300
5,000 to 10,000	5	40	25	0.75	0.55	850	350
10,000 to 20,000	6	45	30	1.00	0.80	850	400
20,000 to 42,000	7	48	35	1.30	1.10	850	600
>42,000	7	48	35	1.30	1.10	1200	700

* Radio altitude for sensitivity levels 2/3, thereafter pressure altitude from the host aircraft's barometric altimeter (Data source: ARINC)

Test your understanding 22.1

What is the difference between TCAS I and TCAS II?

Test your understanding 22.2

What do the abbreviations TA and RA stand for?

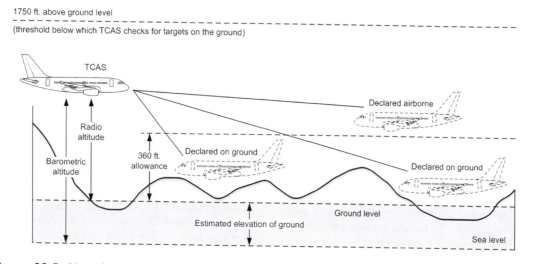

Figure 22.5 Aircraft deemed by TCAS as airborne/declared on the ground (courtesy of ARINC)

boundaries required to trigger a TA or RA provides a further refinement to the calculation of collision avoidance; this distance modification is known as DMOD.

Aircraft that are on the ground are filtered out of the collision avoidance algorithms to reduce nuisance warnings. Other Mode S equipped aircraft are monitored if their altitude is less than 1750 feet above the ground. Referring to Figure 22.5, by using a combination of the host aircraft's barometric and radio altitude, together with the barometric altitude of the target aircraft, any target below 360 feet is deemed to be on the ground.

22.3 TCAS equipment

The system consists of one TCAS computer, two directional antennas, a control panel and two displays, see Figure 22.6. TCAS operates in conjunction with the Mode S surveillance system

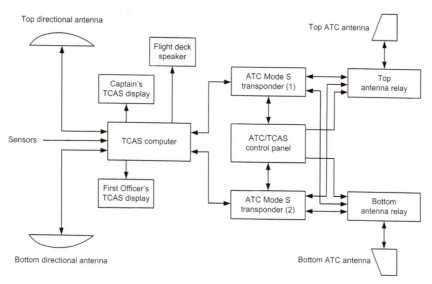

Figure 22.6 TCAS airborne equipment

which includes two transponders, a control panel and antennas, as described in Chapter 21. Visual warnings can be displayed on the instantaneous vertical speed indicator (IVSI) or electronic flight instrument system (EFIS).

22.3.1 Antennas

These are located on the top and bottom of the fuselage as illustrated in Figure 22.7; this provides **antenna diversity**, a technique in which the datalink signal is transmitted along different propagation paths. The upper antenna is directional and is used for tracking targets above the host aircraft; the bottom antenna can either be **omnidirectional**, or **directional** as an operator specified option. Interrogation codes are transmitted via the Mode S transponder on a 1030 MHz carrier wave; reply codes are transmitted on a 1090 MHz carrier wave. The phase array directional antennas are electronically steerable and transmit in four lateral segments at varying power levels. Note that two Mode S transponder antennas (Figure 22.7) are also required for TCAS operation. The latter is suppressed when the Mode S transponder is transmitting so that TCAS does not track the host aircraft.

22.3.2 Computer

This is a combined transmitter, receiver and processor that performs a number of functions including:

- **Monitoring** of the surveillance airspace volume for aircraft
- **Tracking** other aircraft
- **Monitoring** its own aircraft altitude
- Issuing **warnings** for potential flight path conflicts
- Providing recommended **manoeuvres** to avoid potential flight path conflicts.

Inputs to the computer include the host aircraft's heading, altitude and maximum airspeed. Other configuration inputs include landing gear lever position and weight on wheels sensor. Collision avoidance algorithms are used to interpret data from the host aircraft and proximate traffic.

22.3.3 Control panel

This is a combined ATC/TCAS item, see Figure 22.8. The four-digit aircraft identification code is selected by either rotary switches or push buttons, and displayed in a window. (Refer to Chapter 21

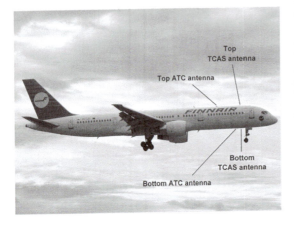

Figure 22.7 Location of ATC/TCAS antennas

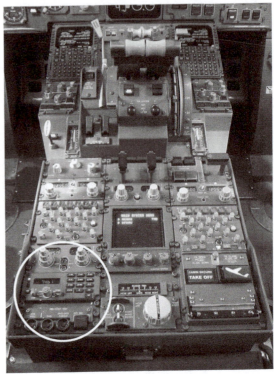

Figure 22.8(a) Location of ATC/TCAS control panel

Figure 22.8(b) ATC/TCAS control panel

for detailed ATC operation.) The mode select switch is used to disable TCAS surveillance, enable traffic advisories only, or both traffic and resolution advisories. With the system in standby, the TCAS transponder is powered but it will not interrogate or reply to interrogations; all surveillance and tracking functions are disabled. The above/below switch (ABV-N-BLW) allows the crew to select three bands of surveillance above or below the aircraft:

- ABV +7000 feet/−2700 feet
- N ±2700 feet
- BLW +2700 feet/−7000 feet.

22.3.4 Displays

The displays used for TCAS advisories vary between aircraft types. These include the instantaneous vertical speed indicator (IVSI) and/ or the electronic flight instrument system (EFIS). In either case, the advisory warnings are based on the same icons. Details of both IVSI and EFIS displays are provided in the description of TCAS system operation that follows in the next section.

22.4 System operation

22.4.1 TCAS compatibility

Referring to Figure 22.4, there will be a combination of aircraft systems in any given airspace. If the host aircraft is fitted with TCAS I equipment, then its computer will provide traffic advisories, regardless of the surrounding aircraft ATC transponder types. (Aircraft not fitted with a transponder are not tracked by TCAS.) When the host aircraft is fitted with TCAS II, but other aircraft have different transponder types, the

advisories provided to the host crew will be as shown in Figure 22.4.

TCAS requires that the aircraft is equipped with a Mode S ATC transponder; the computers in TCAS II equipped aircraft will coordinate their guidance commands such that they provide **complementary** manoeuvres. In the latest versions of TCAS, aircraft performance is taken into account when providing these commands.

22.4.2 Advisory warnings

Traffic icons are shown relative to the host aircraft; these are colour coded to depict their threat level as shown in Figure 22.9. These icons are supplemented by altitude information for the other aircraft: relative altitude (± to depict if the other aircraft is above or below the host aircraft)

Own aircraft
(white or cyan)

Non Intruding Traffic.
Altitude unknown
(white or cyan)

Proximity Traffic 200 ft.
below and descending
(white or cyan)

Traffic Advisory
(Intruder) 700 ft. above
and level (amber)

Resolution Advisory
(Threat) 100 ft. below
and climbing (red)

Figure 22.9 Traffic warning icons

and vertical manoeuvre (climbing or descending indicated by an arrow). These icons are displayed on the IVSI (Figure 22.10) or EHSI (Figure 22.11).

Referring to Figure 22.11, the EHSI also has a display area for TCAS system indications (system off, self testing etc.), traffic icons and TA/RA indications:

- **Off-scale**: intruder aircraft is out of display range
- **Traffic**: TCAS has detected intruder within the protected airspace
- **No bearing**: TCAS cannot determine the bearing of an intruder
- **TA/RA**: TCAS has identified an intruder or threat aircraft.

22.4.3 TCAS guidance

In the event of a resolution advisory, the IVSI will indicate red and green bands around the display to guide the pilot into a safe flight path, see Figure 22.10.

Aircraft with an electronic attitude direction indicator (EADI) have 'fly-out-of' guidance (Figure 22.12) in the form of a red boundary. The pilot has to climb or descend, keeping the aircraft outside of these calculated boundaries until the

---

Key point

TCAS interrogates aircraft within a surveillance volume of airspace and obtains their flight path details. This data is analysed along with the host aircraft's flight path. If there is a potential conflict between flight paths, a visual and audible warning is given to the crew.

Key point

TCAS traffic and resolution advisories are displayed on either the IVSI or EFIS. If the host aircraft is fitted with TCAS II, but other traffic has different TCAS equipment or ATC transponders, the information provided to the host aircraft will vary.

RA is cleared. Note that the latest versions of TCAS take aircraft performance into account when issuing vertical guidance.

22.4.4 TCAS commands

Aural warnings are produced via dedicated TCAS speakers in the cockpit, or through the aircraft's audio system. Traffic advisories (TA) are announced by the words 'traffic, traffic', stated once for each TA. Resolution advisories (RA) are announced as shown in Table 22.2; these are referenced to an IVSI for illustration purposes. Aural warnings are inhibited at altitudes less than 500 feet above ground level.

22.4.5 TCAS surveillance

TCAS has an effective and reliable surveillance range of 14 nm. The host aircraft can simultaneously track at least 30 transponder-equipped aircraft within its surveillance range, with traffic densities of up to 24 aircraft within a five nm radius. The surveillance function transmits interrogations at 1030 MHz; transponders on nearby aircraft reply on 1090 MHz. These replies are decoded into range, altitude, and bearing by the TCAS computer and then analysed for potential conflicts by collision avoidance algorithms in the computer's software. If the ATC transponder is interrogated by a TCAS II-equipped aircraft, it will select the appropriate antenna to transmit the reply. This technique is called **antenna diversity**; this

---

Key point

TCAS I issues traffic advisories (TA); these assist the crew in visually identifying intruder traffic.

Key point

TCAS II issues traffic advisories (TA) and resolution advisories (RA); the latter provides recommended manoeuvres needed to increase or maintain vertical separation.

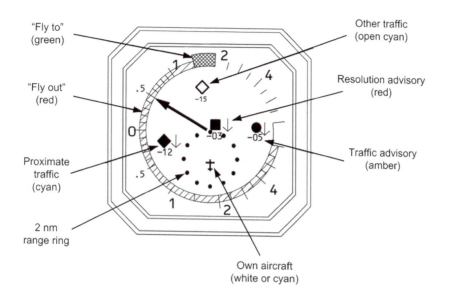

Figure 22.10 IVSI display with TCAS icons

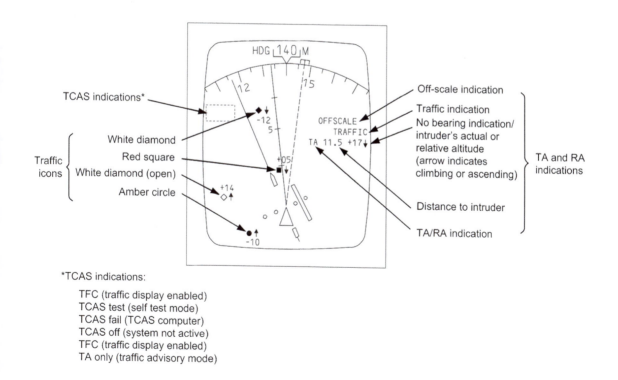

*TCAS indications:

    TFC (traffic display enabled)
    TCAS test (self test mode)
    TCAS fail (TCAS computer)
    TCAS off (system not active)
    TFC (traffic display enabled)
    TA only (traffic advisory mode)

Figure 22.11 EHSI display with TCAS icons and messages

enhances visibility with TCAS-equipped aircraft flying above the host aircraft.

Mode S surveillance

TCAS surveillance of Mode S-equipped aircraft is relatively straightforward because of its inherent selective address feature. TCAS monitors spontaneous broadcast transmissions (referred to as **squitters**) which are generated once per second by the Mode S transponder. In addition to aircraft data, the squitter contains the unique Mode S address of the sending aircraft. Once the squitter message has been received and decoded, TCAS sends a Mode S interrogation to the Mode S address contained in the squitter. The Mode S transponder replies to this interrogation, and the information received is used by TCAS to determine the range, bearing and altitude of the Mode S aircraft.

Limiting the rate at which a Mode S aircraft is interrogated reduces congestion of the 1030/1090 MHz channels. At extended ranges, a target is interrogated at least once every five seconds. As the target aircraft approaches the protected airspace where a TA may be required, the interrogation rate increases to once per second.

Mode A/C surveillance

TCAS uses a Mode C only 'all-call' to interrogate nearby aircraft equipped with Mode A/C transponders. The nominal rate for these interrogations is once per second. Aircraft equipped with Mode A transponders reply to TCAS interrogations with no data contained in the altitude field of the reply. The replies from Mode C transponders are monitored and tracked in range, altitude and bearing.

Surveillance of Mode C transponders by TCAS equipped aircraft is complicated by problems associated with synchronous and non-synchronous garbling (in a similar way to that described for the ATC system, Chapter 21). The length of reply message from a Mode C transponder is 21 μs (refer to Chapter 21). Since the speed of radar pulse propagation is 3×10^{-8} m/s, the distance travelled in 21 μs will be 6300 metres, or 3.4 nm. All Mode C transponders within a range difference of ±1.7 nm from the host aircraft will send replies that overlap when received by TCAS; this is referred to as

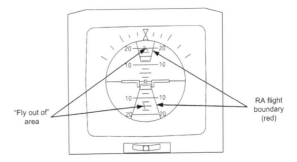

Figure 21.12   EADI vertical guidance

synchronous garbling, see Figure 22.13. A technique known as **whisper-shout** (WS) is used to overcome this. Whisper-shout varies the power level of interrogations on a **progressive** basis; aircraft that are close to the host aircraft send their replies. The next interrogation suppresses the transponders that have already replied, see Figure 22.14, but seeks replies from aircraft that did not reply to the first interrogation. This process is repeated 24 times to ensure that all Mode C transponders in the given airspace provide a reply. Using directional signals as

---

Test your understanding 22.3

What are the two types of TCAS advisory warnings provided to the flight crew in the event of a potential flight path conflict?

Test your understanding 22.4

1. What symbols and colours are used for non-threat and proximity traffic?

2. What symbols and colours are used for TA and RA warnings?

Test your understanding 22.5

What is the purpose of the 'whisper-shout' technique?

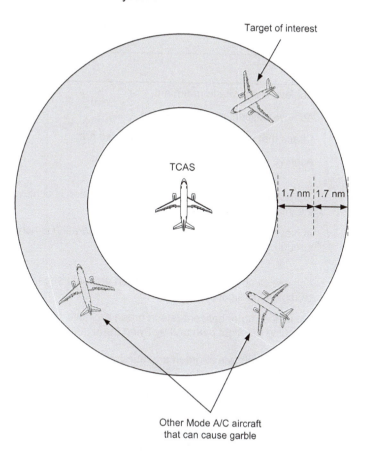

Figure 22.13 Synchronous garble area

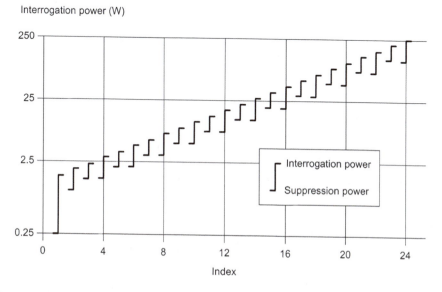

Figure 22.14 Whisper-shout interrogation

Table 22.2 TCAS aural annunciations

Aural warning	Notes
Climb, climb	Achieve climb rate in green arc.
Descend, descend	Achieve descent rate in green arc.
Monitor vertical speed	Check that vertical speed is out of the red arc.
Adjust vertical speed, adjust	Achieve vertical speed within the green arc.
Climb, crossing climb— climb, crossing climb	Achieve climb rate in green arc; safe separation is achieved by flying through the intruder's flight path.
Descend, crossing descend —descend, crossing descend	Achieve descent rate in green arc; safe separation is achieved by flying through the intruder's flight path.
Maintain vertical speed, maintain	Achieve climb or descent rate in green arc; safe separation is achieved by not changing the flight path.
Maintain vertical speed, crossing maintain	Achieve climb or descent rate in green arc; safe separation is achieved by not changing the flight path and by flying through the intruder's flight path.
Increase climb—increase climb	Achieve climb rate in green arc; aural warning has an increased sense of urgency.
Increase descent—increase descent	Achieve descent rate in green arc; aural warning has an increased sense of urgency.
Climb, climb now—climb, climb now	Received after a 'descend' RA has failed to reduce separation, i.e. a change of avoiding manoeuvre is required to achieve safe separation. Aural warning has an increased sense of urgency.
Descend, descend now— descend, descend now	Received after a 'climb' RA has failed to reduce separation, i.e. a change of avoiding manoeuvre is required to achieve safe separation. Aural warning has an increased sense of urgency.
Clear of conflict	Separation of aircraft is now adequate, i.e. the RA has now been removed

illustrated in Figure 22.15 further reduces the number of overlapping replies.

Non-synchronised garble or **false replies from unsynchronised interrogator transmissions** (FRUIT) is caused by undesired transponder replies that were generated in response to interrogations from ground stations or TCAS equipped aircraft. (Note that **FRUIT** is sometimes written as false replies uncorrelated in time.) Since these replies are transitory, algorithms in the TCAS surveillance logic can discard them.

The final consideration in TCAS surveillance is the effect of **multi-path errors**; these are caused by more than one reply being received for a single interrogation. This is a reflected reply, and usually occurs over flat terrain. A technique known as dynamic minimum triggering level (DMTL) is used within the computer to discriminate against delayed and lower power level replies.

Test your understanding 22.6

What is the difference between these two TCAS warnings: 'climb, climb' and 'climb, climb now'?

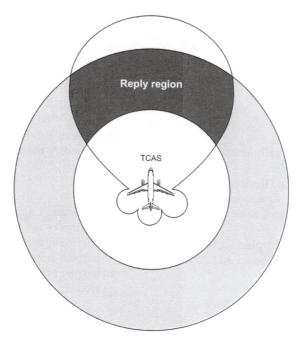

Figure 22.15 Directional transmission of TCAS and reply region

22.5 Multiple choice questions

1. The closest point of approach is determined as a function of:
   (a) range
   (b) time
   (c) closing speed.

2. TCAS interrogation codes are transmitted and received on:
   (a) 1030 MHz and 1090 MHz carrier waves
   (b) 1090 MHz and 1030 MHz carrier waves
   (c) 1030 MHz or 1090 MHz carrier waves.

3. What colour and shape of symbol is used for proximity traffic?
   (a) Solid red square
   (b) Solid cyan diamond
   (c) Solid orange circle.

4. What TCAS equipment is required on host aircraft and threat traffic in order to provide coordinated manoeuvres?
   (a) TCAS I
   (b) Mode C transponder
   (c) TCAS II.

5. TCAS II requires which type of ATC transponder:
   (a) Mode S
   (b) Mode A
   (c) Mode C.

6. The version of TCAS that provides vertical flight manoeuvre guidance to the crew is:
   (a) TAS
   (b) TCAS II
   (c) TCAS I.

7. The directional TCAS antennas transmit in:
   (a) four vertical segments at varying power levels
   (b) four lateral segments at fixed power levels
   (c) four lateral segments at varying power levels.

8. The whisper-shout technique:
   (a) varies the power level of interrogations on a decreasing basis
   (b) varies the power level of interrogations on an increasing basis
   (c) maintains the power level of interrogations on a progressive basis.

9. Recommended manoeuvres needed to
   increase or maintain vertical separation are
   provided by what type of TCAS warning?
   (a) Resolution advisory
   (b) Traffic advisory
   (c) Non-threat traffic.

10. Traffic advisories (TA):
    (a) assist the crew in visually searching and
        identifying an intruder
    (b) provide recommended manoeuvres needed
        to maintain vertical separation
    (c) provide lateral guidance to the crew.

11. TCAS interrogations and replies are sent:
    (a) directly between the onboard ATC
        transponders
    (b) via a ground link
    (c) directly to ATC ground controllers.

12. TCAS warning icons are shown relative to
    the:
    (a) host aircraft
    (b) intruder aircraft
    (c) proximate traffic.

13. Range tau is based on the:
    (a) altitude separation and closing speed of
        traffic
    (b) slant range and closing speed of traffic
    (c) altitude separation and combined vertical
        speed.

14. TCAS is an airborne system based on:
    (a) secondary radar
    (b) primary radar
    (c) DME.

15. An arrow and ± sign combined with a TCAS
    icon indicates the:
    (a) relative altitude information for the
        intruder aircraft
    (b) bearing of the intruder aircraft
    (c) recommended avoidance manoeuvre.

Appendix 1 Abbreviations and acronyms

Abbrev.	Meaning
AA	Aircraft Address
AAIS	Advanced Aircraft Information System
ACARS	Aircraft Communication Addressing and Reporting System
ACAS	Airborne Collision Avoidance System
ACM	Aircraft Condition Monitoring
ACMS	Aircraft Condition Monitoring System
ADAPT	Air Traffic Management Data Acquisition, Processing and Transfer
ADF	Automatic Direction Finder
ADI	Attitude Director Indicator
ADIRS	Air Data/Inertial Reference System
ADIRU	Air Data Inertial Reference Unit
ADM	Air Data Module
ADR	Air Data Reference
ADS	Air Data System
ADS	Automatic Dependent Surveillance
ADS-B	Automatic Dependent Surveillance-Broadcast
AEEC	Airlines Electronic Engineering Committee
AF	Audio Frequency
AFC	Automatic Frequency Control
AFCS	Auto Flight Control System (Autopilot)
AFDX	Avionics Full Duplex
AFIS	Airborne Flight Information Service
AFMS	Advanced Flight Management System
AFS	Automatic Flight System (Autopilot)
AGC	Automatic Gain Control
AGL	Above Ground Level
AHRS	Attitude/Heading Reference System
AHS	Attitude Heading System
AI	Airbus Industries
AIAA	American Institute of Aeronautics and Astronautics
AIDS	Aircraft Integrated Data System
AIMS	Airplane Information Management System
AIS	Aeronautical Information System
AIV	Anti-Icing Valve
ALT	Altitude
AM	Amplitude Modulation
AMD	Advisory Map Display
AMI	Airline Modifiable Information
AMLCD	Active Matrix Liquid Crystal Display
AMSL	Above Mean Sea Level
ANSI	American National Standards Institute

Abbrev.	Meaning
ANSIR	Advanced Navigation System Inertial Reference
ANT	Antenna
AOA	Angle of Attack
AOC	Airline Operational Control
AP	Autopilot
APATSI	Airport Air Traffic System Interface
API	Application Programming Interface
APM	Advanced Power Management
APP	Approach
APR	Auxiliary Power Reserve
APU	Auxiliary Power Unit
ARINC	Aeronautical Radio Incorporated
ARR	Arrival
ARTAS	Advanced Radar Tracker and Server
ARTS	Automated Radar Terminal System
ASAAC	Allied Standard Avionics Architecture Council
ASCB	Aircraft System Common Data Bus
ASCII	American Standard Code for Information Interchange
ASI	Air Speed Indicator
ASIC	Application Specific Integrated Circuit
ASR	Aerodrome Surveillance Radar
ATA	Actual Time of Arrival
ATC	Air Traffic Control
ATCRBS	ATC Radio Beacon System
ATE	Automatic Test Equipment
ATFM	Air Traffic Flow Management
ATI	Air Transport Indicator
ATLAS	Abbreviated Test Language for Avionics Systems
ATM	Air Traffic Management
ATN	Aeronautical Telecommunications Network
ATR	Air Transportable Racking
ATS	Air Traffic Services
ATSU	Air Traffic Services Unit
AVC	Automatic Volume Control
AVLAN	Avionics Local Area Network
AWIN	Aircraft Weather Information
AWLU	Aircraft Wireless Local Area Network Unit
B-C	Back-Course
BCD	Binary Coded Decimal
BCO	Binary Coded Octal

BDS	Comm. B Data Selector
BER	Bit Error Rate
BFO	Beat Frequency Oscillator
BGW	Basic Gross Weight
BIOS	Basic Input/Output System
BIST	Built-In Self-Test
BIT	Built-in Test
BITE	Built-in Test Equipment
BIU	Bus Interface Unit
BLEU	Blind Landing Experimental Unit
BPS	Bits Per Second
CAA	Civil Aviation Authority
CABLAN	Cabin Local Area Network
CADC	Central Air Data Computer
CAI	Computer Aided Instruction
CAN	Controller Area Network
CAS	Collision Avoidance System
CAS	Crew Alerting System
CAT	Clear-Air-Turbulence
CCA	Circuit Card Assembly
CDDI	Copper Distributed Data Interface
CDI	Course Deviation Indicator
CDROM	Compact Disk Read-Only Memory
CDS	Common Display System
CDTI	Cockpit Display of Traffic Information
CDU	Control Display Unit
CEATS	Central European Air Traffic Service
CFDS	Central Fault Display System
CH	Compass Heading
CIDIN	Common ICAO Data Interchange Network
CIDS	Cabin Intercommunication Data System
CIO	Carrier Insertion Oscillator
CLB	Climb
CMC	Central Maintenance Computer
CMOS	Complementary Metal Oxide Semiconductor
CMP	Configuration Management Plan
CMS	Centralized Maintenance System
CMU	Communications Management Unit
CNS	Communications Navigation and Surveillance
COMPAS	Computer Orientated Metering, Planning and Advisory System
COTS	Commercial Off-The-Shelf
CPA	Closest Point of Approach
CPDLC	Controller Pilot Data Link Communications
CPM	Core Processing Module
CPU	Central Processing Unit
CRC	Cyclic Redundancy Check
CRM	Crew Resource Management
CRT	Cathode Ray Tube
CRZ	Cruise
CTO	Central Technical Operations
CVOR	Conventional VOR
CVR	Cockpit Voice Recorder

CW	Continuous Wave
D8PSK	Differential Eight Phase Shift Keying
DA	Drift Angle
DAC	Digital to Analog Converter
DADC	Digital Air Data Computer
DAP	Downlink Aircraft Parameters
DATAC	Digital Autonomous Terminal Access Communications System
DBRITE	Digital Bright Radar Indicator Tower Equipment
DCU	Data Concentrator Unit
DDM	Difference in Depth of Modulation
DDR	Digital Data Recorder
DECU	Digital Engine Control Unit
DEOS	Digital Engine Operating System
DEP	Departure
DES	Descent
DEU	Digital Electronics Unit
DF	Downlink Format
DFDAU	Digital Flight Data Acquisition Unit
DFDR	Digital Flight Data Recorder
DFGC	Digital Flight Guidance Computer
DFGS	Digital Flight Guidance System
DFLD	Database Field Loadable Data
DG	Directional Gyro
DGPS	Differential Global Positioning System
DH	Decision Height
DIR INTC	Direct Intercept
DIS	Distance
DITS	Digital Information Transfer System
DMA	Direct Memory Access
DME	Distance Measuring Equipment
DMEP	Data Management Entry Panel
DMOD	Distance Modification
DO	Design Organisation
DP	Departure Procedures
DP	Decimal Point
DPM	Data Position Module
DPSK	Differential Phase Shift Keying
DPU	Display Processor Unit
DRAM	Dynamic Random Access Memory
DS	Data Segment
DSB	Double Sideband
DSB-SC	Double Sideband Suppressed Carrier
DSP	Display Select Panel
DSP	Digital Signal Processing
DSRTK	Desired Track
DTG	Distance To Go
DTG	Dynamically Tuned Gyroscope
DTOP	Dual Threshold Operation
DU	Display Units
DUATS	Direct User Access Terminal System
DVOR	Doppler VOR
EADI	Electronic Attitude Director Indicator

EARTS	En-route Automated Radar Tracking System		FBL	Fly-By-Light
EAS	Express Air System		FBW	Fly-By-Wire
EAT	Expected Approach Time		FCC	Flight Control Computer
EATMP	European Air Traffic Management Programme		FCC	Federal Communications Commission
			FCGC	Flight Control and Guidance Computer
EATMS	Enhanced Air Traffic Management System		FCS	Flight Control System
EC	European Commission		FCU	Flight Control Unit
ECAM	Electronic Centralized Aircraft Monitoring		FD	Flight Director
ECB	Electronic Control Box		FDAU	Flight Data Acquisition Unit
ECM	Electronic Countermeasures		FDC	Flight Director Computer
ECS	Environmental Control System		FDD	Floppy Disk Drive
ECU	Electronic Control Unit		FDDI	Fibre Distributed Data Interface
EEC	Electronic Engine Control		FDE	Fault Detection and Exclusion
EEPROM	Electrically Erasable Programmable Read-Only Memory		FDM	Frequency Division (Domain) Multiplexing
			FDMU	Flight Data Management Unit
EFCS	Electronic Flight Control System		FDR	Flight Data Recorder
EFIS	Electronic Flight Instrument System		FDS	Flight Director System
EGNOS	European Geostationary Navigation Overlay Service		FET	Field Effect Transistor
			FFS	Full Flight Simulator
			FG	Flight Guidance
EGPWS	Enhanced Ground Proximity Warning System		FGC	Flight Guidance Computer
			FGI	Flight guidance by digital Ground Image
EHF	Extremely High Frequency		FGS	Flight Guidance System
EHS	Enhance Surveillance		FIR	Flight Information Region
EHSI	Electronic Horizontal Situation Indicator		FIS	Flight Information System
EIA	Electronic Industries Association		FIS-B	Flight Information Services-Broadcast
EICAS	Engine Indication and Crew Alerting Systems		FL	Flight Level
			FLIR	Forward Looking Infrared
EIDE	Enhanced Integrated Drive Electronics		FLS	Field Loadable Software
EIS	Electronic Instrument System		FM	Frequency Modulation
EL	Elevation-Station		FMC	Flight Management Computer
ELAC	Elevator and Aileron Computer		FMCDU	Flight Management Control and Display Unit
ELF	Extremely Low Frequency			
eLORAN	Enhanced LORAN		FMCS	Flight Management Computer System
ELM	Extended Length Message		FMGC	Flight Management Guidance Computer
ELS	Elementary Surveillance		FMS	Flight Management System
ELS	Electronic Library System		FOG	Fibre Optic Gyroscope
EMC	Electromagnetic Compatibility		FRUIT	False Replies from Unsynchronised Interrogator Transmissions
EMI	Electromagnetic Interference			
EPROM	Erasable Programmable Read-Only Memory		FSK	Frequency Shift Keying
			FSS	Fixed Satellite Service
EROPS	Extended Range Operations			
ERU	Electronic Routing Unit		G	Giga (10^9 multiplier)
ES	Extended Squitter		G/S	Glide Slope
ESA	European Space Agency		GA	General Aviation
ESD	Electrostatic Discharge		GAT	General Air Traffic
ESD	Electrostatic Sensitive Device		GBST	Ground-Based Software Tool
ETA	Estimated Time of Arrival		GBT	Ground Base Transmitter
ETOPS	Extended Range Twin-engine Operations		GES	Ground Earth Station
EXEC	Execute		GDOP	Geometric Dilution Of Precision
			GHz	Gigahertz (10^9 Hz)
FAA	Federal Aviation Administration		GICB	Ground Initiated Comm. B
FAC	Flight Augmentation Computer		GLONASS	
FADEC	Full Authority Digital Engine Control			Global Navigation Satellite System
FANS	Future Airline Navigation Systems		GLS	GPS Landing System
FAR	Federal Aviation Regulations		GMT	Greenwich Mean Time

GND	Ground
GNSS	Global Navigation Satellite System
GPM	Ground Position Module
GPS	Global Positioning System
GPWS	Ground Proximity Warning System
GRI	Group Repetition Interval
GS	Ground Speed
GW	Gross Weight
HDD	Head Down Display
HDG	Heading
HF	High Frequency
HIRF	High-energy Radiated Field/High-intensity Radiated Field
Hex	Hexadecimal
HFDS	Head-up Flight Display System
HFDL	High Frequency Data Link
HGS	Head-up Guidance System
HIRF	High-Intensity Radiated Field
HIRL	High-Intensity Runway Lights
HM	Health Monitoring
HMCDU	Hybrid Multifunction Control Display Unit
HPA	High Power Amplifier
HSI	Horizontal Situation Indicator
HUD	Head-Up Display
HUGS	Head-Up Guidance System
Hz	Hertz (cycles per second)
I/O	Input/Output
IAC	Integrated Avionics Computer
IAPS	Integrated Avionics Processing System
IAS	Indicated Air Speed
IATA	International Air Transport Association
IC	Interrogator Codes
ICAO	International Civil Aviation Organisation
IDENT	Identification
IF	Intermediate Frequency
IFE	In-Flight Entertainment
IFF	Identification, Friend or Foe
IFOG	Interferometric Fibre Optic Gyro
IFPS	International Flight Plan Processing System
IFR	Instrument Flight Rules
IHF	Integrated Human Interface Function
IHUMS	Integrated Health and Usage Monitoring System
II	Interrogator Identifier
ILS	Instrument Landing System
IM	Inner Marker
IMA	Integrated Modular Avionics
IMU	Inertial Measurement Unit
INS	Inertial Navigation System
IP	Internet Protocol
IPC	Instructions Per Cycle
IPR	Intellectual Property Right
IPX/SPX	Inter-network Packet Exchange/Sequential Packet Exchange

IR	Infra-Red
IRMP	Inertial Reference Mode Panel
IRS	Inertial Reference System
IRU	Inertial Reference Unit
ISA	Inertial Sensor Assembly
ISA	International Standard Atmosphere
ISAS	Integrated Situational Awareness System
ISDB	Integrated Signal Database
ISDU	Inertial System Display Unit
ISO	International Standards Organisation
IVSI	Instantaneous Vertical Speed Indicator
IWF	Integrated Warning Function
JAA	Joint Airworthiness Authority
JAR	Joint Airworthiness Requirement
JEDEC	Joint Electron Device Engineering Council
K	Kilo (10^3 multiplier)
kHz	Kilohertz (10^3 Hz)
KIAS	Indicated Airspeed in Knots
Km	Kilometre
Knot	Nautical Miles/Hour
KT	Knots
LAAS	Local Area Augmentation System
LAN	Local Area Network
LASER	Light Amplification by Stimulated Emission of Radiation
Lat.	Latitude
LATAN	Low-Altitude Terrain-Aided Navigation
LCD	Liquid Crystal Display
LDU	Lamp Driver Unit
LED	Light-Emitting Diode
LF	Low Frequency
LIDAR	Light Radar
LNAV	Lateral Navigation
LO	Local Oscillator
LOC	Localizer
Long	Longitude
LORADS	Long Range Radar and Display System
LOS	Line Of Sight
LRRA	Low Range Radio Altimeter
LRM	Line Replaceable Module
LRU	Line Replaceable Unit
LSAP	Loadable Aircraft Software Part
LSB	Least Significant Bit
LSB	Lower Sideband
LSD	Least Significant Digit
LSI	Large Scale Integration
LSS	Lightning Sensor System
LUF	Lowest Usable Frequency
M	Mega (10^6 multiplier)
MASI	Mach Airspeed Indicator
MAU	Modular Avionics Unit
MCDU	Microprocessor Controlled Display Units

MCP	Mode Control Panel		OBS	Omni Bearing Selector
MCS	Master Control Station		ODS	Operations Display System
MCU	Modular Component Unit		OEI	One Engine Inoperative
MDAU	Maintenance Data Acquisition Unit		OEM	Original Equipment Manufacturer
MEL	Minimum Equipment List		OLDI	On-Line Data Interchange
MF	Medium Frequency		OM	Optical Marker
MFD	Multifunction Flight Display		OMS	On-board Maintenance System
MFDS	Multifunction Display System		OS	Operating System
MHRS	Magnetic Heading Reference System		OSS	Option Selectable Software
MHz	Megahertz (10^6 Hz)			
MLS	Microwave Landing System		PAPI	Precision Approach Path Indicator
MLW	Maximum Landing Weight		PBN	Performance Based Navigation
MM	Middle Marker		PCA	Preconditioned Air System
MMI	Man Machine Interface		PCB	Printed Circuit Board
MMR	Multimode Receiver		PCC	Purser Communication Center
MMS	Mission Management System		PCHK	Parity Check(ing)
MNPS	Minimum Navigation Performance Specification		PDB	Performance Database
			PDL	Portable Data Loader
MOS	Metal Oxide Semiconductor		PFD	Primary Flight Display
MOSFET	Metal Oxide Semiconductor Field Effect Transistor		PIC	Pilot In Command
			PLL	Phase Locked Loop
MRC	Modular Radio Cabinets		PM	Protected Mode
MRO	Maintenance/Repair/Overhaul		PMAT	Portable Maintenance Access Terminal
MSB	Most Significant Bit		PMO	Program Management Organization
MSD	Most Significant Digit		PMS	Performance Management System
MSG	Message		PNF	Pilot Non Flying
MSI	Medium Scale Integration		Pos. init.	Position initialisation
MSK	Minimum Shift Keying		POS	Position
MSL	Mean Sea Level		POST	Power-On Self-test
MSU	Mode Select Unit		PP	Pre-Processor
MSW	Machine Status Word		PPI	Plan Position Indicator
MT	Maintenance Terminal		PPM	Pulse Position Modulation
MTBF	Mean Time Between Failure		PPOS	Present Position
MTBO	Mean Time Between Overhaul		PQFP	Plastic Quad Flat Package
MTC	Mission and Traffic Control systems		PRF	Pulse Repetition Frequency
MTOW	Maximum Takeoff Weight		PRI	Primary
MUF	Maximum Usable Frequency		PROG	Progress
			PROM	Programmable Read-Only Memory
NASA	National Aeronautics and Space Administration		PSK	Phase Shift Keying
			PSM	Power Supply Module
NAVSTAR			PSR	Primary Surveillance Radar
	Navigation System with Timing and Ranging		Q	Quality Factor
NCD	No Computed Data		QAM	Quadrature Amplitude Modulation
ND	Navigation Display		QE	Quadrantal Error
NDB	Navigation Database		QoS	Quality of Service
NDB	Non-Directional Beacon		QPSK	Quadrature Phase Shift Keying
NGATS	Next Generation Air Transport System			
NIC	Network Interface Controller		R/T	Receiver/Transmitter
nm	Nautical mile		RA	Resolution Advisory
NMS	Navigation Management System		RA	Radio Altitude
NMU	Navigation Management Unit		Radar	Radio Direction and Ranging
NVM	Non-Volatile Memory		RAIM	Receiver Autonomous Integrity Monitoring
			RAM	Random Access Memory
OAT	Outside air temperature		RDMI	Radio and Distance Magnetic Indicator
OBI	Omni Bearing Indicator		RF	Radio Frequency

RIMM	RAM Bus In-line Memory Module		TIS-B	Traffic Information Services-Broadcast
RISC	Reduced Instruction Set Computer		TK	Track
RLG	Ring Laser Gyro		TKE	Track Angle Error
RMI	Radio Magnetic Indicator		TRF	Tuned Radio Frequency
RNAV	Area Navigation		TS	Task Switched
RNP	Required Navigation Performance		TTL	Transistor–Transistor Logic
ROM	Read-Only Memory		TTP	Time Triggered Protocol
RTE	Route		TWDL	Two-Way Data Link
RVR	Runway Visual Range			
			UAT	Universal Access Transceiver
SA	Service Availability		UDP	User Datagram Protocol
SAARU	Secondary Attitude/Air Data Reference Unit		UF	Uplink Format
			UHF	Ultra High Frequency
SAT	Static Air Temperature		ULSI	Ultra Large Scale Integration
SATCOM	Satellite Communications		UMS	User Modifiable Software
SC	Suppressed Carrier		USB	Universal Serial Bus
SCMP	Software Configuration Management Plan		USB	Upper Sideband
SDD	System Definition Document		USCG	United States Coast Guard
SDI	Source/Destination Identifier		UTC	Coordinated Universal Time
SEC	Secondary		UTP	Unshielded Twisted Pair
SELCAL	Selective Calling		UV	Ultra-Violet
SHF	Super High Frequency		UVPROM	Ultra Violet Programmable Read-Only Memory
SI	Surveillance Indicator			
SID	Standard Instrument Departure			
SLS	Side Lobe Suppression		VAC	Volts, Alternating Current
SMART	Standard Modular Avionics Repair/Test		VAS	Virtual Address Space
SNMP	Simple Network Management Protocol		VCO	Voltage Controlled Oscillator
SPDA	Secondary Power Distribution Assembly		VDL	Very High Frequency Data Link
SPDT	Single Pole Double Throw		VDC	Volts, Direct Current
SPI	Special Position Identity		VFR	Visual Flight Rules
SRAM	Synchronous Random Access Memory		VG	Vertical Gyro
SRD	System Requirement Document		VHF	Very High Frequency
SROM	Serial Read Only Memory		VHSIC	Very High Speed Integrated Circuit
SSB	Single Sideband		VIA	Versatile Integrated Avionics
SSI	Small Scale Integration		VLF	Very Low Frequency
SSM	Sign/Status Matrix		VLSI	Very Large Scale Integration
SSR	Secondary Surveillance Radar		VME	Versatile Module Eurocard
SSV	Standard Service Volume		VNAV	Vertical Navigation
STAR	Standard Terminal Arrival Routes		VOR	VHF Omnirange
STDMA	Self-organising Time Division Multiple Access		VORTAC	VOR TACAN Navigation Aid
			VPA	Virtual Page Address
STP	Shielded Twisted Pair		VSI	Vertical Speed Indicator
SW	Software			
SWR	Standing Wave Ratio		W/S	Whisper-Shout
			WAAS	Wide Area Augmentation System
T/C	Top of Climb		WAN	Wide Area Network
T/D	Top of Descent		WD	Wind Direction
TA	Traffic Advisory		WS	Wind Speed
TACAN	Tactical Air Navigation		WX	Weather
TAS	True Air Speed		WXP	Weather Radar Panel
TAT	Total Air Temperature		WXR	Weather Radar
Tau	Minimum time to collision threshold			
TAWS	Terrain Awareness Warning System		XTK	Cross Track
TBO	Time Between Overhaul		XTAL	Crystal
TCAS	Traffic Alert Collision Avoidance System			
TDM	Time Division (Domain) Multiplexing		ZFW	Zero Fuel Weight

Appendix 2 Revision papers

These revision papers are designed to provide you with practice for examinations. The questions are typical of those used in CAA and other examinations. Each paper has 20 questions and each should be completed in 25 minutes. Calculators and other electronic aids must not be used.

Revision Paper 1

1. A radio wave is said to be polarised in:
   (a) the direction of travel
   (b) the E-field direction
   (c) the H-field direction.

2. Radio waves tend to propagate mainly as line of sight signals in the:
   (a) MF band
   (b) HF band
   (c) VHF band.

3. An isotropic radiator will radiate:
   (a) only in one direction
   (b) in two main directions
   (c) uniformly in all directions.

4. A vertical quarter wave antenna will have a polar diagram in the horizontal plane which is:
   (a) unidirectional
   (b) omnidirectional
   (c) bi-directional.

5. The attenuation of an RF signal in a coaxial cable:
   (a) increases with frequency
   (b) decreases with frequency
   (c) stays the same regardless of frequency.

6. The method of modulation used for aircraft VHF voice communication is:
   (a) MSK
   (b) D8PSK
   (c) DSB AM.

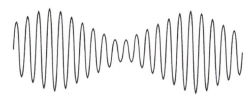

Figure A2.1 See Paper 1, Question 7

7. The type of modulation shown in Figure A2.1 is:
   (a) AM
   (b) FM
   (b) PSK.

8. The standard for ACARS is defined in:
   (a) ARINC 429
   (b) ARINC 573
   (c) ARINC 724.

9. The frequency range currently used in Europe for aircraft VHF voice communication is:
   (a) 88 MHz to 108 MHz
   (b) 108 MHz to 134 MHz
   (c) 118 MHz to 137 MHz.

10. The type of antenna shown in Figure A2.2 is
    (a) a unipole
    (b) a dipole
    (c) a Yagi.

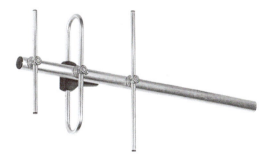

Figure A2.2 See Paper 1, Question 10

11. The typical frequency emitted by a ULB is:
    (a) 600 Hz
    (b) 3.4 kHz
    (c) 37.5 kHz.

12. ELT transmissions use:
    (a) Morse code and high-power RF at HF
    (b) pulses of acoustic waves at 37.5 kHz
    (c) low-power RF at VHF or UHF.

13. Which one of the following gives the function
    of the block marked 'X' in Figure A2.3?
    (a) power amplifier
    (b) matching unit
    (c) SWR detector.

14. The angular difference between magnetic
    north and true north is called the:
    (a) magnetic variation
    (b) great circle
    (c) prime meridian.

15. Morse code tones are used to identify the
    VOR:
    (a) identification
    (b) frequency
    (c) radial.

16. When hovering over water, the 'worst case'
    conditions for Doppler signal to noise ratios
    are with:
    (a) smooth sea conditions
    (b) rough sea conditions
    (c) tidal drift.

17. Once aligned, the inertial navigation system
    is always referenced to:
    (a) magnetic north
    (b) true north
    (c) latitude and longitude.

18. During prolonged periods of poor satellite
    reception, the GPS receiver:
    (a) enters into a dead reckoning mode
    (b) re-enters the acquisition mode
    (c) rejects all satellite signals.

19. What effect will radome delamination have
    on the weather radar signal?
    (a) None
    (b) Decreased attenuation
    (c) Increased attenuation.

20. The purpose of traffic advisories (TA) is to:
    (a) assist the crew in visually searching and
        identifying an intruder
    (b) provide recommended manoeuvres needed
        to maintain vertical separation
    (c) provide lateral guidance to the crew.

Revision Paper 2

1. The HF range extends from:
    (a) 300 kHz to 3 MHz
    (b) 3 MHz to 30 MHz
    (c) 30 MHz to 300 MHz.

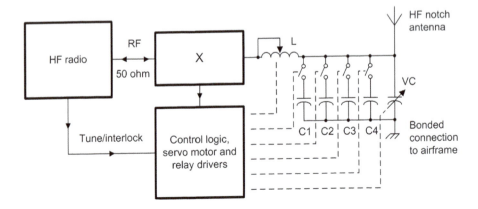

Figure A2.3 See Paper 1, Question 13

2. A transmitted radio wave will have a plane wavefront:
   (a) in the near field
   (b) in the far field
   (c) close to the antenna.

3. When radio waves travel in a cable they travel:
   (a) at the speed of light
   (b) slower than the speed of light
   (c) faster than the speed of light.

4. A radio wave at 11 MHz is most likely to propagate over long distances as:
   (a) a ground wave
   (b) a sky wave
   (c) a space wave.

5. The height of the E-layer is approximately:
   (a) 100 km
   (b) 200 km
   (c) 400 km.

6. The impedance measured at the input of a long length of correctly terminated coaxial cable will be:
   (a) the same as the characteristic impedance of the cable
   (b) zero
   (c) infinite.

7. The ATC transponder code of 7700 is used for:
   (a) general air emergency
   (b) loss of radio
   (c) hijacking.

8. During the alignment mode, a flashing IRS align light indicates:
   (a) the system is ready to navigate
   (b) the aircraft was moved during align mode
   (c) the present position entered agrees with the last known position.

9. The decision height and runway visual range for a Category 2 automatic approach are:
   (a) 100 ft and 300 m respectively
   (b) 200 ft and 550 m respectively
   (c) less than 100 ft and 200 m respectively.

10. The typical pulse rate for a ULB is:
    (a) 0.9 pulses per sec
    (b) 10 pulses per sec
    (c) 60 pulses per sec.

11. The CVR is usually located:
    (a) on the flight deck
    (b) in the avionic equipment bay
    (c) in the ceiling of the aft passenger cabin.

12. Transmission from an ELT is usually initially detected by:
    (a) low-flying aircraft
    (b) one or more ground stations
    (c) a satellite.

13. Which one of the following gives the approximate LOS range for an aircraft at an altitude of 15,000 feet?
    (a) 74 nm
    (b) 96 nm
    (c) 135 nm.

14. The function of the compressor stage in an aircraft VHF radio is:
    (a) to reduce the average level of modulation
    (b) to increase the average level of modulation
    (c) to produce 100% modulation at all times.

15. The radiation efficiency of an antenna:
    (a) increases with antenna loss resistance
    (b) decreases with antenna loss resistance
    (c) is unaffected by antenna loss resistance.

16. Display of the FMCS CDU identification page after power-up confirms the:
    (a) IRS is aligned
    (b) navigation source(s) in use
    (c) FMC has passed its BITE check.

17. Loran-C operates in which frequency band?
    (a) 190–1750 kHz
    (b) 90–110 kHz
    (c) 108–112 MHz.

18. The characteristic impedance of a coaxial cable depends on:
    (a) the ratio of inductance to capacitance
    (b) the ratio of resistance to inductance
    (c) the sum of the resistance and reactance.

19. When a DME indicator is receiving no computed data, it will display:
    (a) dashes
    (b) zeros
    (c) eights.

20. Quadrantal error (QE) for an ADF system is associated with the:
    (a) ionosphere
    (b) physical aspects of terrain
    (c) physical aspects of the aircraft structure.

Revision Paper 3

1. The free-space path loss experienced by a radio wave:
   (a) increases with frequency but decreases with distance
   (b) decreases with frequency but increases with distance
   (c) increases with both frequency and distance.

2. For a given HF radio path, the MUF changes most rapidly at:
   (a) mid-day
   (b) mid-night
   (c) dawn and dusk.

3. In the HF band radio waves tend to propagate over long distances as:
   (a) ground waves
   (b) space waves
   (c) ionospheric waves.

4. A standing wave ratio of 1:1 indicates:
   (a) that there will be no reflected power
   (b) that the reflected power will be the same as the forward power
   (c) that only half of the transmitted power will actually be radiated.

5. An aircraft is flying on heading of 090° to intercept the selected VOR radial of 180°; the HSI will display that the aircraft is:
   (a) right of the selected course
   (b) left of the selected course
   (c) on the selected course.

6. The typical bandwidth of a DSB AM voice signal is:
   (a) 3.4 kHz
   (b) 7 kHz
   (c) 25 kHz.

7. The operational state of an ELT is tested using:
   (a) a test switch and indicator lamp
   (b) immersion in a water tank for a short period
   (c) checking battery voltage and charging current.

8. The air testing of an ELT can be carried out:
   (a) at any place or time
   (b) only after notifying the relevant authorities
   (c) only at set times using recommended procedures.

9. A Type-W ELT is activated by:
   (a) a member of the crew
   (b) immersion in water
   (c) a high G-force caused by deceleration.

10. The angle between north and the flight path of the aircraft is the:
    (a) ground track angle
    (b) drift angle
    (c) heading.

11. An RDMI provides the following information:
    (a) distance and bearing to a navigation aid
    (b) deviation from a selected course
    (c) the frequency of the selected navigation aid.

12. Marker beacon outputs are given by:
    (a) coloured lights and Morse code tones
    (b) deviations from the runway centreline
    (c) deviations from the glide slope.

13. Inertial navigation system errors are a factor of:
    (a) the aircraft's velocity
    (b) how long the system has been in the 'align' mode
    (c) how long the system has been in the 'navigation' mode.

14. The GPS navigation concept is based upon calculating satellite:
    (a) speed
    (b) altitude
    (c) range.

15. The FMC recognises specific aircraft types by:
    (a) CDU entry
    (b) program pins
    (c) the navigation database.

16. Weather radar operates in which bands of radar frequencies?
    (a) C- and X-band
    (b) L-band
    (c) HF.

17. ATC Mode C (altitude information) is derived from an:
    (a) altimeter or air data computer
    (b) ATC transponder
    (c) ATC control panel.

18. TCAS II requires which type of ATC transponder?
    (a) Mode S
    (b) Mode A
    (c) Mode C.

19. The outer marker is displayed on the primary flying display as a coloured icon that is:
    (a) yellow
    (b) white
    (c) cyan.

20. DME is based on what type of radar?
    (a) Primary
    (b) Secondary
    (c) VHF.

Revision Paper 4

1. Which one of the following gives the approximate length of a half-wave dipole for use at 300 MHz?
   (a) 50 cm
   (b) 1 m
   (c) 2 m.

2. In the horizontal plane, a vertical dipole will be:
   (a) bi-directional
   (b) omnidirectional
   (c) unidirectional.

3. The function of the HF antenna coupler is to:
   (a) reduce static noise and interference
   (b) increase the transmitter output power
   (c) match the HF antenna to the HF radio.

4. Another name for a quarter-wave vertical antenna is:
   (a) a Yagi antenna.
   (b) a dipole antenna
   (c) a Marconi antenna.

5. A full-wave dipole fed at the centre must be:
   (a) current fed
   (b) voltage fed
   (c) impedance fed.

6. The antenna shown in Figure A2.4 is used for:
   (a) ILS
   (b) GPS
   (c) VHF communications.

7. The channel spacing currently used in Europe for aircraft VHF voice communication is:
   (a) 8.33 kHz and 25 kHz
   (b) 12.5 kHz and 25 kHz
   (c) 25 kHz and 50 kHz.

Figure A2.4 See Paper 4, Question 6

8. On which frequencies do ELT operate?
   (a) 125 MHz and 250 MHz
   (b) 122.5 MHz and 406.5 MHz
   (c) 121.5 MHz and 406.025 MHz.

9. The DME interrogator is part of the:
   (a) airborne equipment
   (b) DME navigation aid
   (c) VORTAC.

10. Localizer transmitters are located:
    (a) at the threshold of the runway, adjacent to
        the touchdown point
    (b) at the stop end of the runway, on the
        centreline
    (c) at three locations on the extended
        centreline of the runway.

11. With which other system can Loran-C
    systems share their aircraft antennas?
    (a) DME
    (b) VOR
    (c) ADF.

12. Autotuning of navigation aids is used by
    RNAV systems to:
    (a) update the navigation database
    (b) create waypoints in the CDU
    (c) select the best navigation aids for
        optimised area navigation.

13. When hovering directly over an object in the
    sea with a six-knot tide, the Doppler
    navigation system will indicate:
    (a) six knots, drift in the opposite direction of
        the tide
    (b) six knots, drift in the direction of the tide
    (c) zero drift.

14. Errors in an inertial navigation system are:
    (a) random and build up as a function of time
    (b) fixed and irrespective of time
    (c) random and irrespective of time.

15. The purpose of RAIMS in GPS is to:
    (a) identify the selected satellite
    (b) speed up the satellite acquisition process
    (c) provide error detection.

16. FMCS alerting messages require attention
    from the crew:
    (a) before guided flight can be continued
    (b) when time is available
    (c) at the completion of the flight.

17. The most severe weather radar images are
    colour coded:
    (a) black
    (b) magenta
    (c) green.

18. Mode A replies contain the following aircraft
    information:
    (a) identification
    (b) identification and altitude
    (c) identification, altitude and aircraft address.

19. Recommended manoeuvres needed to
    increase or maintain vertical separation are
    provided by what type of TCAS warning:
    (a) resolution advisory
    (b) traffic advisory
    (c) non-threat traffic.

20. During sunrise and sunset, ADF transmissions
    are affected by:
    (a) coastal refraction
    (b) static build-up in the airframe
    (c) variations in the ionosphere.

Appendix 3 Answers

Answers to review questions

Chapter 1 (page 13)

1. B
2. C
3. A
4. A
5. C
6. C
7. C
8. A
9. C
10. B
11. A
12. C
13. C
14. C
15. B
16. C
17. C
18. A
19. B
20. C
21. A
22. C
23. B
24. B
25. C

Chapter 2 (page 39)

1. C
2. C
3. A
4. B
5. B
6. A
7. A
8. B
9. C
10. B

11. C
12. A
13. A
14. C
15. A
16. B

Chapter 3 (page 59)

1. C
2. A
3. A
4. C
5. B
6. B
7. B
8. C
9. C
10. B
11. C
12. A
13. A
14. B
15. B
16. C

Chapter 4 (page 72)

1. A
2. C
3. A
4. C
5. B
6. A
7. B
8. B
9. B
10. C
11. C
12. A

Chapter 5 (page 84)

1. A
2. A
3. A
4. C

5. B
6. A
7. B
8. C
9. A
10. C

Chapter 6 (page 92)

1. B
2. A
3. A
4. B
5. C
6. C
7. C
8. B
9. A
10. A
11. A
12. C
13. C
14. A
15. C
16. C

Chapter 7 (page 100)

1. C
2. B
3. C
4. A
5. C
6. C
7. C
8. C
9. B
10. B
11. B
12. A
13. B

Chapter 8 (page 116)

1. B
2. A
3. A
4. B
5. B

6. C
7. A
8. B
9. B
10. B
11. C
12. B
13. B
14. C
15. C

Chapter 9 (page 125)

1. B
2. A
3. B
4. B
5. C
6. A
7. A
8. B
9. B
10. C
11. A
12. C
13. B
14. A
15. C
16. A

Chapter 10 (page 139)

1. C
2. C
3. B
4. A
5. B
6. B
7. B
8. A
9. A
10. B
11. A
12. A
13. A
14. B
15. A
16. B

Chapter 11 (page 149)

1. B
2. C
3. A
4. A
5. A
6. A
7. B
8. A
9. A
10. A
11. B
12. B
13. A
14. A
15. B
16. C
17. A

Chapter 12 (page 162)

1. C
2. B
3. C
4. C
5. A
6. B
7. A
8. A
9. A
10. A
11. A
12. A
13. C
14. A

Chapter 13 (page 168)

1. A
2. B
3. B
4. B
5. B
6. C
7. A
8. B
9. A
10. C

11. C
12. B

Chapter 14 (page 178)

1. A
2. C
3. C
4. A
5. A
6. B
7. C

Chapter 15 (page 186)

1. A
2. B
3. C
4. C
5. B
6. B
7. A
8. A
9. C
10. C
11. A
12. B
13. B

Chapter 16 (page 199)

1. A
2. C
3. B
4. B
5. B
6. B
7. A
8. B
9. C
10. C
11. B
12. A
13. B
14. B
15. C
16. C

Chapter 17 (page 215)

1. B
2. C
3. A
4. B
5. C
6. A
7. A
8. A
9. B
10. A
11. C
12. C
13. A
14. A
15. B
16. A
17. C

Chapter 18 (page 225)

1. A
2. B
3. A
4. B
5. A
6. C
7. A
8. C
9. C
10. B
11. B
12. B
13. C
14. B
15. C
16. B
17. A
18. B

Chapter 19 (page 237)

1. B
2. A
3. C
4. A
5. B

6. B
7. B
8. A
9. B
10. C
11. B
12. C
13. A
14. A
15. B

Chapter 20 (page 252)

1. A
2. C
3. C
4. B
5. A
6. B
7. A
8. A
9. A
10. A
11. C

Chapter 21 (page 270)

1. B
2. C
3. C
4. C
5. C
6. A
7. B
8. A
9. A
10. A
11. A

Chapter 22 (page 283)

1. C
2. A
3. B
4. C
5. A
6. B

7. B
8. B
9. A
10. A
11. A
12. A
13. B
14. A
15. A

11. C
12. C
13. C
14. B
15. B
16. C
17. B
18. A
19. A
20. C

Answers to revision papers

Revision Paper 1 (page 291)

1. A
2. C
3. C
4. B
5. A
6. C
7. A
8. C
9. C
10. C
11. C
12. C
13. C
14. A
15. A
16. A
17. B
18. B
19. C
20. A

Revision Paper 2 (page 292)

1. B
2. B
3. B
4. B
5. A
6. A
7. A
8. B
9. A
10. A

Revision Paper 3 (page 294)

1. C
2. C
3. C
4. A
5. A
6. B
7. A
8. C
9. B
10. C
11. A
12. A
13. C
14. C
15. B
16. A
17. A
18. A
19. C
20. B

Revision Paper 4 (page 295)

1. A
2. B
3. C
4. C
5. B
6. C
7. A
8. C
9. A
10. B
11. C
12. C

13.   C
14.   A
15.   B
16.   A
17.   B
18.   A
19.   A
20.   C

Decibels (dB) are a convenient means of expressing gain (amplification) and loss (attenuation) in electronic circuits. In this respect, they are used as a *relative* measure (i.e. comparing one voltage with another, one current with another, or one power with another). In conjunction with other units, decibels are sometimes also used as an *absolute* measure. Hence dBV are decibels relative to 1 V, dBm are decibels relative to 1 mW, etc.

The decibel is one-tenth of a bel which, in turn, is defined as the logarithm, to the base 10, of the ratio of output power (P_{out}) to input power (P_{in}).

Gain and loss may be expressed in terms of power, voltage and current such that:

$$A_p = \frac{P_{out}}{P_{in}} \quad A_v = \frac{V_{out}}{V_{in}} \quad \text{and } A_i = \frac{I_{out}}{I_{in}}$$

where A_p, A_v or A_i is the power, voltage or current gain (or loss) expressed as a ratio, P_{in} and P_{out} are the input and output powers, V_{in} and V_{out} are the input and output voltages, and I_{in} and I_{out} are the input and output currents. Note, however, that the powers, voltages or currents should be expressed in the same units/multiples (e.g. P_{in} and P_{out} should both be expressed in W, mW, µW or nW). It is often more convenient to express gain in decibels (rather than as a simple ratio) using the following relationships:

$$A_p = 10\log_{10}\left(\frac{P_{out}}{P_{in}}\right) \quad A_v = 20\log_{10}\left(\frac{V_{out}}{V_{in}}\right)$$

$$\text{and} \quad A_i = 20\log_{10}\left(\frac{I_{out}}{I_{in}}\right)$$

Note that a positive result will be obtained whenever P_{out}, V_{out}, or I_{out} is greater than P_{in}, V_{out}, or I_{out}, respectively. A negative result will be obtained whenever P_{out}, V_{out}, or I_{out} is less than P_{in}, V_{in}, or I_{in}.

A negative result denotes attenuation rather than amplification. A negative gain is thus equivalent to an attenuation (or loss). If desired, the formulae may be adapted to produce a positive result for attenuation simply by inverting the ratios, as shown below:

$$A_p = 10\log_{10}\left(\frac{P_{in}}{P_{out}}\right) \quad A_v = 20\log_{10}\left(\frac{V_{in}}{V_{out}}\right)$$

$$\text{and} \quad A_i = 20\log_{10}\left(\frac{I_{in}}{I_{out}}\right)$$

where A_p, A_v, or A_i is the power, voltage or current gain (or loss) expressed in decibels, P_{in} and P_{out} are the input and output powers, V_{in} and V_{out} are the input and output voltages, and I_{in} and I_{out} are the input and output currents. Again note that the powers, voltages or currents should be expressed in the same units/multiples (e.g. P_{in} and P_{out} should both be expressed in W, mW, µW or nW).

It is worth noting that, for identical decibel values, the values of voltage and current gain can be found by taking the square root of the corresponding value of power gain. As an example, a voltage gain of 20 dB results from a voltage ratio of 10 while a power gain of 20 dB corresponds to a power ratio of 100.

Finally, it is essential to note that the formulae for voltage and current gain are only meaningful when the input and output impedances (or resistances) are identical. Voltage and current gains expressed in decibels are thus only valid for matched (constant impedance) systems. Table A4.1 gives some useful decibel values.

Example A4.1

An amplifier with matched input and output resistances provides an output voltage of 1V for an input of 25 mV. Express the voltage gain of the amplifier in decibels.

Solution

The voltage gain can be determined from the formula:

$$A_v = 20 \log_{10}(V_{out}/V_{in})$$

where $V_{in} = 25$ mV and $V_{out} = 1$ V

Thus:

$$A_v = 20 \log_{10}(1 \text{ V}/25 \text{ mV})$$

$$= 20 \log_{10}(40) = 20 \times 1.6 = \textbf{32 dB}$$

Example A4.2

An audio amplifier provides a power gain of 33 dB. What output power will be produced if an input of 2 mW is applied?

Solution

Here we must rearrange the formula to make P_{out} the subject, as follows:

$$A_v = 10 \log_{10}(P_{out}/P_{in})$$

thus:

$$A_p/10 = \log_{10}(P_{out}/P_{in})$$

or

$$\text{antilog}_{10}(A_p/10) = P_{out}/P_{in}$$

Hence:

$$P_{out} = P_{in} \times \text{antilog}_{10}(A_p/10)$$

Now $P_{in} = 2$ mW $= 20 \times 10^{-3}$ W and $A_v = 33$ dB, thus:

$$P_{out} = 2 \times 10^{-3} \text{ antilog}_{10}(33/10)$$

$$= 2 \times 10^{-3} \times \text{antilog}_{10}(3.3)$$

$$= 2 \times 10^{-3} \times 1.995 \times 10^{-3} = \textbf{4 W}$$

Example A4.3

An antenna has a gain of 7 dB relative to a reference dipole. What power should be applied to the antenna in order to maintain the same signal strength as that produced when 20 W is fed to the dipole?

Solution

The required power can be determined from the formula:

$$A_p = 10 \log_{10}(P_{ant}/P_{ref})$$

from which:

$$P_{ant} = P_{ref}/\text{antilog}_{10}(A_p/10)$$

Now $P_{ref} = 20$ W and $A_p = 7$ dB, thus:

$$P_{ant} = 20/\text{antilog}_{10}(7/10)$$

$$= 20/\text{antilog}_{10}(0.7) = 20/5 = \textbf{4 W}$$

Table A4.1   Decibels and ratios of power, voltage and current

Decibels (dB)	Power gain (ratio)	Voltage gain (ratio)	Current gain (ratio)
0	1	1	1
1	1.26	1.12	1.12
2	1.58	1.26	1.26
3	2	1.41	1.41
4	2.51	1.58	1.58
5	3.16	1.78	1.78
6	3.98	2	2
7	5.01	2.24	2.24
8	6.31	2.51	2.51
9	7.94	2.82	2.82
10	10	3.16	3.16
13	19.95	3.98	3.98
16	39.81	6.31	6.31
20	100	10	10
30	1,000	31.62	31.62
40	10,000	100	100
50	100,000	316.23	316.23
60	1,000,000	1,000	1,000
70	10,000,000	3,162.3	3,162.3

Index